# THERMOELECTRIC MATERIALS AND DEVICES

# THERMOELECTRIC MATERIALS AND DEVICES

Lidong Chen

Ruiheng Liu

Xun Shi

*Shanghai Institute of Ceramics,*
*Chinese Academy of Sciences,*
*Shanghai, China*

ELSEVIER

Elsevier
Radarweg 29, PO Box 211, 1000 AE Amsterdam, Netherlands
The Boulevard, Langford Lane, Kidlington, Oxford OX5 1GB, United Kingdom
50 Hampshire Street, 5th Floor, Cambridge, MA 02139, United States



**British Library Cataloguing-in-Publication Data**
A catalogue record for this book is available from the British Library

**Library of Congress Cataloging-in-Publication Data**
A catalog record for this book is available from the Library of Congress

ISBN: 978-0-12-818413-4

For Information on all Elsevier publications
visit our website at https://www.elsevier.com/books-and-journals

*Publisher:* Matthew Deans
*Acquisitions Editor:* Christina Gifford
*Editorial Project Manager:* Gabriela D. Capille
*Production Project Manager:* Prasanna Kalyanaraman
*Cover Designer:* Greg Harris

Typeset by MPS Limited, Chennai, India

# Contents

CHAPTER

# 1

# General principles of thermoelectric technology

## 1.1 Introduction

The first thermoelectric effect, namely the Seebeck effect, was discovered in 1821, which describes the electromotive force generated by the temperature difference. In the following thirty years or more, Peltier effect and Thomson effect were successively discovered. These effects are the three main physical effects in thermoelectric technology that describe the direct conversion between thermal and electrical energies [1–3]. Although the discoveries of both Seebeck and Peltier effects were made using a circuit composed of two different conductors and the effects were only observed at the junctions between dissimilar conductors, they are actually the bulk properties of the materials involved, not the interfacial phenomena. Solid state physics developed in the following century reveals that all the three thermoelectric effects originate from the energy difference of carriers in different materials and/or in the different parts of materials under different temperatures.

Thomson built the relationship among the three effects, and developed the basic thermodynamic theories for thermoelectric effects [3]. Thomson's work showed that a circuit composed of two conductors with positive and negative Seebeck coefficients (usually called the thermocouple) is a type of heat engine. Such heat engine can generate electrical power by virtue of the temperature difference, or pump heat to realize refrigeration. However, since the reversible thermoelectric effects are always accompanied by the irreversible Joule heat and heat conduction, its energy conversion efficiency is principally low. Thermoelectric effects have been widely used for temperature calibrations as thermocouples, but they had no practical application as heat engine, and there

*Thermoelectric Materials and Devices*
DOI: https://doi.org/10.1016/B978-0-12-818413-4.00001-6

had been no useful theory to guide the design and fabrication of thermoelectric heat engines for a long time. Such situation did not change until 1911, when Altenkirch, for the first time, analyzed the relationship between the energy conversion efficiency and materials' physical parameters (Seebeck coefficient, electrical conductivity, and thermal conductivity) in thermoelectric devices [4]. He pointed out that to enhance the energy conversion efficiency, large Seebeck coefficient and electrical conductivity, and low thermal conductivity are required. This outlines the embryo for the criterion that is nowadays used to judge the thermoelectric performance of materials—figure of merit ($Z$) or dimensionless figure of merit ($ZT$).

This chapter will briefly illustrate the thermoelectric effects and the relationship between the thermoelectric conversion efficiency and the physical properties of materials.

## 1.2 Thermoelectric effects

### 1.2.1 Seebeck effect

The direct conversion from heat to electricity in solid materials was discovered by a German scientist, Thomas Johann Seebeck, in 1821. It is thus named as the Seebeck effect. In the next 20–30 years, the researchers successively discovered the Peltier and Thomson effects. These three effects as well as the Joule effect are the physical foundations during the thermoelectric conversion processes.

Thomas Johann Seebeck connected two different metal wires end-to-end to form a loop, and then he observed a magnetic field around the circuit when heating one junction and holding the other at low temperature, as shown in Fig. 1.1A. He wrote in his paper that "From the above described experiments, it follows that the main and important condition for the emergence of magnetism in these metal circuits is the presence of temperature difference in the circuit links" [1]. Therefore, he named it as thermomagnetism. Soon after that, this phenomenon was reexplained by Hans Christian Oersted. Oersted's experiment demonstrated that the magnetic field around the circuit was not directly contributed by the temperature difference. Instead, the temperature difference generated a voltage $V_{ab}$ and thus an electric current in the circuit to provide magnetism observed in experiment. Accordingly, he proposed the concept of thermoelectricity. Nonetheless, since the phenomenon was firstly discovered by Seebeck, it was named as the Seebeck effect until today.

As shown in Fig. 1.1B, when two conductors (a) and (b) connect each other with cold-end temperature $T$ and hot-end temperature $T + \Delta T$, the

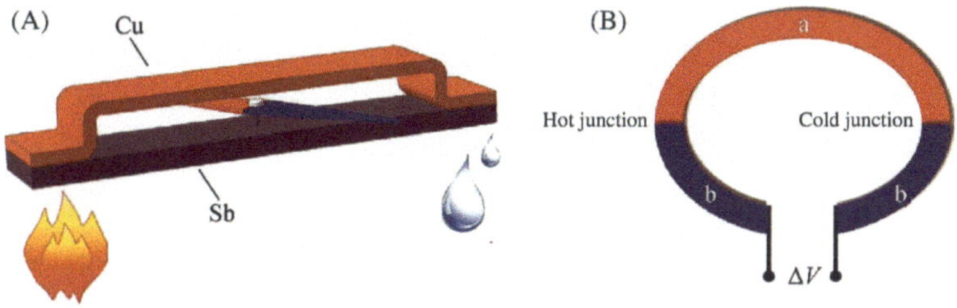


**FIGURE 1.1** (A) Experimental phenomenon and (B) equivalent diagram of Seebeck effect.

electrical potential difference $V_{ab}$ in the circuit can be measured at the free ends (having the same temperature) of (b), which is expressed as

$$V_{ab} = S_{ab}\Delta T \tag{1.1}$$

where $S_{ab}$ is the differential Seebeck coefficient of the two conductors with the unit of $\mu V/K$. $V_{ab}$ is directional, which depends on the intrinsic properties of two constituting materials and the direction of temperature gradient. $S_{ab}$ is defined as positive when the thermoelectric current flowing from hot end to cold end in conductor (a). Seebeck coefficient is also called thermopower or thermal EMF coefficient.

The generation of thermoelctric potential can be simply but principally explained by the fluctuation of charge distribution under a temperature gradient. As shown in Fig. 1.2, taking the $p$-type semiconductor (holes are majority carriers) as an example, when the temperature field is uniform, the distribution of carriers (concentration, energy, and velocity) is also uniform and the material as a whole is electrically neutral. When there is a temperature difference between the two ends of material, the hole carriers at the hot end (the temperature is $T+\Delta T$) gain higher energy ($E+\Delta E$) than the cold end (the temperature is $T$), and thus become more prone to diffuse toward the cold end. Driven by the temperature/energy difference, more holes diffuse to and accumulate at the cold end, and the distribution of charges becomes nonuniform anymore, forming an inner electric field. The inner electric field yields a reversed drift charge current. When a dynamic equilibrium is established between the thermal activated diffuse and inner field driven drift charge flows, a steady voltage $V$ is formed.

Based on the definition of thermoelectric potential described earlier, the absolute Seebeck coefficient of a material at temperature $T$ is defined as

$$S = \lim_{\Delta T \to 0} \frac{V}{\Delta T} \tag{1.2}$$

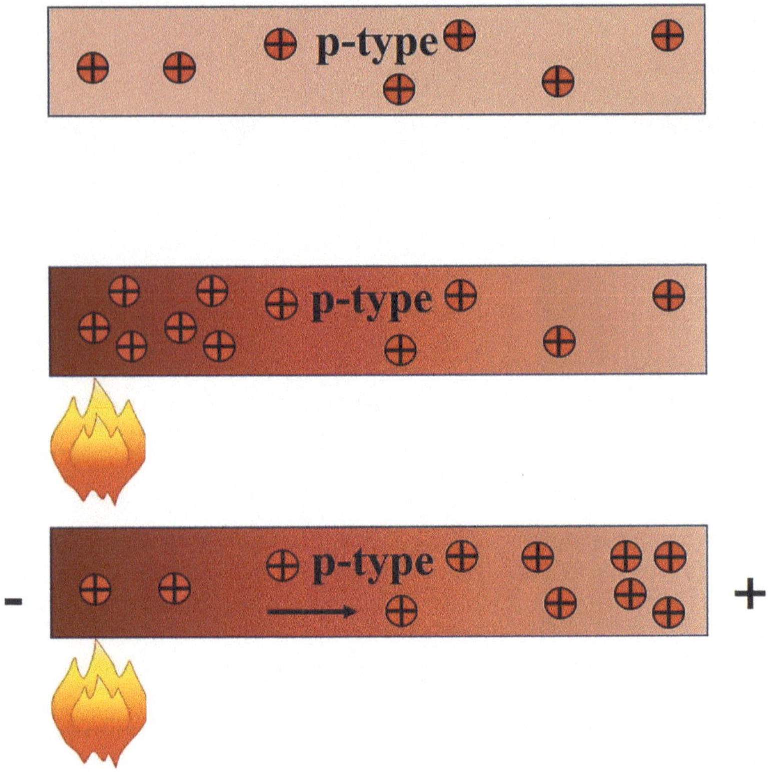


FIGURE 1.2 Schematic depiction of Seebeck effect.

The relation between the differential Seebeck coefficient $S_{ab}$ and the absolute Seebeck coefficients $S_a$, $S_b$ is

$$S_{ab} = S_a - S_b \tag{1.3}$$

The absolute Seebeck coefficient is independent with the direction of temperature field and thus it is an intrinsic property of material. In *p*-type semiconductors, the majority charge carriers (holes) diffuse from hot end to cold end driven by the energy gradient, which has the same direction as thermoelectric potential inside the material. According to Eqs. (1.1)–(1.3), the absolute Seebeck coefficient is positive. Correspondingly, the direction of the thermoelectric potential in *n*-type semiconductors is from cold end to hot end, and the absolute Seebeck coefficient is negative. Normally, the Seebeck coefficient of metals is very small with the values about several microvolts per Kelvin (μV/K), while the Seebeck coefficient of semiconductors can reach several tens or hundreds of microvolts per Kelvin (μV/K).

### 1.2.2 Peltier effect

Peltier effect is the inverse process of Seebeck effect, which describes the phenomenon of directly pumping heat by carriers (holes and/or electrons). When applying a current in the circuit composed of two different conductors, in addition the generated Joule heat, extra heat will be released or absorbed at these two junctions (Fig. 1.3). This effect was firstly discovered by a French scientist, J.C.A. Peltier, in 1834 and thus was named as the Peltier effect. Peltier connected Bi and Sb wires and observed the freezing of the water droplets in one of the junctions of the two metals when applying an electric current on the circuit. After the current was reversed, the ice was melted (Fig. 1.3A).

As shown in Fig. 1.3C, when two pieces of conductors with different Fermi levels are connected and if an electric current is applied on this link, the electrons will jump either from the high energy level to the low energy level or in the opposite direction across the interface potential barrier, and therefore either release heat or absorb heat at the junctions. For example, in the metal/*n*-type semiconductor link, when the electrons flow from latter to the former driven by the electric field, the electrons jump from high energy level to low energy level accompanied with heat release at the junction. Experimental results show that the heat absorbed or released per unit time is proportional to the electric current

$$\frac{dQ}{dt} = \pi_{\mathrm{ab}} I \tag{1.4}$$

when current flows from a to b. Here $\pi_{\mathrm{ab}}$ is the differential Peltier coefficient with the unit of $V$, $t$ is the time, and $I$ is the current. When current

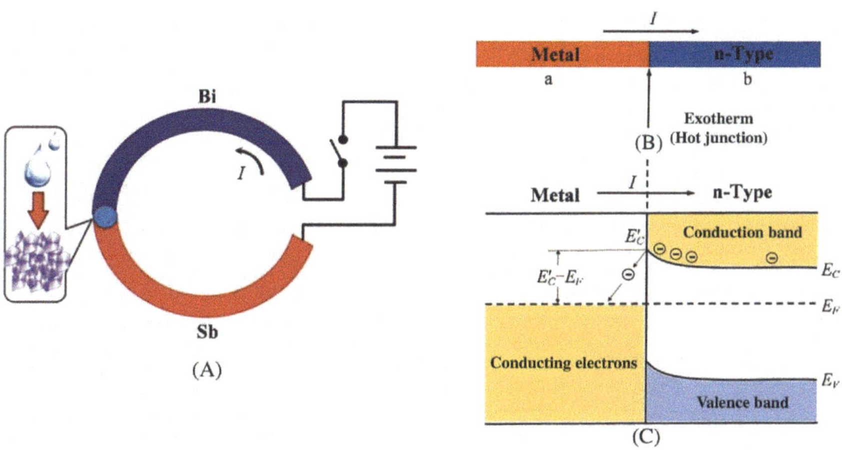


FIGURE 1.3 (A) Experimental phenomenon and (B) schematic depiction of Peltier effect.

flows from metals to $p$-type materials (electron flows from low-energy-level conductor to the high one), heat is absorbed, and $\pi_{ab}$ is negative. Apparently,

$$\pi_{ab} = -\pi_{ba} \tag{1.5}$$

In analogy to Seebeck coefficient, the differential Peltier coefficient at the junctions is related to the absolute Peltier coefficients of the two constituting materials via

$$\pi_{ab} = \pi_a - \pi_b \tag{1.6}$$

### 1.2.3 Thomson effect

The fact that both the Seebeck and Peltier effects occur only at junctions between different conductors might suggest that they are interfacial phenomena, but they are really dependent on the bulk properties of the materials involved. It is known nowadays that these two effects stem from the different properties of the materials connected together, that is, the difference in electron energies between the two conductors. The correlation between Seebeck and Peltier effects had not been realized until William Thomson (later became Lord Kelvin) recognized this issue in 1855. He analyzed the relationship between Seebeck effect and Peltier effect and then proposed that there must be the third effect, that is, when an electric current passes through a piece of uniform conductor with a temperature gradient, reversible heat absorption or release should occur through the whole piece beside the Joule heat. This effect was experimentally verified in 1867 and was termed as Thomson effect.

When an electric current $I$ passes through a piece of conductor with temperature difference $\Delta T$ along the current direction, the heat released or absorbed per unit time is

$$\frac{dQ}{dt} = \beta \Delta T I \tag{1.7}$$

where $\beta$ is the Thomson coefficient with the unit of V/K. If the direction of current is consistent with that of the temperature gradient (from cold side to hot side) and the conductor absorbs heat, $\beta$ is positive and vice versa. Considering the analogy of this expression with the definition of material's specific heat, Thomson vividly called $\beta$ as "specific heat of the current." The origin of the Thomson effect is similar to the Peltier effect. The difference is that the potential difference in the Peltier effect comes from that of the carriers in different conductors, while it is caused by the temperature gradient in a single conductor for the Thomson effect. Compared to the aforementioned two effects, the

Thomson effect contributes little to thermoelectric conversion and is therefore often neglected in the analysis of energy conversion processes and device design.

### 1.2.4 Relations between thermoelectric effects and coefficients

The Seebeck, Peltier, and Thomson effects are the intrinsic properties of bulk materials and these three coefficients are related to each other. Thomson derived the relations among these three coefficients according to equilibrium thermodynamics [3]

$$\pi_{ab} = S_{ab}T \tag{1.8}$$

$$\beta_a - \beta_b = \frac{TdS_{ab}}{dT} \tag{1.9}$$

These are called the Kelvin relations. The exact derivation of the Kelvin relations should rely on irreversible thermodynamics [5]. These two equations are verified by the experimental investigations on numerous metals and semiconductors. For a single conductor, Eq. (1.9) can be rewritten as

$$\beta = \frac{TdS}{dT} \tag{1.10}$$

It can be also rewritten as

$$S = \int_0^T \frac{\beta}{T} dT \tag{1.11}$$

The thermoelectric coefficients in Eqs. (1.8) and (1.9) are the differential values of two conductors. As demonstrated in Eqs. (1.3) and (1.6), the absolute Seebeck (or Peltier) coefficient becomes equal to the differential Seebeck (or Peltier) coefficient if the second material in the circuit is regarded as having zero Seebeck (or Peltier) coefficient. This can be realized in practice by using a superconductor as the second material, because both the Seebeck and Peltier coefficients are zero at the superconducting state. Generally, the absolute Seebeck coefficient of lead is calibrated by measuring the differential Seebeck coefficient in the circuit composed of lead and superconductor. If the absolute Seebeck coefficient of a material at low temperature is determined by connecting a circuit using a superconductor as the reference material, by using the Eq. (1.11), one can find the values at higher temperatures above the critical superconducting temperature after measuring the Thomson coefficient [6,7]. Absolute Seebeck coefficients of other materials can be calibrated by measuring the differential Seebeck coefficients in the circuit composed of lead and the target materials. The Peltier coefficient is difficult to measure in the experiment, and therefore it is often calculated via the Kelvin relation by using the measured Seebeck coefficient.

It is clear that, the Thomson effect is a spontaneous phenomenon as the Seebeck coefficient changes along a temperature gradient inside a conductor. Obviously, all the thermoelectric effects take place throughout the whole material caused by temperature gradients and/or electric current, though the Seebeck and Peltier effects are observed macroscopically at the junctions.

## 1.3 Theory of thermoelectric power generation and refrigeration

A practical thermoelectric device is usually constituted by *n*- and *p*-type materials (legs) connected electrically in series and thermally in parallel. A pair of *n*- and *p*-type legs, conveniently called as $\pi$-shape element, is the basic unit of a thermoelectric device. Usually, a number of $\pi$-shape elements make up a practical thermoelectric module via a parallel or series connection. The working principle of thermoelectric power generation and refrigeration can be schematically shown in Fig. 1.4.

Thermoelectric devices can be designed into several configurations such as plate-like, cascaded, film, and ring-shaped devices for different applications and/or working environment. Among them, the plate-like device is the most typical one (Fig. 1.5), which has been widely used as

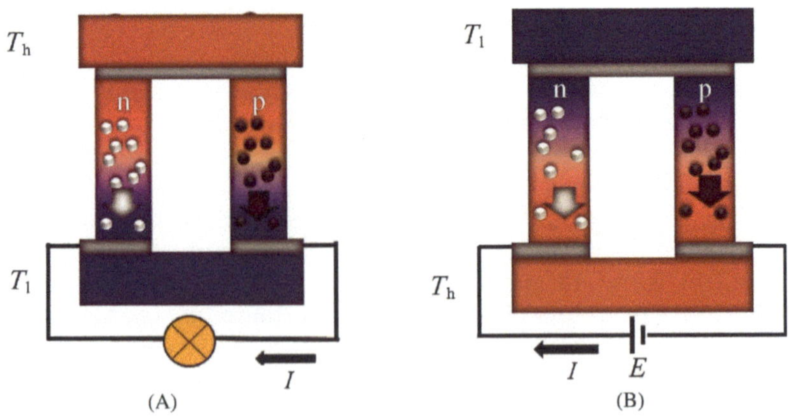


**FIGURE 1.4** Schematic depiction of thermoelectric (A) power generation and (B) refrigeration.

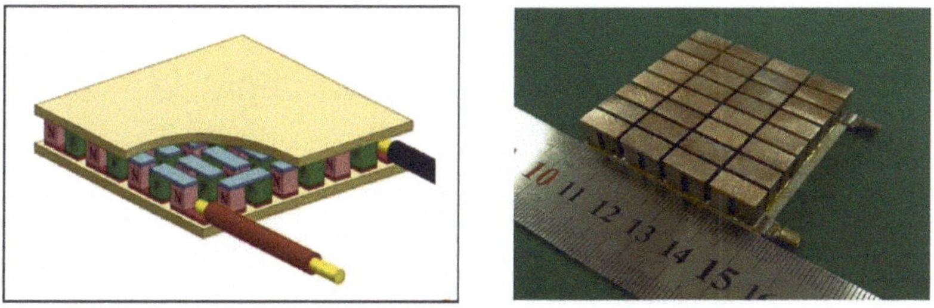

FIGURE 1.5 Structure and photo of plate-like thermoelectric devices.

power generation and refrigeration. Taking this type of device as an example, this chapter will describe the relationship between energy conversion efficiency and material performance (thermoelectric coefficients). To simplify the model and obtain a concise relation, the materials' physical parameters are taken as temperature-independent constants. In addition, it is assumed that heat flows in the one way from the hot end to the cold end through thermoelectric legs, and there is no heat exchange (such as thermal radiation, conduction, and convection) between the legs and the surrounding medium. However, practically, the physical parameters (thermal conductivity, electrical conductivity, and the Seebeck coefficient) of thermoelectric materials are usually dependent on temperature, and it is not easy to keep unidirectional heat flow from heat source and sink because the heat exchange between thermoelectric legs and surrounding medium cannot be completely prevented. The performance prediction and design of practical devices are much more complex and will be discussed in detail in Chapter 7, Design and Fabrication of Thermoelectric Devices.

## 1.3.1 Thermoelectric power generation

### *1.3.1.1 Efficiency* η

Thermoelectric devices can generate electric power and drive the load when there is a temperature gradient between the two ends. The energy conversion efficiency is the most important performance indicator for thermoelectric devices. Giving the temperatures at the hot and cold ends of the $\pi$-shaped device as $T_h$ and $T_l$ (shown in Fig. 1.6), respectively, the thermoelectric energy conversion efficiency ($\eta$) is defined as

$$\eta = \frac{P}{Q_h} \tag{1.12}$$

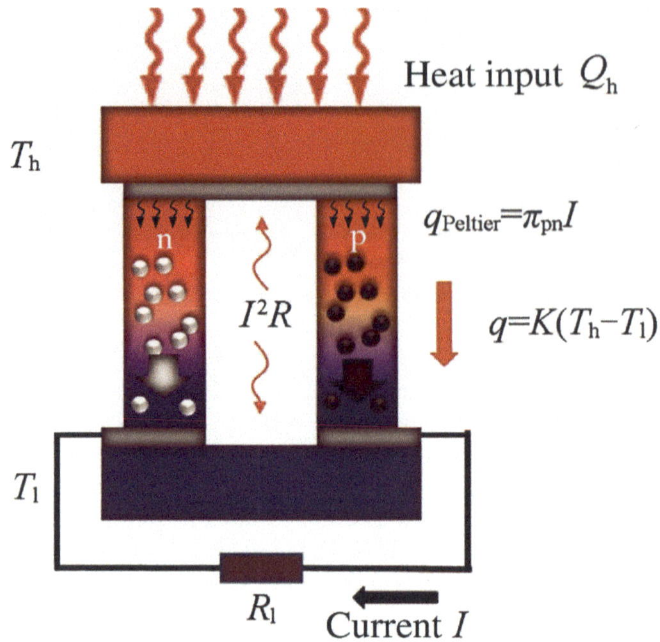


**FIGURE 1.6** Thermoelectric power generation.

where, $P$ is the output power on the load, and $Q_h$ is the heat input at the hot end (supplied by heat source). Here we do not consider the thermal and electrical resistances at the interfaces as well as the Thomson heat within the legs. On the assumption of unidirectional heat flow without side heat dissipation, the net income heat at the hot junction will be transferred from hot end to cold end by thermal conduction [$K(T_h - T_l)$] and Peltier pump. According to the Peltier effect, when taking a p-type conductor as example, heat will be absorbed at the current-in end (hot end in Fig. 1.6) and be released at the current-out end. The amount of Peltier pumped heat from hot end to cold end in Fig. 1.6 is $\pi_{pn}I$, where $I$ is the current and $\pi_{pn}$ is the total Peltier coefficient of the two legs. On the other hand, the net heat income at the hot junction is composed of two parts, the heat input (supplied from heat source) at the hot end ($Q_h$) and the Joule heat ($I^2R/2$, where $R$ is the total electrical resistance of the two legs). Here, it is reasonable to assume that the Joule heat $I^2R$ transfers equally to the hot end and cold end, therefore only half of the Joule heat ($I^2R/2$) reaches to the hot end. Then, we can obtain the following equation

$$\pi_{pn}I + K(T_h - T_l) = Q_h + \frac{1}{2}I^2R \tag{1.13}$$

Combining Eq. (1.8), the heat input is

$$Q_h = S_{pn}T_hI - \frac{1}{2}I^2R + K(T_h - T_l) \tag{1.14}$$

where $S_{pn}$ is the total Seebeck coefficient of $n$ and $p$ legs. The Seebeck voltage in the circuit is

$$V = S_{pn}(T_h - T_l) \tag{1.15}$$

Giving the resistance of the load as $R_l$, the loop current and output power are

$$I = \frac{S_{pn}(T_h - T_l)}{R + R_l} \tag{1.16}$$

$$P = \left[\frac{S_{pn}(T_h - T)_l}{R + R_l}\right]^2 R_l \tag{1.17}$$

Eq. (1.12) can be refined as

$$\begin{aligned}\eta = \frac{P}{Q_h} &= \frac{I^2R_l}{S_{pn}T_hI - \frac{1}{2}I^2R + K(T_h - T_l)} \\ &= \frac{S_{pn}^2(T_h - T_l)R_l}{\frac{1}{2}S_{pn}^2R(T_h + T_l) + S_{pn}^2T_hR_l + K(R + R_l)^2}\end{aligned} \tag{1.18}$$

Here it is convenient to define a parameter $Z$, called thermoelectric figure of merit, as

$$Z = \frac{S_{pn}^2}{RK}$$

where $Z$ is mainly determined by the properties of the thermoelectric legs; $R = \frac{l_n}{A_n}\rho_n + \frac{l_p}{A_p}\rho_p$, $K = \frac{A_n}{l_n}\kappa_n + \frac{A_p}{l_p}\kappa_p$ in which $\rho$ and $\kappa$ are the electrical resistivity and thermal conductivity, respectively; $A$ and $l$ are the sectional area and length of the thermoelectric legs. Then the efficiency can be expressed as

$$\eta = \frac{T_h - T_l}{T_h}\frac{R_l/R}{(1 + R_l/R) - \frac{T_h - T_l}{2T_h} + \frac{(1 + R_l/R)^2}{ZT_h}} \tag{1.19}$$

where $\frac{T_h - T_l}{T_h}$ is the Carnot cycle efficiency. One can define $\varepsilon = R_l/R$ and obtain the maximum value of $\eta$ when $\frac{\partial\eta}{\partial\varepsilon} = 0$. In Eq. (1.19), when $\frac{\partial\eta}{\partial\varepsilon} = 0$, $\varepsilon = R_l/R = \sqrt{1 + Z\overline{T}}$ ($\overline{T}$ is the average temperature of the hot and cold ends, $\overline{T} = (T_h + T_l)/2$). Then the thermoelectric generator exhibits the maximum energy conversion efficiency

$$\eta_{max} = \frac{T_h - T_l}{T_h} \frac{\sqrt{1 + Z\overline{T}} - 1}{\sqrt{1 + Z\overline{T}} + T_l/T_h} \tag{1.20}$$

It is seen from Eq. (1.20) that the maximum efficiency is only related to the temperature difference and the $ZT$ value. Like other heat engines, thermoelectric generator takes the Carnot cycle efficiency as the up limit of its energy conversion efficiency.

As discussed earlier, the parameter $ZT$ of a device is a dimensionless value, which is determined by the properties of the thermoelectric material. It is conventionally to define the dimensionless figure of merit ($ZT$) of a material to evaluate its thermoelectric performance

$$ZT = \frac{S^2\sigma}{\kappa}T \tag{1.21}$$

Clearly, for a given operation temperature range, a larger $ZT$ will produce higher efficiency. Fig. 1.7 demonstrates the relationship between the efficiency of thermoelectric power generation and the average $ZT$ of the material for a given $T_l = 300K$ and different $T_h$. For example, if we want to obtain a 25% efficiency using thermoelectric technology, a comparable level to the conventional heat engine, the average $ZT$ of the constituent materials should be larger than 2.0 even under a hot side temperature of 1000K. Nowadays, for most of the state-of-the-art thermoelectric materials, the average $ZT$ over wide temperature range is smaller than unity, and thus the conversion efficiency of practical device is much inferior to the conventional heat engines (Fig. 1.8). Therefore, enhancing materials' $ZT$ takes always priority in field of thermoelectricity.

#### 1.3.1.2 *Output power*

The Seebeck voltage produced in a device under a temperature gradient ($T_h - T_l$) is $V = S_{pn}(T_h - T_l)$ as shown in Eq. (1.15). Obviously, this voltage is the sum of the internal voltage drop and the voltage drop on the road. The latter is the output voltage $V_0$

$$V_0 = S_{pn}(T_h - T)_l \frac{R_l}{R_l + R} \tag{1.22}$$

The current is

$$I_0 = \frac{S_{pn}(T_h - T_l)}{R_l + R} \tag{1.23}$$

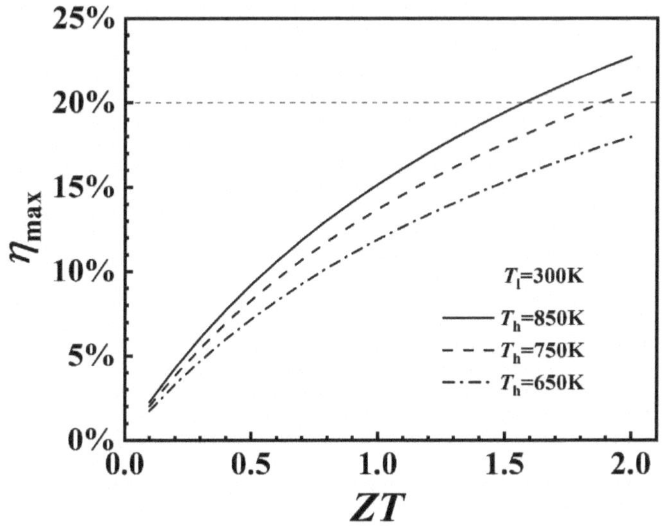


FIGURE 1.7 Dependence of $\eta_{max}$ on $ZT$ when the temperature of the cold end is fixed at 300K.

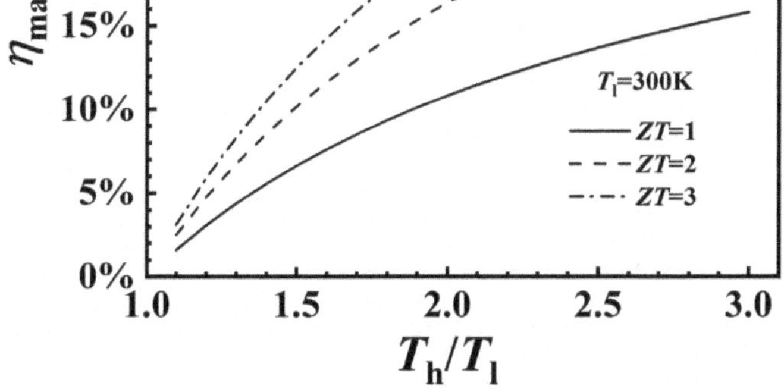


FIGURE 1.8 Dependence of $\eta_{max}$ on the temperature difference when the temperature of cold end is fixed at 300K.

Therefore, the output power $P_0$ is derived as

$$P_0 = \frac{S_{pn}^2 (T_h - T_l)^2 R_l}{(R_l + R)^2} \tag{1.24}$$

or

$$P_0 = \frac{\varepsilon}{(\varepsilon + 1)^2} \frac{S_{pn}^2 (T_h - T_l)^2}{R} \tag{1.25}$$

When the load resistance $R_l$ is equal to the internal resistance $R$, that is, $\varepsilon = R_l/R = 1$, the output power reaches the maximum value $P_{max}$

$$P_{max} = \frac{S_{pn}^2 (T_h - T_l)^2}{4R} = \frac{S_{pn}^2 \Delta T^2}{4R} \tag{1.26}$$

The corresponding efficiency at the maximum output is

$$\eta = \frac{\left(\frac{T_h - T_l}{T_h}\right)}{\left[2 - \frac{1}{2}\left(\frac{T_h - T_l}{T_h}\right) + \frac{4}{ZT_h}\right]} \tag{1.27}$$

Let $A_l = A_n + A_p$, the sum of the sectional areas of the two thermoelectric legs. And then the output power per unit area is

$$\frac{P_0}{A_l} = \frac{\varepsilon}{(\varepsilon + 1)^2} \frac{S_{pn}^2 (T_h - T_l)^2}{(A_n + A_p)(\rho_n l_n / A_n + \rho_p l_p / A_p)} \tag{1.28}$$

## 1.3.2 Thermoelectric refrigeration

### *1.3.2.1 Coefficient of performance*

Fig. 1.9 shows the working principle of thermoelectric refrigeration. The main parameters describing the refrigeration performance include coefficient of performance (COP), maximum refrigerating capacity, and maximum temperature difference. COP is defined as

$$\text{COP} = \frac{Q_c}{P} \tag{1.29}$$

where $Q_c$ is the amount of heat absorbed at the cold end (refrigeration capacity) and $P$ is the input electric power. As described in Fig. 1.3, when a current $I$ flows from the $n$-type leg to the $p$-type one at the upper junction in Fig. 1.9, heat absorption and heat release will take place at the upper junction and the lower junction, respectively, resulting in the formation of a temperature drop ($\Delta T = T_h - T_l$) between the

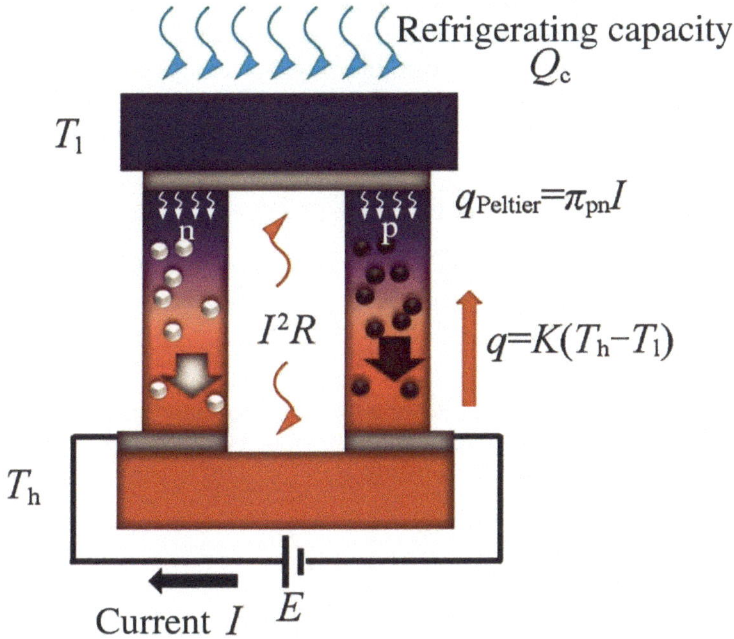


FIGURE 1.9 Working principle of thermoelectric refrigeration.

two ends. The heat pumped from the upper to the lower end per unit time by Peltier effect is $\pi_{pn}I$. At the same time, thermal conduction is inevitable against the Peltier heat pumping due to the temperature gradient. Giving the total thermal conductance of the legs $K$, the heat flow driven by temperature difference is $K\Delta T$. Similar as in power generator, Joule heat will be generated when the current is passed through the refrigeration device. The Joule heat $I^2R$ transfers equally to the hot end and cold end, therefore only half of the Joule heat ($I^2R/2$) reaches the cold end. Based on the aforementioned analyses and assumptions, the thermal balance equation at the cold end can be established:

$$Q_C + \frac{1}{2}I^2R = \pi_{pn}I - K(T_h - T_l) \tag{1.30}$$

The refrigerating capacity per unit time at the junction $Q_c$ is

$$Q_C = \pi_{pn}I - \frac{1}{2}I^2R - K(T_h - T_l) \tag{1.31}$$

or

$$Q_C = S_{pn}T_lI - \frac{1}{2}I^2R - K(T_h - T_l) \tag{1.32}$$

where $R = \frac{l_n}{A_n}\rho_n + \frac{l_p}{A_p}\rho_p$, $K = \frac{A_n}{l_n}\kappa_n + \frac{A_p}{l_p}\kappa_p$. Here the subscripts $n$ and $p$ represent $n$- and $p$-type thermoelectric legs, respectively.

The voltage applied on the two thermoelectric legs is the sum of the internal voltage drop ($V_R = IR$) and the "anti-Seebeck voltage" to overcome the thermoelectric voltage induced by temperature difference $[V_S = S_{pn}(T_h - T_l)]$:

$$V = V_R + V_S = IR + S_{pn}(T_h - T_l) \tag{1.33}$$

Therefore, the input power $P$ (the rate of expenditure of electrical energy by the thermoelectric legs) is

$$P = IV = I^2R + S_{pn}(T_h - T_l)I \tag{1.34}$$

where the first term is the internal resistive loss and the second term is the rate of working to overcome the thermoelectric voltage. Substituting Eqs. (1.32) and (1.34) into the definition of COP, Eq. (1.29), we obtain

$$\mathrm{COP} = \frac{S_{pn}T_lI - \frac{1}{2}I^2R - K(T_h - T_l)}{I^2R + S_{pn}(T_h - T_l)I} \tag{1.35}$$

Apparently, if fixing temperature difference $T_h - T_l$, the COP varies with the applied current. Let $d(\mathrm{COP})/dI$ be zero, one can find the optimum current $I_{COP}$ at which the COP reaches the maximum value

$$I_{COP} = \frac{(S_p - S_n)(T_h - T_l)}{R\left[\sqrt{1 + Z\overline{T}} - 1\right]} \tag{1.36}$$

or

$$I_{COP} = \frac{(S_p - S_n)(T_h - T_l)}{\left(\frac{l_n}{A_n}\rho_n + \frac{l_p}{A_p}\rho_p\right)\left[\sqrt{1 + Z\overline{T}} - 1\right]} \tag{1.37}$$

where $Z$ holds the same definition as that in Eq. (1.20). Under this optimum current, the maximum efficiency $\mathrm{COP}_{max}$ is

$$\mathrm{COP}_{max} = \frac{T_l}{T_h - T_l}\frac{\sqrt{1 + Z\overline{T}} - \frac{T_h}{T_l}}{\sqrt{1 + Z\overline{T}} + 1} \tag{1.38}$$

Correspondingly, the applied voltage and input power are

$$(V_{CD})_{COP} = S_{pn}\frac{(T_h - T_l)\sqrt{1 + Z\overline{T}}}{\sqrt{1 + Z\overline{T}} - 1} \tag{1.39}$$

$$P_{COP} = \frac{\sqrt{1 + Z\overline{T}}}{R}\left[S_{pn}\frac{(T_h - T_l)}{\sqrt{1 + Z\overline{T}} - 1}\right]^2 \tag{1.40}$$

respectively, where $\overline{T} = (T_h + T_l)/2$ is the average temperature of the thermoelectric elements. Further, one can obtain the released heat per unit time at the hot end

$$Q = Q_c + P = S_{pn}T_hI + \frac{1}{2}I^2R - K(T_h - T_l) \tag{1.41}$$

Similarly, the COP of a thermoelectric device at heating mode is

$$\mathrm{COP} = \frac{S_{pn}T_hI + \frac{1}{2}I^2R - K(T_h - T_l)}{I^2R + S_{pn}(T_h - T_l)I} \tag{1.42}$$

#### 1.3.2.2 *Maximum depression of temperature* $\Delta T_{max}$

Another key parameter of thermoelectric refrigerator is the temperature difference that can be established between the two ends $\Delta T = T_h - T_l$. Obviously, this temperature difference is related to the refrigerating capacity and current. Based on the thermal balance equation [Eq. (1.30)], it is obtained that

$$\Delta T = \frac{S_{pn}T_hI - \frac{1}{2}I^2R - Q_c}{K} \tag{1.43}$$

When the cold end is adiabatic, that is, $Q_c = 0$, let $d\Delta T/dI$ be 0 under which $\Delta T$ reaches the maximum value, and the optimum current $I_T$ is

$$I_T = \frac{S_{pn}T_l}{R} \tag{1.44}$$

The corresponding maximum temperature depression $\Delta T_{max}$ is

$$\Delta T_{max} = \frac{1}{2}ZT_l^2 \tag{1.45}$$

This can also be used to know the minimum temperature that can be reached at the cold end of the refrigerator

$$(T_l)_{min} = \frac{\left[\sqrt{1 + 2ZT_h} - 1\right]}{Z} \tag{1.46}$$

$\Delta T_{max}$ or $(T_l)_{min}$ is usually used to evaluate the performance of thermoelectric device as refrigerator. They are determined by the figure of merit ($Z$) of the constituent thermoelectric materials and the temperature of heat sink ($T_h$).

#### 1.3.2.3 *Maximum refrigerating capacity* $Q_{c,max}$

The heat absorbed per unit time by the refrigerator (or called the refrigerating capacity) is given by Eq. (1.31). Obviously, for giving thermoelectric materials, $Q_c$, is related to the current ($I$) and the temperatures at the two ends ($T_h$, $T_l$). Similarly, let $dQ_c/dI$ be 0, one can obtain the optimum current $I_q$ which satisfies $Q_c$ reaching the maximum value at a certain temperature,

$$I_q = I_T = \frac{S_{pn}T_l}{R}, \tag{1.47}$$

The corresponding $Q_c$ is

$$Q_c = \frac{S_{pn}^2 {T_l}^2}{2R} - K(T_h - T_l) \tag{1.48}$$

Next, the maximum refrigerating capacity $Q_{c,max}$ is obtained when the temperature difference is 0, that is,

$$Q_{c,max} = \frac{S_{pn}^2 {T_l}^2}{2R} \tag{1.49}$$

Apparently, $Q_{c,max}$ is independent of the thermal conducting properties of the thermoelectric material. In addition, based on Eqs. (1.48) and (1.49) as well as (1.43) and (1.45), the relationship between the refrigerating capacity and temperature difference is

$$Q_c = K(\Delta T_{max} - \Delta T) \tag{1.50}$$

$$\Delta T = \frac{1}{K}(Q_{c,max} - Q_c) \tag{1.51}$$

It is seen that $Q_c$ varies with $\Delta T$ in a linear way. To plot $Q_c$ against $\Delta T$, one can easily find $\Delta T_{max}$ and $Q_{c,max}$ from the intercepts on the two axes, and the total thermal conductance $K$ $\left(K = \frac{A_n}{l_n}\kappa_n + \frac{A_p}{l_p}\kappa_p\right)$ of the thermocouple from the slop. It is convenient to estimate the thermal conductance of the constituting thermoelectric materials in a device from the experimental variation of $Q_c$ with $\Delta T$.

## References

[1] J. Seebeck, Magnetische Polarisation der Metalle und Erze durch Temperatur-Differenz Abh, Akad. Wiss. Berl. (1822) 289–346.

[2] J.C.A. Peltier, Nouvelles Expériemences sur la Caloricite descourants électriques, Ann. Chim. Phys. 56 (1834) 371–386.

[3] W. Thomson, On a mechanical theory of thermo-electric currents, Proc. R. Soc. Edinb. (1851) 91–98.

[4] E. Altenkirch, Elektrothermische Kalteerzeugung, Phys. Z. 12 (1911) 920.

[5] I. Muller, Thermodynamics of Irreversible Processes, North-Holland Pub.Co-distributors for USA, Interscience Publishers, New York, 1951.

[6] G. Borelius, W.H. Keesom, C.H. Johansson, J.O. Linde, Establishment of an absolute scale for the thermo-electric force, Proc. K. Akad. Wetensch. Amst. 35 (1932) 10–14.

[7] J.W. Christian, J.P. Jan, W.B. Pearson, I.M. Templeton, Proceedings of the royal society of London series a-mathematical and physical sciences (1958) 213–245.

CHAPTER

# 2

# Strategies to optimize thermoelectric performance

## 2.1 Introduction

The dimensionless figure of merit ($ZT$) is the key criterion that quantifies thermoelectric (TE) performance of materials. According to the definition $ZT = S^2\sigma T/\kappa$, high-performance TE materials should possess a large Seebeck coefficient $S$, a high electrical conductivity $\sigma$ ($S^2\sigma$ is termed as power factor), and a low thermal conductivity $\kappa$. Fig. 2.1 qualitatively outlines the dependence of the three TE parameters on carrier concentrations. $\sigma$ and $S$ change in crosscurrent with the increase of carrier concentrations. For most semiconductors, the optimal carrier concentration corresponding to the highest power factor usually lies within the range of $10^{19}$–$10^{20}\ \mathrm{cm}^{-3}$, a value in the heavily doped or degenerate semiconductors. The thermal conductivity is majorly composed of two parts: lattice thermal conductivity ($\kappa_L$) and electronic thermal conductivity ($\kappa_e$). The latter is proportional to $\sigma$, so high electrical conductivity will directly result in a high $\kappa_e$, which is unfavorable for high $ZT$. Due to the correlation and interaction of carriers and phonons, the strategies aiming at reducing $\kappa_L$ by intensifying phonon scattering usually also influence $S$ and $\sigma$. Therefore the three parameters $S$, $\sigma$, and $\kappa$ that determine TE performance are basically dependent on each other. This is the primary cause of why enhancing TE performance is extremely difficult. The independent and synergetic regulation of electrical and thermal transports has been the long-term goal in thermoelectrics.

Based on semiconductor physics and transport theory in solids, this chapter will outline the basic theory on tuning transport properties of TE materials and focus on the correlation among the key transport

*Thermoelectric Materials and Devices*
DOI: https://doi.org/10.1016/B978-0-12-818413-4.00002-8

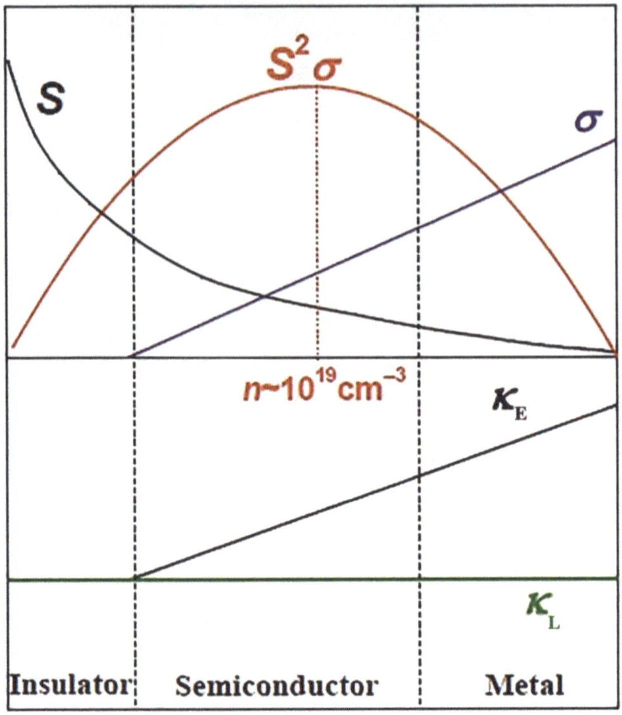


FIGURE 2.1 Dependence of electrical conductivity, Seebeck coefficient, power factor, and thermal conductivity on carrier concentrations.

parameters. The strategies to explore and design high-performance TE materials based on the developed theoretical models and proposed multiscale structure modulation in the recent decades will also be summarized.

## 2.2 Basic theory for transports in thermoelectrics

### 2.2.1 Band model for carrier transport

Most of the TE materials are semiconductors. The electrical conductivity and Seebeck coefficient are mainly determined by material's intrinsic parameters, such as carrier concentration, mobility, and scattering processes. According to the band structure theory of semiconductors, as considering a single-carrier system with a parabolic band, carriers are in the equilibrium state and obey the Fermi–Dirac distribution when the external electrical field is zero. The carrier concentration is the integral of the density of state (DOS) $g(E)$ at the band edge multiplied by occupancy probability $f(E)$, while $g(E)$ and $f(E)$ can be expressed as [1–4]

$$g(E) = \frac{4\pi(2m^*)^{3/2}}{h^3} E^{1/2} \tag{2.1}$$

$$f(E) = \frac{1}{1 + \exp\left(\frac{E - E_F}{k_B T}\right)} \tag{2.2}$$

where $m^*$ is the effective mass of carriers, $h$ is the Planck's constant, $E_F$ is the Fermi level, $k_B$ is the Boltzmann constant, and $f$ is the Fermi–Dirac distribution function.

When considering the movement of electrons under external electrical field and temperature gradient, the steady state is statistically described by the Boltzmann equation

$$\frac{1}{\hbar}(\Delta_k E \cdot \Delta T)\frac{\partial f}{\partial T} - \frac{e}{\hbar}\varepsilon \cdot \Delta_k f = -\frac{f - f_0}{\tau} \tag{2.3}$$

where $f$ is the distribution function at the nonequilibrium state, $f_0$ is the distribution function at the equilibrium state, $\varepsilon$ is the external field, $\nabla T$ is the temperature gradient, $\tau$ is the relaxation time, and $\hbar$ is the reduced Planck's constant. When the deviation from the equilibrium state is small, the distribution function is

$$\frac{df(E)}{dt} = \frac{f(E) - f_0(E)}{\tau} \tag{2.4}$$

For a certain scattering process, a corresponding scattering parameter $\lambda$ is introduced, so the relaxation time is related to the carrier energy via

$$\tau = \tau_0 E^{\lambda - 1/2} \tag{2.5}$$

In the one-dimensional (1D) case, it is derived that

$$\frac{f(E) - f_0(E)}{\tau} = U_x\left\{e\varepsilon_x + \left[T\frac{d}{dT}\left(\frac{E_F}{T}\right) + \frac{E}{T}\right]\frac{dT}{dx}\right\} \tag{2.6}$$

where $U_x$ is the drift velocity along $x$-direction. Considering the definition of the current density and the features of parity function integral, the 1D current density (charge flux) is expressed as

$$\begin{aligned} i_x &= \pm \int_0^\infty eU_x \cdot f(E)g(E)dE \\ &= \pm e\left\{e\varepsilon_x + T\frac{d}{dT}\left(\frac{E_F}{T}\right)\frac{dT}{dx}\right\}K_1 \pm \frac{e}{T}\frac{dT}{dx}K_2 \end{aligned} \tag{2.7}$$

in which + and − correspond to holes and electrons, respectively. The heat flux contributed by carriers is

$$j_x = \int_0^\infty (E - E_F)U_x g(E)\frac{\partial f_0}{\partial E}\varphi(E)d(E)$$
$$= \left[e\varepsilon_x + T\frac{d}{dT}\left(\frac{E_F}{T}\right)\frac{dT}{dx}\right]K_2 + \frac{1}{T}\frac{dT}{dx}K_3 - \frac{E_F}{e}i_x \tag{2.8}$$

where

$$K_m = \int_0^\infty \tau U_x^2 g(E)E^{m-1}\frac{\partial f_0}{\partial E}dE \quad (m = 1 \sim 3) \tag{2.9}$$

When $\nabla T = 0$, the electrical conductivity is

$$\sigma = \frac{i_x}{e_x} = e^2 K_1 \tag{2.10}$$

Based on Eqs. (2.7) and (2.8), the Peltier coefficient can be derived based on its definition

$$\pi = \frac{j_x}{i_x} = \mp\frac{1}{e}\left[\frac{K_2}{K_1} - E_F\right] \tag{2.11}$$

The Seebeck coefficient is

$$S = \frac{\pi}{T} = \mp\frac{1}{eT}\left[\frac{K_2}{K_1} - E_F\right] \tag{2.12}$$

In fact, Seebeck coefficient can be also derived directly using Eq. (2.7) under the condition $\varepsilon_x = 0$ and $i_x = 0$. The result is the same with Eq. (2.12).

The carrier thermal conductivity can be calculated based on the definition under the condition $i_x = 0$

$$\kappa_e = -\frac{j_x}{dT/dx} = \frac{1}{T}\left(K_3 - \frac{K_2^2}{K_1}\right) \tag{2.13}$$

where the minus sign means that heat transports from high temperature to low temperature.

Now we have derived three parameters related to $ZT$ by solving the Boltzmann equation. All of them contain the integral $K_m$ that is related to the distribution of carriers, physical properties of semiconductors, and the relaxation time. For the spherical iso-energy surface, the drift mobility is

$$U_x^2 = \frac{2E}{3m^*} \tag{2.14}$$

Then we can arrive at

$$K_m = \frac{8\pi}{3}\left(\frac{2}{\hbar^2}\right)^{3/2}(m^*)^{1/2}\tau_0(\lambda+m)\cdot(k_BT)^{\lambda+m}F_{\lambda+m}(\eta) \tag{2.15}$$

where

$$F_n(\eta) = \int_0^\infty \frac{x^n dx}{1+\exp(x-\eta)} \tag{2.16}$$

Eq. (2.16) is called the Fermi integral. $\eta = E_F/k_BT$ is the reduced Fermi level; $x = E/k_BT$ is the reduced carrier energy; $n$ is an integer or a half-integer. The Fermi integrals possess only numerical solutions. For most practical TE materials, $\eta$ falls in the range of $-2.0$ to 5.0.

Using Eqs. (2.11)–(2.13), the basic parameters (Seebeck coefficient, carrier concentration, mobility, electrical conductivity, and Lorenz number) related to carrier transport can be expressed as

$$S = \mp\frac{k_B}{e}\left(\eta - \frac{(\lambda+2)F_{\lambda+1}(\eta)}{(\lambda+1)F_\lambda(\eta)}\right) \tag{2.17}$$

$$n = 4\pi\left(\frac{2m^*k_BT}{h^2}\right)^{3/2}F_\lambda(\eta) \tag{2.18}$$

$$\mu = \frac{2e}{3m^*}\tau_0(\lambda+1)(k_BT)^{\lambda-1/2}\frac{F_\lambda(\eta)}{F_{1/2}(\eta)} \tag{2.19}$$

$$\sigma = ne\mu \tag{2.20}$$

$$L = \left(\frac{k_B}{e}\right)^2\left\{\frac{(\lambda+3)F_{\lambda+2}(\eta)}{(\lambda+1)F_\lambda(\eta)} - \left[\frac{(\lambda+2)F_{\lambda+1}(\eta)}{(\lambda+1)F_\lambda(\eta)}\right]^2\right\} \tag{2.21}$$

Based on Eqs. (2.17)–(2.21), the key TE transport parameters can be expressed as the functions of the basic parameters of the material such as the Fermi level, effective mass, and scattering parameters. In principle, the three TE transport parameters can be obtained by solving Eqs. (2.17)–(2.21). However, the Fermi integrals have only numerical solutions, so the TE parameters only have numerical expressions, which is not convenient for materials design and optimization. Therefore approximation and simplification are needed for Eqs. (2.17)–(2.21) to obtain analytical solutions. Taking n-type semiconductors as an example, setting the conduction band minima (CBM) as the base energy level, when the Fermi level is significantly higher than $k_BT$ (degenerate state) or lower than $-k_BT$ (nondegenerate state), the Fermi–Dirac distribution can be expressed in a more simplified format.

#### 2.2.1.1 Nondegenerate limit ($E_F << -k_BT$)

In this case, $\eta << -1$, the Fermi level situates 2–3 $k_BT$ below CBM for $n$-type materials, and the Fermi–Dirac distribution can be approximated by the Boltzmann distribution. The Fermi integral can be expressed as

$$F_n(\eta) = \exp(\eta)\int_0^\infty x^n \exp(-x)dx = \exp(\eta)\Gamma(n+1) \tag{2.22}$$

where $\Gamma(n+1) = n\Gamma(n)$ is the $\Gamma$ function, and $\Gamma(1/2) = \pi^{1/2}$. For an integer $n$, we have

$$\Gamma(n) = (n-1)! \tag{2.23}$$

The Seebeck coefficient, carrier concentration, mobility, and Lorenz number in the nondegenerate case can be obtained

$$S = \mp\frac{k_B}{e}[\eta - (\lambda+2)] \tag{2.24}$$

$$n = 2\left(\frac{2\pi m^* k_B T}{\hbar^2}\right)^{3/2}\exp(\eta) \tag{2.25}$$

$$\mu = \frac{4}{3\pi^{1/2}}\Gamma(\lambda+2)\frac{e\tau_0(k_BT)^{\lambda-1/2}}{m^*} \tag{2.26}$$

$$L = \frac{\lambda T}{\sigma} = \left(\frac{k_B}{e}\right)^2(\lambda+2) \tag{2.27}$$

#### 2.2.1.2 Degenerate limit ($E_F >> k_BT$)

In this case, $\eta >> 1$, the Fermi level has well crossed CBM for n-type semiconductors, which is analogous to the bands of metals. The Fermi integrals can be expressed as a series that rapidly converges

$$F_n(\eta) = \frac{\eta^{n+1}}{n+1} + n\eta^{n-1}\frac{\pi^2}{6} + n(n-1)(n-2)\eta^{n-3}\frac{7\pi^4}{360} + \cdots\cdots \tag{2.28}$$

We can pick up limited terms to obtain a finite (nonzero) solution as a reasonable approximation. Considering the first term only, the electrical conductivity can be solved

$$\sigma = \frac{8\pi}{3}\left(\frac{2}{\hbar^2}\right)e^2(m^*)^{1/2}\tau_0 {E_F}^{\lambda+1} \tag{2.29}$$

For Seebeck coefficient and Lorenz number, at least two first terms have to be calculated to obtain nonzero solutions

$$S = \mp \frac{\pi^2}{3} \frac{k_B}{e} \frac{(\lambda + 1)}{\eta} \tag{2.30}$$

$$L = \mp \frac{\pi^2}{3} \left(\frac{k_B}{e}\right)^2 \tag{2.31}$$

Eq. (2.31) clearly shows that all the metals have the same Lorenz number that is independent of carrier concentrations and scattering mechanism.

The Eqs. (2.14)–(2.31) are derived under the assumption of a single parabolic band with isotropic properties and are applicable to material systems with simple band structures. For multiband systems, these equations can be employed after the modification of effective mass. For example, there are six energy minima along the [100] direction for the conduction band of silicon. The dispersion relation shall be

$$E(k_x, k_y, k_z) = \frac{\hbar^2 k_x^2}{2m_x} + \frac{\hbar^2 k_y^2}{2m_y} + \frac{\hbar^2 k_z^2}{2m_z} \tag{2.32}$$

where $x$, $y$, and $z$ are the projection of the iso-energy surface on the principal axes. The DOS effective mass for a single electron can be regarded as the average value of those along three directions $m^* = \left(m_x^* m_y^* m_z^*\right)^{1/3}$. Considering the multivalley effect, the total effective mass can be given as:

$$m_{\mathrm{DOS}}{}^* = N_{\mathrm{v}}^{2/3} \left(m_x m_y m_z\right)^{1/3} \tag{2.33}$$

where $N_{\mathrm{v}}$ is the number of equivalent valleys. Things are even more complicated for the nonparabolic band, and the treatment details can be found in books on semiconductor physics [1–3].

All of the earlier discussions have been focused on the single-carrier system. However, most of TE materials are narrow-gap semiconductors in which bipolar effect, that is, mixed conduction of electrons and holes, is anticipated due to thermal excitation. In such case, the electrical conductivity and Seebeck coefficient can be written as [4]

$$\sigma_{\mathrm{total}} = \sigma_{\mathrm{e}} + \sigma_{\mathrm{h}} = \mathrm{e}(n_{\mathrm{e}}\mu_{\mathrm{e}} + n_{\mathrm{h}}\mu_{\mathrm{h}}) \tag{2.34}$$

$$S_{\mathrm{total}} = \frac{S_{\mathrm{e}}\sigma_{\mathrm{e}} + S_{\mathrm{h}}\sigma_{\mathrm{h}}}{\sigma_{\mathrm{e}} + \sigma_{\mathrm{h}}} \tag{2.35}$$

where subscripts e and h denote the contributions by electrons and holes, respectively. As seen from Eqs. (2.34) and (2.35), mixed conduction enhances the electrical conductivity, but significantly decreases the value of Seebeck coefficient and the power factor due to the opposite signs of the two types of carriers.

## 2.2.2 Scattering of carriers

The Seebeck coefficient and electrical conductivity can be expressed using several basic parameters within the framework of SPB, including the effective mass $m^*$, reduced Fermi level $\eta$, carrier concentration $n$, scattering parameter $\lambda$, and the energy-independent relaxation time $\tau_0$. The effective mass is determined by the band structure, the Fermi level and carrier concentration can be tuned via element doping, and the scattering parameter is related to the scattering mechanisms of materials. Carriers are mostly scattered by lattice vibrations, ionized impurities, neutral impurities, large defects, etc.

At 0K, carriers in a perfect lattice will not be scattered. In any crystal above 0K, however, there must exist lattice vibrations. Carriers in the crystal will deviate from the periodicity and thus be scattered. This effect can be ascribed to the scattering by lattice waves, including acoustic waves and optical waves. In semiconductors, long acoustic waves with low frequencies and energies have the wavelength several times larger than the atomic distance, which contribute dominantly to carrier scattering. Particularly, longitudinal waves are dilatational ones and cause the dilatation or compression of the lattice and thus the fluctuation of bands in reciprocal space. This effect exerts an extra potential to the original uniform periodic potential field and hence scatters carriers. Bardeen and Shockley proposed the concept of deformation potential based on the characters of acoustic phonons, and the relaxation time limited by acoustic branches is [5]

$$\tau = \frac{h^4 v^2 \rho_{\mathrm{d}}}{8\pi k_{\mathrm{B}} T \psi^2 (2m^*)^{3/2}} E^{-1/2} \tag{2.36}$$

where $\Psi$ is the deformation potential coefficient, $\rho_{\mathrm{d}}$ is the density, and $v$ is the sound velocity. As seen from Eqs. (2.5) and (2.36), the scattering parameter $\lambda$ is 0 for acoustic phonon scattering.

Element doping is widely employed to tune the carrier concentrations and Fermi levels in semiconductors. When the dopant atoms are ionized to generate holes or electrons, they are also charged. Carriers will be scattered by the Coulomb force when approaching these ionized impurity atoms. This effect is equivalent to adding a local Coulomb field to the lattice potential. Conwell and Weisskopf derived the relaxation time limited by ionized impurity scattering [6]

$$\tau_i = \frac{\xi^2 (2m^*)^{1/2} E^{3/2}}{ze^4 N_i \pi} \left[ \ln\left(1 + \frac{\xi E}{ze^2 N_i^{1/2}}\right)^2 \right] \tag{2.37}$$

**TABLE 2.1** Temperature dependence of relaxation time and mobility for different scattering mechanisms.

| Scattering mechanism | Relaxation time ($\tau$) | | Mobility Nondegenerate | Degenerate |
|---|---|---|---|---|
| Acoustic phonon scattering | $E^{-1/2}$ | $T^{-1}$ | $T^{-3/2}$ | $T^{-1}$ |
| Optical phonon scattering | $E^{1/2}$ | $T^{-1}$ | $T^{-3/2}$ | $T^{-1}$ |
| Ionized impurity scattering | $E^{3/2}$ | $T^{0}$ | $T^{3/2}$ | $T^{0}$ |
| Alloying scattering | $E^{1/2}$ | $T^{0}$ | $T^{0}$ | $T^{0}$ |

where $N_i$ is the concentration of ionized impurities, $\xi$ is the dielectric constant, and $z$ is the effective charge. According to Eq. (2.37), the scattering parameter $\lambda$ is 2 for ionized impurity scattering.

In addition to acoustic phonon scattering and ionized impurity scattering, carriers can also be scattered by alloying (due to compositional fluctuations), other carriers, neutral impurities, and grain boundaries. The dependence of relaxation time and mobility on energy and temperature for various scattering mechanisms is listed in Table 2.1. If more than one processes are involved, the reciprocal of the relaxation time is the sum of individual ones

$$\frac{1}{\tau} = \sum_{i}^{n} \frac{1}{\tau_i} \tag{2.38}$$

### 2.2.3 Thermal transport and phonon scattering in solids

Heat conduction in solids is strongly correlated to the constituent particles and their interactions, or simply saying, the constituent particles transfer thermal energies through their collisions [7]. Heat carriers in solids are mainly phonons, electrons, and photons. Photonic thermal conduction comes from the radiation of high-frequency electrons and this process usually occurs at high temperature, usually beyond the temperature range where TE materials work. In a single-carrier system, the thermal conductivity ($\kappa$) is usually the sum of carrier thermal conductivity ($\kappa_c$) and lattice thermal conductivity ($\kappa_L$).

$$\kappa = \kappa_c + \kappa_L \tag{2.39}$$

According to the Wiedemann–Franz law, the carriers' contribution to thermal conduction is

$$\kappa_c = L_0 \sigma T \tag{2.40}$$

where $L_0$ is the Lorenz number. For metals, $L_0$ is a constant with the value of $2.45 \times 10^{-8}\ V^2/K^2$.

In the mixed conduction system, the Seebeck coefficient shall be deteriorated and the thermal conductivity shall be increased, which is unfavorable to the enhancement of TE performance. Electron–hole pairs are generated by intrinsic excitation, and they can be recombined with a certain probability. This recombination will release a certain amount of heat that is larger than the bandgap, thus inducing extra heat conduction. This phenomenon is termed as bipolar diffusion, and its contribution ($\kappa_B$) to thermal conductivity is [4]

$$\kappa_B = \frac{\sigma_h \sigma_e}{(\sigma_h + \sigma_e)}(S_h - S_e)^2 T \tag{2.41}$$

After considering this effect, Eq. (2.39) becomes

$$\kappa = \kappa_c + \kappa_L + \kappa_B \tag{2.42}$$

Bipolar effect increases the thermal conductivity and significantly worsens the TE performance. As a result, a good TE material is usually heavily doped semiconductors in which single-carrier transport is roughly maintained to avoid the bipolar diffusion.

Lattice thermal conduction is rather independent, so regulation on this parameter has become a key strategy for performance optimization. By introducing the concept of phonons, heat conduction by lattice vibration from the high-temperature end to the low-temperature end can be regarded as the transport of heat-carrying phonons. In analogy to carrier transport in the lattice, studies on heat conduction can turn to the understanding of phonon collision process. Supposing the mean free path (MFP) between two collisions is $l$ and employing the heat conduction model of the ideal gas kinetics, lattice thermal conductivity in solids can be expressed as

$$\kappa_L = \frac{1}{3} C_V v l \tag{2.43}$$

where $C_v$ is the volumetric heat capacity and $v$ is the velocity of phonon propagation. The MFPs ($l$) are determined by the scattering processes on phonons.

The accurate description of lattice thermal conductivity can be obtained by solving the Boltzmann equation, which is quite complicated. Conventionally we employ the Debye model and employ the "relaxation time approximation" for analysis. In this framework, the lattice thermal conductivity is expressed as [7,8]

$$\kappa_L = \frac{k_B}{2\pi^2 \nu}\left(\frac{k_B T}{\hbar}\right)^3 \int_0^{\theta_D/T} \frac{x^4 e^x}{\tau_c^{-1}(e^x - 1)^2} dx \tag{2.44}$$

where $x = \hbar\omega/k_B T$ is the reduced phonon energy, $\omega$ is the phonon frequency, $\theta_D$ is the Debye temperature, $\tau_c$ is the relaxation time, and $k_B$ is the Boltzmann constant. In TE materials, various phonon scattering mechanisms exist as shown in Fig. 2.2.

The total relaxation time $\tau_c$ is the collective effect of various scattering processes

$$\frac{1}{\tau_c} = \frac{1}{\tau_B} + \frac{1}{\tau_U} + \frac{1}{\tau_D} + \frac{1}{\tau_r} + \cdots \tag{2.45}$$

with grain-boundary scattering (B), phonon–phonon Umklapp scattering (U), point-defect scattering (D), and resonant scattering (r). $\tau_B$ is determined by the average phonon velocity $v_s$ and the grain size $L$

$$\frac{1}{\tau_B} = \frac{v_s}{L} \tag{2.46}$$

The second term on the right of Eq. (2.45) is also called the Umklapp process. There are mainly two ways of phonon interaction. One is the normal (N-) process that usually makes an effect at low temperatures. In this process, the wave vector of newly generated phonons after collision still falls in the first Brillouin zone, that is, the main direction of heat transport is maintained producing only little thermal resistance. The second is the Umklapp (U-) process, in which the wave vector of newly generated phonons goes beyond the first Brillouin zone. The relaxation time is

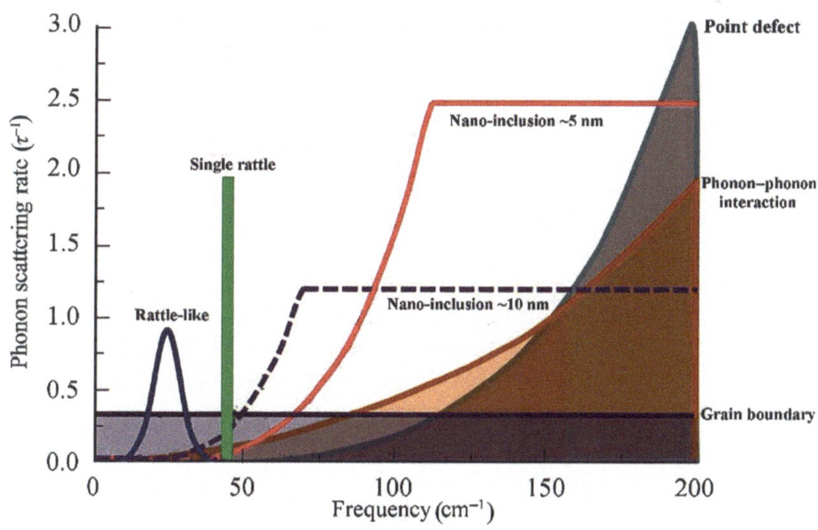


FIGURE 2.2 Frequency-dependent phonon scattering rate [9].

$$\tau_{\mathrm{U}}^{-1} = \frac{\hbar\gamma^2}{M{\nu_s}^2\theta_{\mathrm{D}}}\omega^2 T\exp\left(-\frac{\theta_{\mathrm{D}}}{T}\right) \tag{2.47}$$

where $M$ is the average atomic mass and $\gamma$ is the Grüneisen parameter. At sufficiently high temperatures (higher than the Debye temperature), $\omega$ is a constant and nearly all the phonons shall gain enough energy to participate in this process. Therefore the scattering rate of U-process is mainly dependent on the number of phonons, which is in proportion to the temperature. Based on Eqs. (2.44) and (2.47), at high temperatures where U-process is dominant, the lattice thermal conductivity is inversely proportional to the temperature.

The third term on the right of Eq. (2.45) is the point-defect scattering. This process involves two parts: the mass and strain fluctuations. When lattice distortion induced by element doping is very small, the effect of the strain fluctuations is far less than the mass fluctuations. According to the studies by Callaway et al. [10,11] in the 1960s on thermal conductivity of solid solutions, the relaxation time determined by mass fluctuations is

$$\tau_{\mathrm{D}}^{-1} = \frac{V}{4\pi v_{\mathrm{s}}^3}\omega^4\sum f_{\mathrm{i}}(\overline{m} - m_i\overline{m})^2 \tag{2.48}$$

where $V$ is the average atomic volume, $f_{\mathrm{i}}$ is the atomic ratio of the atom with the mass of $m_i$, and $\overline{m}$ is the average atomic mass. Point-defect scattering is effective in suppressing heat transport and enhancing TE performance.

$1/\tau_{\mathrm{r}}$ in Eq. (2.45) represents the resonant scattering of phonons. In filled skutterudites and clathrates, loosely bonded atoms resonantly scatter phonons and reduce the lattice thermal conductivity. Based on the empirical formula proposed by Pohl [12], the relaxation time is

$$\tau_{\mathrm{r}}^{-1} = \frac{C\omega^2}{(\omega^2 - \omega_0^2)^2} \tag{2.49}$$

where $C$ is a constant that is proportional to the density of resonant defects, $\omega$ is the phonon frequency, and $\omega_0$ is the local resonant vibration frequency of the loosely bonded atoms. Nonetheless, there is still some controversy on the quantitative description of this mechanism.

In addition to the typical scattering mechanisms mentioned earlier, there are other processes such as charge-carrier scattering and nanoparticle scattering, and the latter will be discussed in Section 2.4.

Thermal transport in real crystals is a collective effect of various processes, and each one delivers different contributions to the scattering for phonons having different frequencies and/or at different temperatures. For example, grain-boundary scattering is dominant at low temperatures; point-defect and resonant scattering contribute significantly at mid-to-high temperatures; while the effect of U-process is significant and prevalent at high temperatures. In order to intensify phonon scattering over the whole temperature range and reduce thermal conductivity to the largest extent, the collective or synergetic effects are expected by introducing multiple scattering processes.

Roufosse and Klemens [13] firstly proposed the concept of minimum lattice thermal conductivity. They claimed that, in an amorphous which is fully disordered in atomic scale, the MFP of phonons approaches to the phonon wavelength or the half of phonon wavelength, and the minimum thermal conductivity shall be obtained. In 1979, Slack et al. developed the theory defining the minimum thermal conductivity in solids [14]. Based on computational and experimental work, Slack [14] and Cahill [7,15] found that in some crystalline materials, even if the long-range order structure, the MFP could reach the minimum values if the constituent atoms vibrate in a nearly independent Einstein mode to provide enough intense scattering to phonons. Numerous experimental data revealed that the magnitude of minimum MFP is around the minimum atomic distance in the crystal lattice. Therefore it is an important way to introduce various phonon scattering processes approaching the minimum lattice thermal conductivity for enhancing TE performance.

### 2.2.4 $\beta$ factor as a performance indicator for thermoelectric materials

As mentioned earlier, studies on the optimization of TE performance has been focused on tuning carrier concentrations by element doping. This strategy has brought about a great enhancement of $ZT$. Furthermore, some basic features of performance indicators for TE materials have been concisely summarized. According to the SPB model, the power factor $S^2\sigma$ is determined by the effective mass and mobility [16,17]

$$S^2\sigma \propto \mu m_{\mathrm{DOS}}^{*3/2} \tag{2.50}$$

Thus a high power factor calls for a large effective mass and a high mobility of carriers. Meanwhile, the effective mass is proportional to the change rate of the DOS near the Fermi level. If additional carrier pockets can be introduced around the Fermi level, DOS shall be increased, leading to an enhancement of the Seebeck coefficient and power factor. It is noted that the effective mass here embraces the contribution of band effective mass ($m^* = d^2E/dk^2$) and valley degeneracy as shown in Eq. (2.33). In solids, the mobility is inversely proportional to the band's effective mass but not affected by band degeneracy. Therefore, multivalley band structures are favorable to high electrical performance.

Based on the electrical parameters under SPB at the nondegenerate limit, $ZT$ can be written as

$$ZT = \frac{[\eta-(\lambda+2)]^2}{[\beta\exp(\eta)]^{-1}+(\lambda+2)} \tag{2.51}$$

where $\beta$ is the key parameter that determines the maximum $ZT$. This parameter is first proposed by Chasmar and Stratton [18]

$$\beta = \left(\frac{k}{e}\right)^2 \frac{\sigma_0 T}{\kappa_L} = \left(\frac{k}{e}\right)^2 \frac{2e(k_B T)^{5/2}}{(2\pi)^{3/2}\hbar^3} \frac{\mu_0 m_{DOS}^{*3/2}}{\kappa_L} \tag{2.52}$$

Although this formula is derived for the nondegenerate system, the fundamental relationship between $\beta$ and transport parameters still holds for the degenerate case, that is,

$$\beta \propto \frac{\mu_0 m_{DOS}^{*3/2}}{\kappa_L} \tag{2.53}$$

when the scattering parameter $\lambda$ is set to be 0, the relationship between $ZT$ and $\beta$ is shown in Fig. 2.3. The optimal $\eta$ varies within only 1 $k_B T$ for a certain material, and ranges between $-2$ and 1 for various materials with different $\beta$ values. The larger the $\beta$ factor, the larger the upper limit of $ZT$, and the optimal $\eta$ tends to be negative values approaching

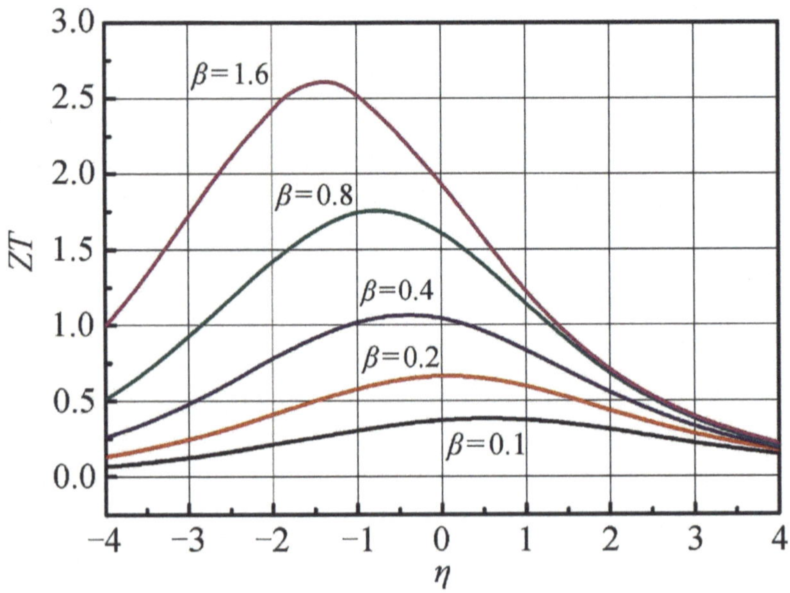


FIGURE 2.3 Dependence of $ZT$ on $\beta$ and reduced Fermi level $\eta$ [4].

nondegenerate intrinsic semiconductors. Most of the state-of-the-art TE materials are weakly degenerate semiconductors with $\beta$ values around 0.4.

Generally, materials with large $\beta$ and high TE performance are featured with [4,6,19,20]: (1) the crystal structures adopt a high symmetry that ensures a high band degeneracy or valley degeneracy near the Fermi level; (2) the compounds are constituted by heavy elements via covalent bonding, which maximizes the product of the effective mass and mobility; (3) the band gaps are around 10 $k_BT$; (4) a complex crystal structure is desired for an intrinsically low lattice thermal conductivity. These characters and the $\beta$ factor have become an effective criterion for searching new TE materials, and several typical approaches standing on the basic transport theories have been developed to optimize TE performance.

## 2.3 Approaches to optimize thermoelectric performance

It is difficult to meet all the requirements in one material for high $ZT$ as discussed in the Section 2.2. It is why high-performance TE materials have been limited to few classic systems despite the existence of vast semiconductors. Fortunately, the basic understanding of the principle transport theory described earlier really provides practical ways toward higher TE performance and exploring new TE materials. At first, we can achieve a high $\mu m_{DOS}^{*3/2}$ through optimizing the DOS around the Fermi level without introducing extra scattering processes. For example, it is easy to realize band degeneracy or introduce resonant level by element alloying and/or doping. Furthermore, we can introduce multiple phonon scattering processes to suppress the lattice thermal conductivity. Some typical approaches will be raised and discussed in detail in this section.

### 2.3.1 Band convergence

According to Wilson and Mott's derivation of Eqs. (2.17)–(2.20), the Seebeck coefficient and electrical conductivity are correlated as

$$S = \frac{\pi^2 k_B^2 T}{3q}\left\{\frac{d[\ln\sigma(E)]}{dE}\right\}_{E=E_F} = \frac{\pi^2 k_B^2 T}{3q}\left\{\frac{dn(E)}{ndE} + \mu\frac{d\mu(E)}{\mu dE}\right\}_{E=E_F} \tag{2.54}$$

For the most cases, improving Seebeck coefficient is more effective to obtain high $ZT$. Eq. (2.54) indicates two ways to enhance the Seebeck coefficient. One is to increase $d\mu(E)/dE$ at Fermi level, which can be realized via intensifying carrier scattering process. The other one is to enhance $dn(E)/dE$ at $E_F$, which can be obtained by increasing the local DOS. Increasing the valley degeneracy $N_v$ around the Fermi level can enhance the DOS effectively, and therefore it is a good way to optimize the power factor.

Although the relationship between the power factor and the band structure can be obtained via combining first-principle calculations and SPB model, it is still rather difficult to provide an accurate analytical description of the power factor. For multiband systems, Goldsmid proposed that TE performance can be enhanced when the bands converge, which can be explained by the equations $\sigma = \sigma_1 + \sigma_2$ and $S = (S_1\sigma_1 + S_2\sigma_2)/(\sigma_1 + \sigma_2)$ [4]. For two valence or conduction bands with different effective masses and mobilities, the power factor reaches the maximum value when the contribution from the two bands is comparable. Many classic TE materials exhibit band convergence characters. For example, the energies of L and Σ bands of PbTe shift with temperature and converge near the Fermi level at 800K (Fig. 2.4A), leading to an enhanced TE performance. In addition to the convergence of bands at different positions of the Brillouin zone, the high degeneracy of Σ bands can also enhance the power factor [21]. Temperature-induced band convergence in PbTe is a special case, and it is more common to approach band convergence by tuning composition in solid solutions. Typically, this is easily applicable to material systems with a large solubility whose band structures are sensitive to the composition, such as $Mg_2Si_{1-x}Sn_x$ and half-Heusler compounds [22].

PbTe, $Mg_2Si$, and half-Heusler alloys possess a cubic structure, and the high symmetry directly brings about the band convergence or multivalley bands. In contrast, in noncubic systems, this feature can be also realized by tuning the lattice parameters. Taking the diamond-like compounds as an example [23], when the distortion parameter $\delta$ $(=c/2a)$

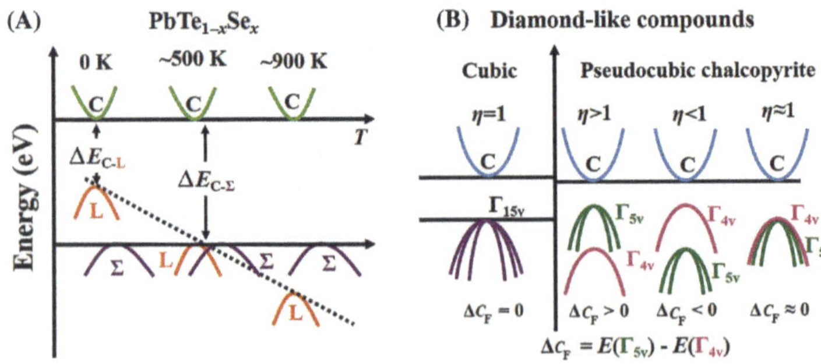


**FIGURE 2.4** (A) Convergence of L and Σ bands induced by temperature in $PbTe_{1-x}Se_x$; (B) schematic relationship between the band split parameter $\Delta C_F$ and structure distortion parameter $\delta$ in tetrahedral diamond-like compounds.

approaches 1, the split bands due to the crystal field effect will converge, leading to high power factors and $ZT$ values.

### 2.3.2 Electron resonant states

According to Eq. (2.54), band convergence can be realized not only by tuning the structure and composition but also by introducing additional energy levels to increase the DOS and Seebeck coefficient. For example, Ga, In, and Tl are reported to introduce resonant levels around the Fermi level of PbTe [24,25]. The Seebeck coefficient was doubled in $1-2$ at% Tl-doped PbTe. Appreciable enhancement in Seebeck coefficient was also observed in Al-doped PbSe [26] and In-doped SnTe [27] due to the resonant state. However, the resonant state will also strongly scatter carriers and deteriorate the electrical conductivity, which must be considered and balanced when employing this strategy. Fig. 2.5 shows the schematic depiction of DOS where the sharp increase near the Fermi level is favorable to a high Seebeck coefficient.

### 2.3.3 Alloying

Alloying refers to introducing isoelectronic defects to the lattice. Due to the difference in mass (mass fluctuations), size, and bonding (strain fluctuations) between the guest and host atoms, phonons are strongly scattered by these point defects, thus lowering the lattice thermal conductivity. This is a conventional yet effective approach to optimize TE performance.

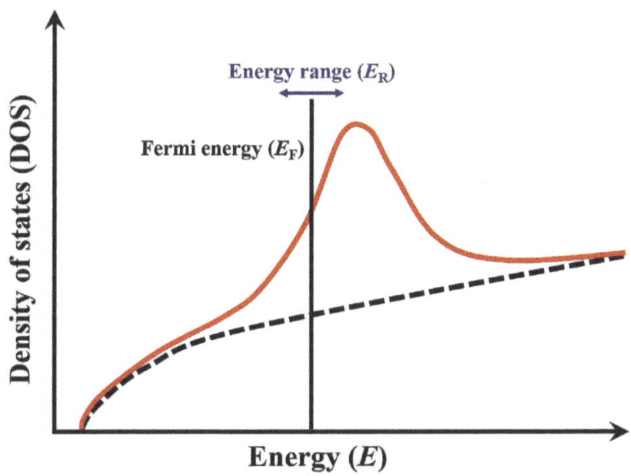


FIGURE 2.5 Schematic density of state of electrons.

In the mid-20th century, several models were established to describe the effect on lattice thermal conductivity of alloys ($\kappa_L$) [10,11,28], among which Callaway and Von Baeyer's [10] model has been widely adopted thereafter. Considering only the Umklapp and point-defect scattering of phonons, lattice thermal conductivities of the real crystals ($\kappa_L$) and ideally pure crystals ($\kappa_L{}^P$) have the following relation:

$$\frac{\kappa_L}{\kappa_L^P} = \frac{\tan^{-1} u}{u} \tag{2.55}$$

$$u^2 = \frac{\pi^2 \theta_D \Omega}{h v^2} \kappa_L^P \Gamma_{expt} \tag{2.56}$$

where $u$, $\theta_D$, $\Omega$, $v$, $h$, and $\Gamma_{expt}$ are disorder parameter, Debye temperature, average atomic volume, Planck's constant, and the experimental disorder scattering parameter, respectively.

According to Slack [28] and Abeles [11], $\Gamma = \Gamma_m + \Gamma_s$ has two parts: the mass fluctuation ($\Gamma_m$) and strain fluctuation ($\Gamma_s$). The average atomic mass and radius can be expressed as

$$\overline{M_i} = \sum_k f_i^k M_i^k \tag{2.57}$$

$$\overline{r_i} = \sum_k f_i^k r_i^k \tag{2.58}$$

The total mass fluctuation can be defined as the sum of the mass fluctuation of each lattice site

$$\Gamma_M = \frac{\sum_{i=1}^{n} c_i \left(\frac{\overline{M_i}}{\overline{\overline{M}}}\right)^2 \Gamma_M^i}{\sum_{i=1}^{n} c_i} \tag{2.59}$$

where $c_i$ is the number of the $i$th sublattice site. $\Gamma_M^i$ is related to the mass and fraction of this sublattice.

$$\Gamma_M^i = \sum_k f_i^k \left(\frac{M_i^k}{\overline{M_i}}\right)^2 \tag{2.60}$$

The average atomic mass is

$$\overline{\overline{M}} = \frac{\sum_{i=1}^{n} c_i \overline{M_i}}{\sum_{i=1}^{n} c_i} \tag{2.61}$$

For the two different atoms at the $i$th sublattice with the mass of $M_i^1$ and $M_i^2$ and fraction of $f_i^1$ and $f_i^2$, the following equations hold:

$$f_i^1 + f_i^2 = 1 \text{ and } \overline{M_i} = f_i^1 M_i^1 + f_i^2 M_i^2 \tag{2.62}$$

Then Eq. (2.59) becomes

$$\Gamma_M = \frac{\sum_{i=1}^{n} c_i \left(\frac{\overline{M_i}}{\overline{\overline{M}}}\right)^2 f_i^1 f_i^2 \left(\frac{M_i^1 - M_i^2}{\overline{M_i}}\right)^2}{\sum_{i=1}^{n} c_i} \tag{2.63}$$

Impurity atoms are different from host atoms in size, coupling force, disorder in strain field, etc. Using the continuous elastic medium approach, Steigmeier [29] and Abeles [11] proposed a quantitative model to describe the change of rigid constants induced by impurity atoms that have different atomic sizes with the host ones and derived a simplified expression of $\Gamma_s$

$$\Gamma_s = \frac{\sum_{i=1}^{n} c_i \left(\frac{\overline{M_i}}{\overline{\overline{M}}}\right)^2 f_i^1 f_i^2 \varepsilon_i \left(\frac{r_i^1 - r_i^2}{\overline{r_i}}\right)^2}{\sum_{i=1}^{n} c_i} \tag{2.64}$$

where $\overline{r_i} = f_i^1 r_i^1 + f_i^2 r_i^2$; $\varepsilon_i$ is the tunable parameter of the $i$th sublattice, which is mainly determined by the Grüneisen parameter $\gamma$ and reflects the effect of anharmonicity [12]. The value of $\varepsilon$ ranges from 10 to 100. This model has successfully explained the experimental lattice thermal conductivity of many solid solutions such as $PbTe_{1-x}Se_x$ [30], $CuGa_{1-x}In_xTe_2$ [31], $Mg_2Si$-based solid solutions [32], and $CoSb_3$-based alloys [33]. As shown in Fig. 2.6 as for the $PbTe_{1-x}Se_x$ system, the lattice thermal conductivity is lowered by 45% at room temperature and 20% at 800K when compared to pristine PbTe.

## 2.3.4 Phonon resonant scattering

In crystalline materials, low-frequency acoustic phonons usually possess high group velocity. Although their DOS is not large, they carry a majority of the vibration energy. As shown in Fig. 2.7B, if a series of independent resonant vibration modes are introduced at the low-frequency range, phonons around this frequency range will be intensely scattered, leading to a great suppression of lattice thermal conductivity. Typical examples of this mechanism include caged compounds such as filled skutterudites and clathrates. In 1996 Sales [34] filled La and Ce into the voids of $(Fe_xCo_{1-x})_4Sb_{12}$ and effectively lowered the lattice thermal conductivity. As shown in Fig. 2.7, the unit cell of skutterudites has two large voids, and the chemical formula can be written as $\square_2M_8X_{24}$ (M = Co, Rh, Ir; X = P, As, Sb; $\square$ = void). The unit cell of the type-I clathrate has eight large voids, and the formula is $\square_8E_{46}$ (E = Si, Ge, Sn; $\square$ = void). Filled atoms (mainly metals) can be introduced into these

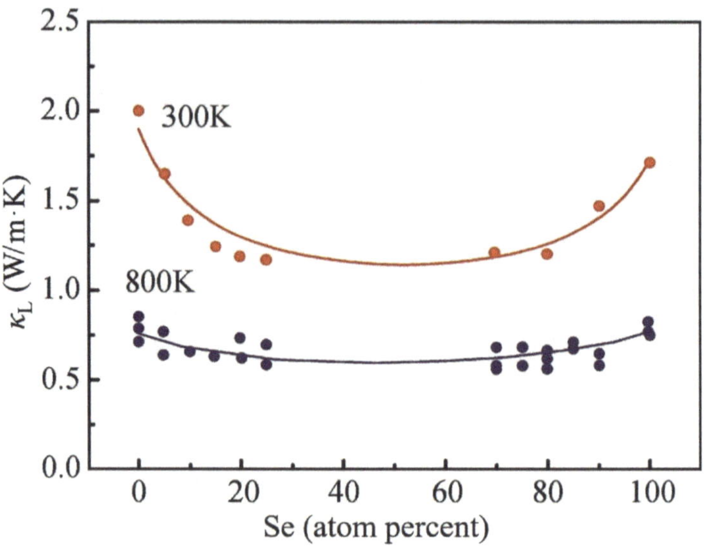


**FIGURE 2.6** Lattice thermal conductivity varying with Se content in $PbTe_{1-x}Se_x$ solid solutions.

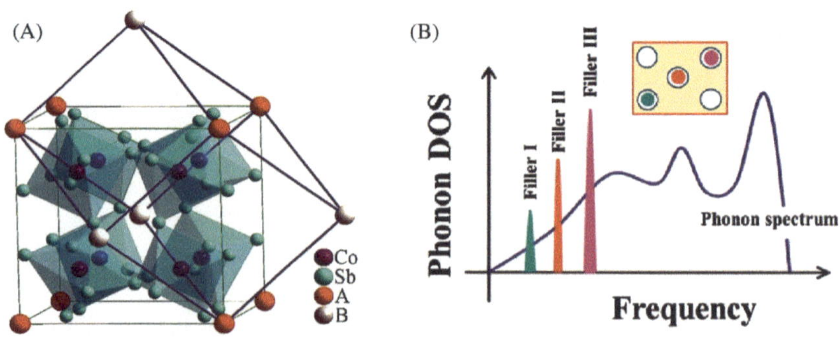


**FIGURE 2.7** (A) Crystallographic unit cell (*thin black line*) and physical primitive cell (*wide blue line*) of filled $CoSb_3$; (B) schematic phonon DOS of caged compounds.

voids. Since the size of these voids is pretty large, the filled cations exhibit a large vibration amplitude compared to the frame host atoms. Later, Shi et al. [35] illustrated the factors influencing the stability and filling limit of the cations and calculated the effect of vibration frequency on phonon scattering. The vibration of these filled atoms in the voids is often considered as "rattlers" contributing to low-frequency resonant modes. This low-frequency localized resonant vibration disturbs the original phonon modes, thus significantly reducing the lattice thermal conductivity. The resonant frequency of the filling atom is determined by its mass, radius, and charge, which is different for different filling atoms. Based on the understanding of the resonant behavior of filled atoms, Yang et al. [36] proposed the broadband phonon scattering mechanism, that is, filling multiple atoms with different resonant frequencies can more effectively scatter phonons with a wide range of frequencies and reduce the lattice thermal conductivity. Following this principle, several high-performance dually and multiply filled skutterudites have been successfully synthesized to show much enhanced *ZT*s [37–40].

Considering the features in electrical and thermal transports for high-performance TE materials, Slack proposed in the 1990s the design strategy of "phonon glass and electron crystal (PGEC)" [19], that is, ideal TE materials should possess glass-like phonon transport properties and crystal-like charge transport properties. The discovery of a series of compounds with local vibrations such as filled skutterudites and clathrates has corroborated the PGEC idea as an important strategy for the exploration of high-performance TE materials.

## 2.3.5 Liquid-like thermoelectric materials

The earlier approaches to suppress lattice thermal conductivity are majorly focused on reducing the MFPs of phonons via multiscale structure regulation following the concept of PGEC. In solids, however, as discussed in Section 2.2.3, the structure-induced phonon scattering shall make the MFPs approaching to the minimum value in the fully disordered glass and thus determines minimum lattice thermal conductivity. As discussed earlier, the lattice thermal conductivity in solids can be expressed as $\kappa = 1/3C_V vl$. In addition to reducing the MFPs via structure regulation such as alloying, resonant scattering, and grain-boundary scattering, $\kappa_L$ can also be reduced by lowering $C_v$ and $v$. Although the two parameters are usually constants in solids, recent research [41] found that the migration of ions in ionic conductors can reduce the heat capacity. In conventional solids, the lattice vibration is composed of both longitudinal and transverse modes, and the volumetric heat capacity is $3Nk_B$. In contrast, only longitudinal waves can propagate in liquids due to the weak molecular bonding and the transverse modes cannot propagate, so the theoretical volumetric heat capacity is $2Nk_B$, which is 2/3 of the solid's value. In $Cu_{2-\delta}X$ (X = S, Se, Te) materials [41–43], there are two sublattices. The rigid FCC X-framework provides the channel for carrier transport, and Cu ions situate at the interstitial sites of X-framework and can freely move at high temperatures. The migration of Cu ions not only strongly scatters the phonons to lower the MFPs but also partially reduces transverse modes.

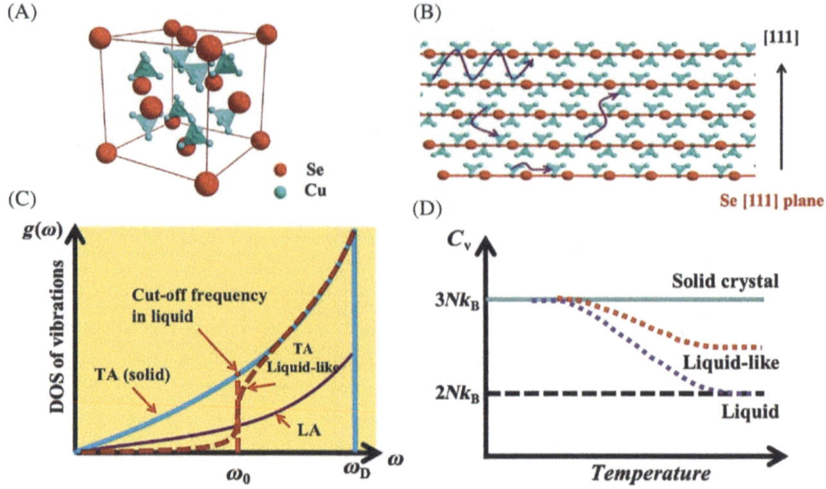


FIGURE 2.8 (A) Unit cell of cubic β-$Cu_2Se$ where only 8c and 32f sites are shown with Cu atoms; (B) projected plane representation of the crystal structure along the cubic direction. The arrows indicate that the Cu ions can freely migrate among the interstitial sites; (C) phonon DOS and (D) heat capacity in liquid-like materials.

As shown in Fig. 2.8, when the phonon frequency is lower than the cut-off frequency, transverse phonons will be softened or even diminished, which leads to an obvious decrease in the heat capacity with temperature, even lower than the Dulong–Petit limit at high temperatures. The damping effect of the shear waves of Cu ion sublattice breaks the limitation of minimum lattice thermal conductivity in amorphous state. In addition to $Cu_{2-\delta}X$, extremely low thermal conductivities and ultrahigh $ZT$ values have also been found in many other Cu- or Ag-containing materials. In analogy to PGEC, the coexistence of solid and liquid-like sublattices endows these materials with the feature of "phonon-liquid electron-crystal" that has turned a new strategy to develop TE materials with low lattice thermal conductivity and high performance.

## 2.4 Thermoelectricity in nanoscale and nano-thermoelectric materials

In the 1990s, Hicks and Dresselhaus theoretically predicted that [44–46] when the geometry of the material reaches nanoscale, obvious dimensional and size effects shall be present in the transports of electrons and phonons, which provides more freedom for performance tuning, thus going beyond the realm of the traditional strategies. It is found by calculation that when the size of the material on a certain dimension is as low as comparable to the MFPs of electrons and phonons, $ZT$ shall greatly increase. The followed experiments were presented one decade after this theoretical work, and since then the concept of nano-thermoelectrics has substantially come to establishment and widely adoptable. This section will give a brief discussion on the transport behavior in nanoscale and strategies for developing nanostructure and nanocomposite TE materials. The fabrication and performance of nanostructured TE materials will be demonstrated in detail in Chapter 5, Low-Dimensional and Nanocomposite Thermoelectric Materials.

### 2.4.1 Carrier transport in nanoscale

According to the carrier transport theory described in Section 2.2, the dispersion relation of carriers in three-dimensional (3D) bulk materials satisfies the parabolic band model [Eq. (2.32)]. When the material is compressed along $z$-axis, the system becomes a two-dimensional (2D) quantum well. Suppose the width of the well is a constant $a$, and then the carrier energy along $z$-axis can be approximated as a constant. Now the dispersion relation [Eq. (2.32)] transforms into

$$\varepsilon(k_x, k_y) = \frac{\hbar^2 {k_x}^2}{2m_x} + \frac{\hbar^2 {k_y}^2}{2m_y} + \frac{\hbar^2 \pi^2}{2m_z a^2} \tag{2.65}$$

If two or three axes are compressed simultaneously, that is, the system evolves into a 1D nanowire or 0D nanodot, the dispersion relation can be simplified in a similar way as 2D quantum well. The DOS curves for different dimensionality are shown in Fig. 2.9.

The DOS of 2D materials is obviously different from that of 3D materials; the power on energy changes from 1/2 to 0, that is, $g(E) = (m_x m_y)^{1/2}/\pi\hbar^2 E^0$. If we still suppose that the carriers are mainly scattered by acoustic phonons, the TE parameters in 2D materials are [44,45]

$$S = \mp \frac{k_B}{e}\left(\eta - \frac{2F_1(\eta^*)}{F_0(\eta^*)}\right) \tag{2.66}$$

$$\sigma = \frac{1}{2\pi a}\left(\frac{2k_B T}{\hbar^2}\right)(m_x m_y)^{1/2} F_0(\eta^*) e\mu \tag{2.67}$$

$$\kappa = \frac{\tau\hbar^2}{4\pi a}\left(\frac{2k_B T}{\hbar^2}\right)^2 \left(\frac{m_y}{m_x}\right)^{1/2} k_B\left(3F_2(\eta^*) - \frac{4{F_1}^2(\eta^*)}{F_0(\eta^*)}\right) \tag{2.68}$$

Here $F$ is still the Fermi integral function, and $a$ is the width of the quantum well with

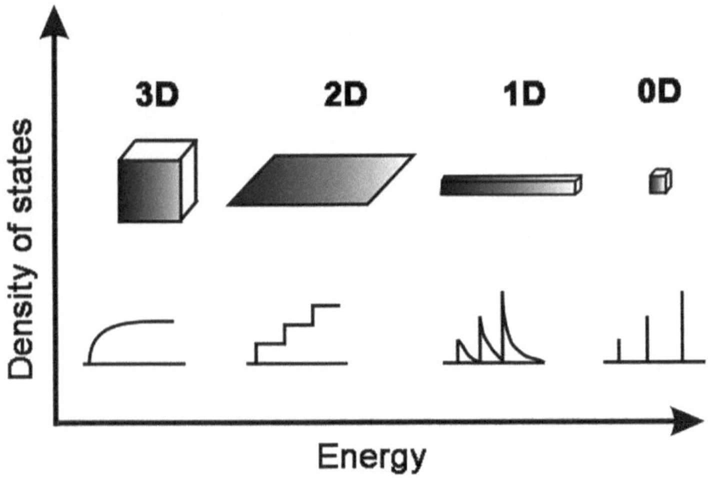


FIGURE 2.9 DOS versus energy for materials with different dimensionalities.

$$\eta^* = \frac{\left(\eta - \frac{\hbar^2\pi^2}{2m_z a^2}\right)}{k_B T} \tag{2.69}$$

The performance of 2D materials, $Z_{2D}T$ is

$$Z_{2D}T = \frac{\left(\frac{2F_1}{F_0} - \eta^*\right)^2 F_0}{\frac{1}{\beta'} + 3F_2 - \frac{4F_1^2}{F_0}} \tag{2.70}$$

where

$$\beta' = \frac{1}{2\pi a}\left(\frac{2k_B T}{\hbar^2}\right)(m_x m_y)^{1/2}\frac{k_B^2 T\mu}{e\kappa_L} \tag{2.71}$$

It is seen that $Z_{2D}T$ is determined by the reduced Fermi level and $\beta'$ factor. Compared to 3D materials, there is an additional degree of freedom for regulation in 2D materials: the width of the quantum well that can significantly affect the value of $\beta'$. When the width is in nanoscale, $\beta'$ of 2D materials is obviously higher than $\beta$ of 3D counterparts, yielding an even higher $Z_{2D}T$.

Similarly, when the dimensionality further decreases to 1, the dispersion relation is confined to constants along two directions and only varies with the wave factor along one certain direction:

$$\varepsilon(k_x) = \frac{\hbar^2 k_x^2}{2m_x} + \frac{\hbar^2\pi^2}{2m_y a^2} + \frac{\hbar^2\pi^2}{2m_z a^2} \tag{2.72}$$

Here the power index of DOS on energy is further reduced to $-1/2$, showing a jagged shape. The TE parameters can be then expressed as

$$S = \mp \frac{k_B}{e}\left(\eta - \frac{3F_{1/2}(\eta^*)}{F_{-1/2}(\eta^*)}\right) \tag{2.73}$$

$$\sigma = \frac{1}{\pi a^2}\left(\frac{2k_B T}{\hbar^2}\right)^{1/2}(m_x)^{1/2}F_{-1/2}e\mu \tag{2.74}$$

$$\kappa = \frac{2\tau}{\pi a^2}\left(\frac{2k_B T}{\hbar^2}\right)^{1/2}(m_x)^{1/2}k_B^2 T\left(\frac{5}{2}F_{3/2} - \frac{9F_{1/2}^2}{2F_{-1/2}}\right) \tag{2.75}$$

with

$$\eta^* = \frac{\left(\eta - \frac{\hbar^2\pi^2}{2m_Y a^2} - \frac{\hbar^2\pi^2}{2m_z a^2}\right)}{k_B T} \tag{2.76}$$

In a similar way, the performance of 1D materials, $Z_{1D}T$, can be written as

$$Z_{1D}T = \frac{\frac{1}{2}\left(\frac{3F_{1/2}}{F_{-1/2}} - \eta^*\right)^2 F_{-1/2}}{\frac{1}{\beta''} + \frac{5}{2}F_{3/2} - \frac{9F_{1/2}^2}{2F_{-1/2}}} \tag{2.77}$$

where

$$\beta'' = \frac{2}{\pi a^2}\left(\frac{2k_B T}{\hbar^2}\right)^{1/2} {m_x}^{1/2}\frac{k_B^2 T\mu}{e\kappa_L} \tag{2.78}$$

It is noted that the size effect on $\beta''$ is more remarkable in 1D materials than 2D materials, which is reversely proportional to the square of the scale of the quantum well. If taking into account the size effect on phonon transport, $ZT$ would be considerably enhanced in low-dimensional materials. Fig. 2.10 shows the calculated relationship between $ZT$ and size for 2D and 1D $Bi_2Te_3$ materials [45]. For 2D materials, when the scale of the quantum well is below 40 Å (1 Å = $10^{-10}$ m), $ZT$ values along all the directions are higher than bulks. When the size is reduced to 10 Å, $ZT$ of the 2D films would reach 1.5 or even 2.3, which is three or five times the value in 3D bulks. In 1D nanowire, when the size is lowered to 10 Å, $ZT$ would be enhanced to above 6, being 10 times higher than bulks. Because the phonon MFP in $Bi_2Te_3$ is around 10 Å, the performance enhancement benefits mainly from the improvement of electrical properties by quantum effect. If the size of quantum well can be further reduced below the MFP of

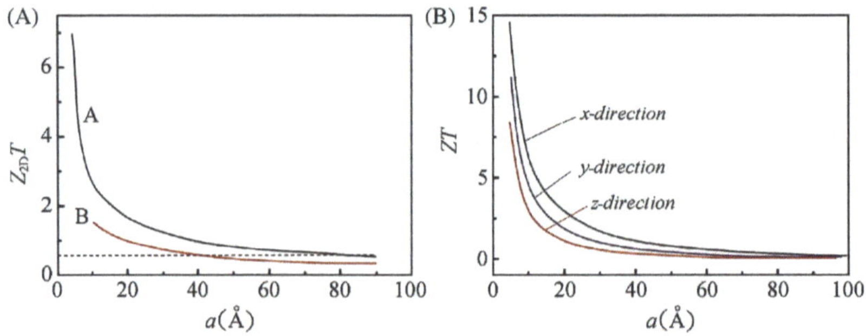


FIGURE 2.10 *ZT* varying with the width of the quantum well along different directions for (A) $Bi_2Te_3$ 2D films (letter A denotes the $a_0-b_0$ plane orientation, and letter B denotes the $a_0-c_0$ plane orientation) and (B) 1D nanowires.

phonons, the predicted enhancement of *ZT* would be much significant, as shown in Fig. 2.10.

## 2.4.2 Heat transport in nanoscale

When the material's size is lowered to the nanoscale or when nanostructures are introduced in bulk materials, thermal transport will be appreciably influenced. The phonon scattering shall be significantly strengthened when the size of materials or secondary/interface phases is comparable to that of MFPs of phonons, thus greatly reducing the lattice thermal conductivity.

Dealing with the thermal conductivity in low-dimensional materials, Ren and Dow [47] simulated the thermal transport in ideal superlattice structures using the Boltzmann quantum transport equations as early as in 1982. They predicted that thermal conductivity in superlattice will be 20% lower than that in the bulks. Phonon transport in 2D thin films is schematically shown in Fig. 2.11. In 2000, Venkatasubranmanian [48] experimentally found that the lattice thermal conductivity of $Bi_{0.5}Sb_{1.5}Te_3$ superlattice with the period of 50Å is as low as 0.22 W/mK. Nowadays, several theoretical models have been adopted to describe thermal transports in superlattice including phonon confinement theory by Balandin [49] and phonon-interface scattering theory by Chen [50,51]. The latter treats the phonon transport in low-dimensional materials as the combined effects of scattering and transmission at the interface, which is a relatively mature theory and thus be discussed briefly here in this section.

Take thermal transports in a single-layer 2D material as an example as shown in Fig. 2.11. If the interactions at the boundaries are primarily

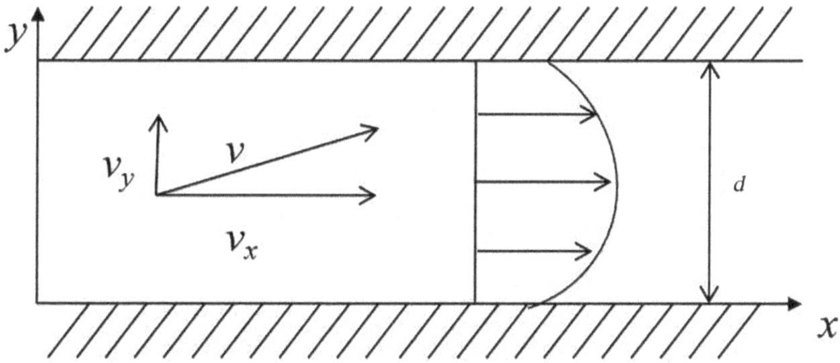


FIGURE 2.11 Phonon propagation in 2D thin films.

specular, elastic scattering, the in-plane lattice thermal conductivity is derived by Chen as follows:

$$\frac{\kappa}{\kappa_B} = 1 - \frac{3}{8\xi}[1 - 4(\Psi_3(\xi) - \Psi_5(\xi))] \tag{2.79}$$

where $\kappa_B$ is the thermal conductivity of bulk materials; $\xi = d/l$, where $d$ is the film thickness, and $l$ is the MFP of phonons; $\psi$ is the series integral.

$$\Psi_n(x) = \int_0^1 \mu^{n-2} e^{-x/\mu} d\mu$$

If there is only partially specular scattering that is inelastic, Eq. (2.79) can be further modified as

$$\frac{\kappa}{\kappa_B} = 1 - \frac{3(1-p)}{2\xi}\int_0^1 (\mu - \mu^3)\frac{1 - e^{-\xi/\mu}}{1 - pe^{-\xi/\mu}} d\mu \tag{2.80}$$

where $p$ is the specularity, and $p = 1$ means the totally specular. Since there always exists inelastic scattering at real interfaces, thermal conductivity depends not only on the film thickness but also on the interface properties. If the interface scattering of phonons is specular, the thermal conductivity of the superlattice is close to that of the bulks; if there exists some diffuse scattering, the thermal conductivity will sharply decrease. In superlattices composed of multiple thin films, thermal conduction within the plane can be regarded as the combined effect of each layer, and the thermal conductivity can be expressed as

$$\kappa = \sum_{i=1}^{n} x_i \kappa_i \left[1 - 1.5p\left(1 - \frac{a_j}{a_i}\right)\frac{A_{si}}{\xi_i} - 1.5(1-p)\frac{A_{di}}{\xi_i}\right] \tag{2.81}$$

Along the normal direction, the effect of interface on phonon scattering and diffuse is more intense. Combining the Fourier's law and the transmission effect, Chen proposed the formula of $\kappa$ in dual superlattice thin films:

$$\frac{\kappa}{\kappa_B} = \frac{3d/(4l)}{\frac{1-(\tau'_{12}+\tau'_{21})/2}{\tau''_{21}} + \ldots + \frac{3d}{4l}} \tag{2.82}$$

where $\tau'_{12}$ and $\tau'_{21}$ are phonon transmission coefficients at the interface between different superlattice layers, $d$ is the film thickness, and $l$ is the MFP of phonons. Therefore, in superlattice films or bulks with superlattice structures, phonon transport is related to both interface scattering and transmission between the layers. When the transmission coefficient is 1 for all the interfaces, phonons can completely transmit the interface without scattering or attenuation, and the thermal conductivity is equal to that of the bulks. In reality, the transmission coefficient of phonons at the interface is always smaller than 1. The smaller the coefficient, the stronger interrupt of the phonon transport by the layers. Also, the interface is not perfectly specular, and diffuse scattering is always present. All these factors will reduce the thermal conductivity. The key to the above model on 2D film thermal conductivity is the description of the interfaces. Therefore the model can be further modified and extended to 1D nanowires, nanoparticle-dispersing, or nanocrystal systems with nanoscale interfaces. Comprehensive theories on phonon-interface scattering can be referred to refs. [50,51].

### 2.4.3 Nanocrystalline and nanocomposite thermoelectric materials

When the grain size approaches nanoscale or nanoparticles are dispersed in bulk materials, the density of grains or phase boundaries is significantly increased. The originally periodical potential field in a grain will be interrupted at the boundaries, and thus both the transport of carriers and phonons will be disturbed or scattered. For example, the distribution of electrons at boundaries is different from that within the grains. Such a difference introduces potential barriers at the interface, which exerts extra scattering to carriers, impedes charge transport, and lowers the mobility [52–54]. This process will strengthen the dependence of carrier relaxation time on energy, which therefore enhances the scattering parameter and the Seebeck coefficient [19,55,56]. In addition, the barrier can scatter low-energy carriers and enhance the average energy of carriers. Such an energy-filtering effect shall also enhance the Seebeck coefficient [57–60]. Nowadays, the description on the effect of nanophase and nanoscale grain boundary on electrical transport is still qualitative, and a mature quantitative description is to be developed.

The understanding about the low-dimensional interface thermal transport can be implemented into nano-bulk materials in principle. The influence of nanocrystals and nanoparticle dispersion is more appreciable and comprehensible on thermal conductivity than on electrical transport. As discussed in Section 2.4.2, nanocrystals and nano-size secondary phase can introduce extra phonon scattering and therefore reduce the lattice thermal conductivity. In solids, heat is conducted mainly via the propagation and interaction of lattice vibrations (phonons), electrons, and photons [61–63]. At high temperatures, photon propagation is the main source of heat conduction, while the acoustic phonons make the dominant contribution at low and medium temperatures. At low temperatures, the average wavelength of phonons is pretty large and phonons can easily bypass the defects without being scattered. With increasing temperatures, the density of phonons increases and the MFP decreases dramatically. In addition, phonons are scattered by the impurities and interfaces because of the nonequilibrium potential around such defects. Since phonons disperse in a wider wavelength than carriers, the scattering to phonon by nanoparticles and/or nano-inclusions is more remarkable than to carriers. In fact, the major role of nanostructuring in TE materials is to selectively scatter the mid-to-long wavelength phonons to reduce the lattice thermal conductivity.

In analogy to point-defect scattering, phonon-drag effect, and resonant scattering, phonon scattering by nano-dispersive phases can also be described by the relaxation time $\tau_{\text{nano}}$ [9]

$$\tau_{\text{nano}}^{-1} = \frac{3xv_{\text{s}}}{(2R)} \tag{2.83}$$

where $x$ and $R$ are the content and the size of nanophases, respectively, and $v_{\text{s}}$ is the phonon velocity in the matrix. The effective lattice thermal conductivity, $\kappa_{\text{eff}}$, around the nanoparticles is derived as

$$\frac{\kappa_{\text{eff}}}{\kappa_{\text{B}}} = \frac{3d/(4l)}{1 + 3d/(4l)} \tag{2.84}$$

It is seen that $\kappa_{\text{eff}}$ depends on the size of nanophases. When the radius of the dispersed particle is comparable to the MFPs of phonons, $\kappa_{\text{eff}}$ can be reduced to 43% of the bulk's.

Since the nano-effect came into wide attention in thermoelectrics, great successes have been obtained in nanostructuring bulk materials: reducing the lattice thermal conductivity via the boundary or interface scatterings and enhancing the Seebeck coefficient via the energy-filtering effect. The selection of suitable nanophases and their dispersions are the top concerns for TE composites. Typically, developing smart fabricating methods to realize controllable dispersion of desired nano-size particles and/or nano-inclusions without agglomeration is the key issue to achieve the synergetic effect of reducing thermal conductivity and optimizing electrical performance.

## References

[1] Z. Wang, Introduction to Statistical Physics, Higher Education Press, Beijing, 1965.
[2] K. Huang, Solid State Physics, Higher Education Press, Beijing, 1966.
[3] P.Y. Yu, M. Cardona, Fundamentals of Semiconductors, Springer, Berlin, 2010.
[4] H.J. Goldsmids, Introduction to Thermoelectricity, Springer, 121 (16) (2016) 339–357.
[5] J. Bardeen, W. Shockley, Deformation Potentials and Mobilities in Non-Polar Crystals, Phys. Rev 80 (1) (1950) 72–80.
[6] E. Conwell, V.F. Weisskopf, Theory of Impurity Scattering in Semiconductors, Physical Review 77 (3) (1950) 388–390.
[7] D.G. Cahill, S.K. Watson, R.O. Pohl, Lower Limit to the Thermal Conductivity of Disordered Crystals, Physical Review B 46 (10) (1992) 6131–6140.
[8] C. Kittel, Introduction to Solid State Physics, John Wiley & Sons Inc, New York, 1996.
[9] J. Yang, L. Xi, W. Qiu, et al., On the tuning of electrical and thermal transport in thermoelectrics: an integrated theory–experiment perspective, NPJ Computational Materials 2 (2016) 15015.
[10] J. Callaway, H.C. Von Baeyer, Effect of point imperfections on lattice thermal conductivity, Physical Review, 120 (120) (1960) 1149–1154.
[11] B. Abeles, Lattice thermal conductivity of disordered semiconductor alloys at high temperatures, Physical Review, 131 (5) (1963) 1906–1911.
[12] R.O. Pohl, Thermal Conductivity and Phonon Resonance Scattering, Physical Review Letter 8 (12) (1962) 481–483.
[13] M. Roufosse, P.G. Klemens, Lattice thermal conductivity of minerals at high temperatures, Journal of Geophysical Research 79 (5) (1974) 703–705.
[14] G.A. Slack, Thermal-conductivity of nonmetals, Bulletin of The American Physical Society 24 (1979) 281.
[15] D.G. Cahill, R.O. Pohl, Lattice vibrations and heat transport in crystals and glasses, Annual Review of Physical Chemistry 39 (1) (1988) 93–121.
[16] G.S. Nolas, J. Sharp, H.J. Goldsmid, Thermoelectrics: Basic Principles and New Materials Developments, Springer, New York, 2001.
[17] H.J. Goldsmid, Applications of Thermoelectricity, Methuen, London, 1960.
[18] R.P. Chasmar, R. Stratton, The thermoelectric figure of merit and its relationship to thermoelectric generator, Journal of electronics and control 7 (1959) 52.
[19] D.M. Rowe, CRC handbook of Thermoelectrics, CRC press, Boca Raton, 1995.
[20] G.D. Mahan, Figure of merit for thermoelectrics, Journal of Applied Physics 65 (4) (1989) 1834.
[21] Y. Pei, X. Shi, A. LaLonde, et al., Convergence of electronic bands for high performance bulk thermoelectrics, Nature 473 (7345) (2011) 66.
[22] W. Liu, X.J. Tan, K. Yin, et al., Convergence of Conduction Bands as a Means of Enhancing Thermoelectric Performance of n-Type $Mg_2Si_{1-x}Sn_x$ Solid Solutions, Physical Review Letters 108 (16) (2012) 166601.
[23] J. Zhang, R. Liu, N. Cheng, et al., High-Performance Pseudocubic thermoelectric materials from non-cubic chalcopyrite compounds, Advanced Materials 26 (23) (2014) 3848–3853.
[24] K. Hoang, S.D. Mahanti, Electronic structure of Ga-, In-, and Tl-doped PbTe: A supercell study of the impurity bands, Physical Review Letters 78 (8) (2008) 085111.

[25] J.P. Heremans, V. Jovovic, E.S. Toberer, et al., Enhancement of thermoelectric efficiency in PbTe by distortion of the electronic density of states, Science 321 (5888) (2008) 554–557.
[26] Q.Y. Zhang, H. Wang, W.S. Liu, et al., Enhancement of Thermoelectric Figure-of-Merit by Resonant States of Aluminum Doping in Lead Selenide, Energy & Environmental Science 5 (1) (2012) 5246–5251.
[27] Q. Zhang, B. Liao, Y. Lan, et al., High thermoelectric performance by resonant dopant indium in nanostructured SnTe, Proceeding of the National Academy of Sciences of the United States of America 110 (33) (2013) 13261–13266.
[28] G.A. Slack, Effect of isotopes on low-temperature thermal conductivity, Physical. Review, 105 (3) (1957) 829–831.
[29] E.F. Steigmeier, The Debye temperatures of III–V compounds, Applied Physics Letters, 3 (1) (1963) 6–8.
[30] H. Wang, A.D. Lalonde, Y.Z. Pei, et al., The criteria for beneficial disorder in thermoelectric solid solutions, Advanced Functional Materials, 23 (12) (2013) 1586–1596.
[31] Y.T. Qin, P.F. Qiu, R.H. Liu, et al., Optimized thermoelectric properties in pseudocubic diamond-like $CuGaTe_2$ compounds, Journal of Material Chemistry A 4 (4) (2016) 1277–1289.
[32] M.B.A. Bashir, S.M. Said, M.F.M. Sabri, et al., Recent advances on $Mg_2Si_{1-x}Sn_x$ materials for thermoelectric generation, Renewable and Sustainable Energy Reviews 37 (3) (2014) 569–584.
[33] G.P. Meisner, D.T. Morelli, S. Hu, et al., Structure and Lattice Thermal Conductivity of Fractionally Filled Skutterudites: Solid Solutions of Fully Filled and Unfilled End Members, Physical Review Letters 80 (16) (1998) 3551–3554.
[34] B.C. Sales, D. Mandrus, R.K. Williams, Filled Skutterudite Antimonides: A New Class of Thermoelectric Materials, Science 272 (5266) (1996) 1325.
[35] X. Shi, W. Zhang, L.D. Chen, et al., Filling Fraction Limit for Intrinsic Voids in Crystals: Doping in Skutterudites, Physical Review Letters 95 (18) (2005) 185503.
[36] J. Yang, W.Q. Zhang, S.Q. Bai, et al., Dual-frequency resonant phonon scattering in $Ba_xR_yCo_4Sb_{12}$ (R = La, Ce, and Sr), Appl. Phys. Lett. 90 (19) (2007) 192111.
[37] H. Shi, Kong, C.P. Li, et al., Low thermal conductivity and high thermoelectric figure of merit in n-type $Ba_xYb_yCo_4Sb_{12}$ double-filled skutterudites, Applied Physics Letters 92 (18) (2008) 182101–182103.
[38] X. Shi, J.R. Salvador, J. Yang, et al., Thermoelectric properties of n-Type multiple-filled skutterudites, International Conferernce on Thermoelectrics 38 (2009) 930.
[39] L. Zhang, A. Grytsiv, P. Rogl, et al., High thermoelectric performance of triple-filled n-type skutterudites (Sr, Ba, Yb)$_y$$Co_4Sb_{12}$, Journal of Physics D: Applied Physics 42 (22) (2009) 225405.
[40] X. Shi, J. Yang, J.R. Salvador, et al., Multiple-Filled Skutterudites: High Thermoelectric Figure of Merit through Separately Optimizing Electrical and Thermal Transports, Journal of American Chemical Society 134 (5) (2012). 2842-2842.
[41] H. Liu, X. Shi, F. Xu, et al., Copper ion liquid-like thermoelectrics, Nature Materials 11 (5) (2012) 422.
[42] Y. He, T. Day, T.S. Zhang, et al., High thermoelectric performance in non-toxic earth-abundant copper sulfide, Advanced Materials, 26 (23) (2014) 3974–3978.
[43] Y. He, T. Zhang, X. Shi, et al., High thermoelectric performance in copper telluride, NPG Asia Materials, 7 (8) (2015) e210.
[44] L.D. Hicks, M.S. Dresselhaus, Thermoelectric figure of merit of a one-dimensional conductor, Physical Review B 47 (24) (1993) 16631–16634.
[45] L.D. Hicks, M.S. Dresselhaus, Effect of quantum-well structures on the thermoelectric figure of merit, Physical Review B 47 (19) (1993) 12727.

[46] L.D. Hicks, T.C. Harman, M.S. Dresselhaus, Use of quantum-well superlattices to obtain a high figure of merit from nonconventional thermoelectric materials, Applied Physics Letters 63 (1993) 3230–3232.
[47] S.Y. Ren, J.D. Dow, Thermal-conductivity of super-lattices, Physics Review B 25 (6) (1982) 3750–3755.
[48] R. Venkatasubramanian, Lattice thermal conductivity reduction and phonon localizationlike behavior in superlattice structures, Physical Review B 61 (4) (2000) 3091–3097.
[49] A.A. Balandin, O.L. Lazarenkova, Mechanism for thermoelectric figure-of-merit enhancement in regimented quantum dot superlattices, Applied Physics Letters 82 (3) (2003) 415–417.
[50] G. Chen, Thermal conductivity and ballistic-phonon transport in the cross-plane direction of superlattices, Physical Review B 57 (23) (1998) 14958–14973.
[51] G. Chen, D. Borca-Tasciuc, R.G. Yang, Nanoscale Heat Transfer//Nalwa H S, Encyclopedia of Nanoscience and Nanotechnology, American Scientific Publishers, 2004.
[52] S.A. Yamini, H. Wang, D. Ginting, et al., Thermoelectric performance of n-type $(PbTe)_{0.75}(PbS)_{0.15}(PbSe)_{0.1}$ composites, ACS Applied Materials and Interfaces 6 (14) (2014) 11476–11483.
[53] J.R. Sootsman, H. Kong, C. Uher, et al., Large enhancements in the thermoelectric power factor of bulk PbTe at high temperature by synergistic nanostructuring, Angewandte Chemie-International Edition, 47 (2008) 8618.
[54] A.F. Ioffe, H.J. Goldsmid, Semiconductor thermoelements and thermoelectric cooling, Inforesearch, London, 1957.
[55] J.P. Heremans, C.M. Thrush, D.T. Morelli, Thermopower enhancement in PbTe with Pb precipitates, Journal of Applied Physics, 98 (6) (2005) 063703.
[56] J. Martin, L. Wang, L.D. Chen, et al., Enhanced Seebeck coefficient through energy-barrier scattering in PbTe nanocomposites, Physical Review B 79 (11) (2009) 5311.
[57] S.V. Faleev, F. Léonard, Theory of enhancement of thermoelectric properties of materials with nanoinclusions, Physical Review B 77 (21) (2008) 214304.
[58] J.P. Heremans, B. Wiendlocha, A.M. Chamoirea, Resonant levels in bulk thermoelectric semiconductors, Energy and Environmental Science, 5 (2) (2012) 5510–5530.
[59] J.H. Yang, H.L. Yip, A.K.Y. Jen, Rational design of advanced thermoelectric materials, Advanced Energy Materials, 3 (5) (2013) 549–565.
[60] J. Zhou, X.B. Li, G. Chen, et al., Semiclassical model for thermoelectric transport in nanocomposites, Physical Review B 82 (82) (2010) 2431–2433.
[61] J.M. Ziman, Electrons and Phonons, Clarendon Press, Oxford, 1960.
[62] J.H. Davies, The Physics of Low-Dimensional Semiconductors: An Introduction, Cambridge University Press, Cambridge, 1998.
[63] D. Vashaee, A. Shakouri, Improved thermoelectric power factor in metal-based superlattices, Physical Review Letters 92 (10) (2004) 106103.

CHAPTER

# 3

# Measurement of thermoelectric properties

## 3.1 Introduction

The performance of thermoelectric materials is principally evaluated by the dimensionless figure of merit, $ZT = S^2\sigma T/\kappa$, where $S$ is the Seebeck coefficient, $\sigma$ is the electrical conductivity, and $\kappa$ is the thermal conductivity. Measuring these three parameters, $S$, $\sigma$ and $\kappa$, becomes the most essential and frequent for characterizing the properties of thermoelectric materials. Although the basic principles of the measurement techniques for these parameters are well established, the complexity of measurement error sources makes it still difficult to accurately determine these basic parameters. In addition, recently, a number of new methods and techniques for measuring nanoscale transport properties have been developed as rapid development of low-dimensional and nanothermoelectrics. The measurement accuracy of nanoscale transport is more challenging than that of the bulk properties. This chapter will give a general description on the basic principles and methods for the measurement of Seebeck coefficient, electrical conductivity, and thermal conductivity, with providing some worth-attention points to eliminate the measurement errors. Although other related physical parameters (such as Hall coefficient, heat capacity, sound velocity, etc.) are also important for evaluating thermoelectric materials, they are not involved in this chapter. The introduction for the related measurements on these properties can be referred to other expert books.

## 3.2 Measurement for bulk materials

### 3.2.1 Electrical conductivity

The principle and technique for measuring electrical conductivity of a bulk material are well established. For a piece of material with

*Thermoelectric Materials and Devices*
DOI: https://doi.org/10.1016/B978-0-12-818413-4.00003-X

uniform composition, the electric field inside the material is evenly distributed when a constant current is passed through. The electric field strength inside the material is $E = V/l$, and the current density flowing through the material is $J = I/A$ ($V$ is the potential difference between two points with distance $l$ along current direction in the material; $I$ is the total intensity of the current passing through the material with cross-sectional area $A$). Then the expression of electrical conductivity is

$$\sigma = \frac{J}{E} = \frac{Il}{VA} \tag{3.1}$$

Or the resistivity ($\rho$) is

$$\rho = \frac{1}{\sigma} = \frac{RA}{l} \tag{3.2}$$

Here, $R = V/I$ is the resistance of the piece of material. According to the above expressions, the current and voltage drop can be measured by either two-probe method or four-probe method [1]. The four-probe method is commonly used for thermoelectric materials in which the influence of Seebeck effect is not ignorable. The schematic configuration of the standard four-probe method is shown in Fig. 3.1. The electric current terminals are usually contacted on the two end faces and the two voltage probes are located between the two current terminals. The samples used in the measurement are usually regular rectangular parallelepiped or cylinder shape and are required to be uniform in both composition and cross section along the longitude direction. The contact face of the current terminals onto the sample is required as large as possible to ensure equipotential on the contact surface, and the distance between the current terminal and the voltage probe on the same side should be enough large so that the electric field inside the sample is evenly distributed. Ryden et al. [2] pointed out that the size factors should be satisfied with the relation of Eq. (3.3), when considering the

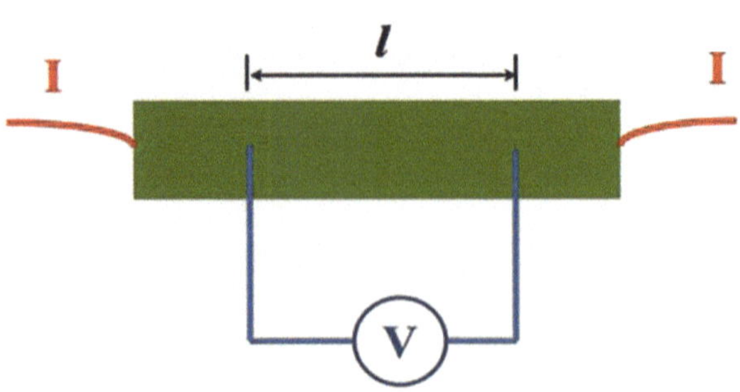


**FIGURE 3.1** The schematic diagram of four-probe method for electrical conductivity measurement.

uncertainty of the contact and uniformity of electric field within the sample to be measured.

$$L - l = 2\zeta \tag{3.3}$$

where $L$ is the distance between the two faces of the sample (i.e., the distance between the electric current terminals); $l$ is the distance between the two voltage probes; $\zeta$ is the thickness of the sample. Therefore, in practical, it is better to use the sample with a large aspect ratio. The voltage probes are required to be point-contacted with the sample, and the contact area is as small as possible to reduce the influence of the electric field distribution in the sample.

The interfacial resistance is unavoidable in the contacts between thermoelectric material and metal probes. For example, the formation of oxides on the surface will introduce extra interfacial resistance. In some cases, the formation of $p-n$ junction between thermoelectric material and probe (or contacting paste) would also produce an additional nonohmic voltage that will influence the measurement accuracy. To eliminate the influence of contacts on the measurement accuracy, it is convenient to apply a series of currents and collect the voltage signal and then obtain the electrical resistance from the slope of $I-V$ curve. In addition, the Seebeck and Peltier effects would also give influence on the measurement accuracy. Because of the Peltier effect, a temperature difference would be produced in the sample between the two ends when a current applied on the sample for voltage measurement. The temperature difference produced by the Peltier effect will result in an additional Seebeck voltage that is collected as a total voltage together with the resistance produced voltage drop. A common method for eliminating the additional Seebeck voltage is to use a DC pulse current to measure the resistance by intermittently introducing a small instantaneous DC current. Since the thermal response is not rapid as electric field, when the current switching speed is higher than 40 Hz, it is effective to eliminate the formation of temperature difference by the Peltier effect. It is also possible to measure the electrical resistance using a forward-reverse current or an alternating current method. Furthermore, because electrical conductivity is usually temperature-dependent, small current is preferred to prevent the temperature increment due to Joule heating.

### 3.2.2 Seebeck coefficient

According to the definition of Seebeck coefficient, the Seebeck coefficient of material is expressed as

$$S = \lim_{\Delta T \to 0} \frac{\Delta V}{\Delta T} \tag{3.4}$$

It is easy to obtain the Seebeck coefficients only by measuring the potential difference ($\Delta V$) generated by the temperature drop ($\Delta T$) at different temperatures. Fig. 3.2 shows a schematic diagram of the Seebeck coefficient measurement. A temperature gradient on the sample can be generated by heating one end by a heater chip while cooling the other end by a heat sink. Usually, two thermocouples (the same type) contacting on two different points of the sample are used to simultaneously collect the temperatures ($T_1$ and $T_2$, respectively) at the two points and the potential difference ($V_x$) between the two points. It is convenient to use one leg from each thermocouple as the voltage probe. Here, the two thermocouples should be electrically and thermally contacted on the sample, and the voltage probes should be the same constituent leg. Keeping the sample around a certain temperature $T_0$ and then changing the heating power of the side heater, a small temperature difference can be produced on the sample. Collecting the temperature differences ($\Delta T$) and the corresponding potential differences ($V_x$) between the two measuring points, one can obtain differential Seebeck coefficient $S_{sr}$ ($S_{sr} = S - S_{ref}$) using the fitted relation $V_x = f(\Delta T)$ or simply by taking the slope of $\Delta V - \Delta T$ when $\Delta T$ is small enough. The absolute Seebeck coefficient $S$ of the sample is obtained by removing the reference material's contribution ($S_{ref}$) (the contribution by the probe wire, that is, the thermocouple leg in the above described configuration).

In practical, copper-constantan, NiCr-NiSi, and Pt-PtRh thermocouples are commonly used for the measurement of low, medium, and high-temperature ranges, respectively. And copper, Ni-Cr alloy wires, and platinum are usually used as the Seebeck voltage probes. These thermocouple wires exhibit relatively small Seebeck coefficients, for example, the absolute Seebeck coefficients of NiCr alloy and platinum at room temperature are 21.5 and 4.9 μV/K, respectively. The absolute

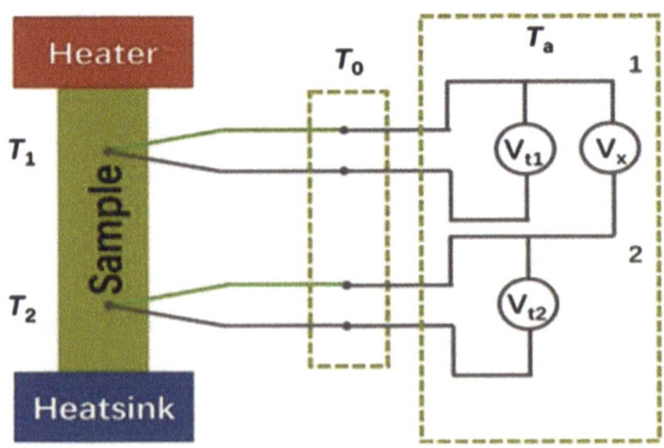


FIGURE 3.2 Schematic diagram for Seebeck coefficient measurement.

Seebeck coefficients of these typical reference materials can be calibrated by measuring Thomson coefficient [3–5]. Roberts calibrated the absolute Seebeck coefficients of several commonly used reference materials, including lead (7K–600K), copper (300K–873K), platinum (900K–1600K), and tungsten (900K–1600K). According to the experimental results, the absolute Seebeck coefficient of these reference materials in a wide temperature range can be obtained. For example, the absolute Seebeck coefficient of platinum (70K–1500K) and copper (70K–1000K) can be expressed as

$$S_{Pt}(T) = 0.186T\left[\exp\left(-\frac{T}{88}\right) - 0.0786 + \frac{0.43}{1+\left(\frac{T}{84.3}\right)^4}\right] - 2.57 \tag{3.5}$$

$$S_{Cu}(T) = 0.041T\left[\exp\left(-\frac{T}{93}\right) + 0.123 - \frac{0.442}{1+\left(\frac{T}{172.4}\right)^4}\right] + 0.804 \tag{3.6}$$

The method of measuring Seebeck coefficient by directly calculating the slope of $\Delta V - \Delta T$ is often called the differential method, which is also the most common method for Seebeck coefficient measurement.

In order to minimize the measurement error, the measuring process should satisfy with: (1) The temperatures ($T_1$, $T_2$) are kept stable during measuring the thermoelectric voltage; (2) the voltage is detected simultaneously with temperature and the voltage probes are located at the same position with the thermocouples; (3) sample composition and geometry are uniform; (4) all the connecting joints in the measurement system (except the contact points of thermocouple on the sample) are kept at the same temperature to avoid the production of any noise electromotive voltage. Furthermore, selecting a proper range of temperature differences is also important to improve the measurement accuracy. According to Eq. (3.4), the Seebeck coefficient needs to be measured under a temperature difference as small as possible, while the Seebeck coefficient measured under a large temperature difference is the average value rather than the exact value at certain temperature. However, if the temperature difference is excessively small, both the temperature difference and the electromotive force signal will be too small and the relative measurement error will be too large. For most thermoelectric semiconductors, the temperature difference $\Delta T$ is generally set from 4K to 10K to balance the two influence factors. It is also preferred to set the temperature difference from a negative value to a positive value (reverse the hot and cold ends) for eliminating the effects of noisy voltages.

It is also needed to pay attention on the chemical reactions inside the sample and at the interface between thermocouples and sample when measuring Seebeck at high temperatures. The chemical reactions would produce extra electromotive force as a measurement noise. The polluted thermocouples and voltage probes will produce great measurement uncertainty of both temperature and electromotive force. Therefore keeping the thermocouples and voltage probes cleaning without reactant is also very important. Measuring Seebeck coefficient across phase transition temperature is more challenging because the structure uniformity is not easy to be guaranteed. A low heating temperature, a small temperature difference, and a short collecting time are benefit to reliable measurement of Seebeck coefficient during phase transition.

## 3.2.3 Thermal conductivity

### *3.2.3.1 Steady-state method*

The principle of measuring thermal conductivity is well established, and there are several typical methods frequently used for determining thermal conductivity of materials. However, performing reliable measurement is still a great challenge. The key issue is how to ensure the heat flow one-directional along the measuring direction and uniform on the cross-section of sample. Because of the existence of various heat exchange such as radiation, conduction, and convection, it is difficult to achieve real thermal insulation condition. The methods for thermal conductivity measurement mainly include steady-state and nonsteady-state ones. The schematic diagram of steady-state method is shown in Fig. 3.3. The sample is placed between a heater and a heat sink, and a certain heating power is applied by the heater to supply a stable heat flux. Assuming that a stable temperature gradient is maintained between the two ends and there is no heat exchange between the measuring sample and surrounding mediums except the one-directional heat flow from the heat-source end to the heat-sink end. Then the

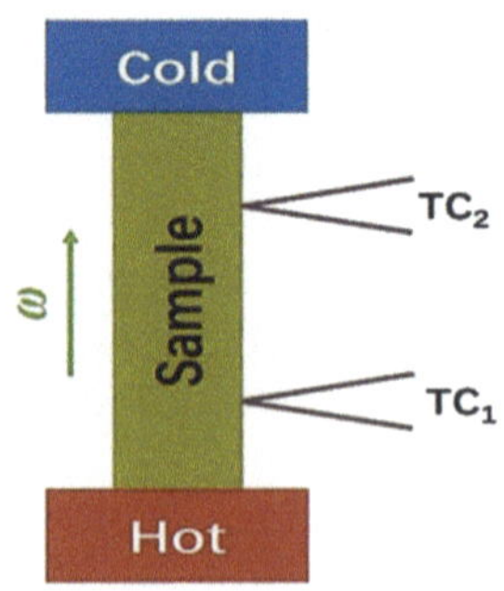


FIGURE 3.3 Schematic diagram of direct steady-state method for thermal conductivity measurement. $\omega$ is the heat flux through the sample.

thermal conductivity is obtained by measuring temperature gradient between both ends when the heat flux flows through the sample [6–9].

According to Fourier's law of heat conduction, the heat flux through a material is expressed as

$$\omega = -\kappa \frac{dT}{dx} \tag{3.7}$$

where $\omega$ is the heat flux (heat flow per area); $\kappa$ is the thermal conductivity of the material; $dT/dx$ is the temperature gradient between both ends. For a sample with a regular shape (such as plate), the heat flux is calculated from the heater power divided by cross-sectional area of the sample on the assumption that the electrical power supplied on the heater is all transferred through the sample. Then, the thermal conductivity of the material can be calculated from the heat flux and the temperature gradient ($\Delta T/\Delta x$) between the two ends because the temperature distribution is linear in a regular-shaped and homogenous sample.

In practice, it is hard to completely eliminate the heat exchanges (through radiation, convection, and heat conduction) between the measuring sample or heater and the surrounding mediums. Furthermore, there are the thermal resistances in all contacts such as that between the measuring sample and the heater and/or heat sink. These will yield measurement uncertainty especially to the heat flux. The radiation heat loss from the sample surface is the main cause of measurement errors. The heat lost through the surface of the sample can be expressed as

$$Q_{\text{rad}} = \varepsilon \sigma_{\text{SB}} A \left(T_0^4 - T_s^4\right) \tag{3.8}$$

where $Q_{\text{rad}}$ is the heat from radiation; $\varepsilon$ is the emissivity ($0 < \varepsilon < 1$), $\sigma_{\text{SB}}$ is the Stefan-Boltzmann constant ($\sigma_{\text{SB}} = 5.7 \times 10^{-8}\,\text{W}/(\text{m}^2 \cdot \text{K}^4)$); $A$ is the surface area; $T_0$ and $T_s$ are the temperature of the sample and the surrounding medium, respectively. When the temperature difference ($\Delta T = T_0 - T_s$) between the sample and the surrounding medium is very small, the heat lost by radiation can be approximated as $Q_{\text{rad}} = \varepsilon A \Delta T T_s^3$ [10]. Keeping the surrounding medium at the same or close temperature with the sample is very important to reduce radiation heat loss. For example, setting a heat shelter wall surrounding the "heater-sample-heat sink" series and keeping the same or close temperature gradient with the series are commonly used as the effective approach to minimize the radiation heat loss. It is also seen that the heat loss caused by the radiation is related to the sample temperature. At low temperatures (less than 200K), the influence of radiant heat loss is not obvious, while it becomes unignorable as the measuring temperature increases. The real heat flux can be estimated by subtracting the radiation heat loss. In addition, the steady-state direct measurement is required to be conducted under a high vacuum condition (usually $10^{-4}-10^{-5}$ torr) to eliminate the heat loss caused by convection. Furthermore, fine thermocouple wires are used to minimize the conduction heat loss through the thermocouples.

In practical experiment, heat flux is often calibrated by a modified steady-state method (called comparative steady-state method). In the comparative steady-state method, two cupper blocks (references) with well-characterized thermal conductivity are set in a sandwich-like structure with the sample (Fig. 3.4), and the real heat flux through the reference/sample series is obtained by averaging the heat flux through the upper and lower references.

The size and shape of sample also have significant impact on the measurement accuracy. In order to minimize the heat exchange between the sample and surrounding medium, the size and specific surface area of the sample should be minimized. When the sample size is too small, the ratio of the contribution by the contact thermal resistance in the

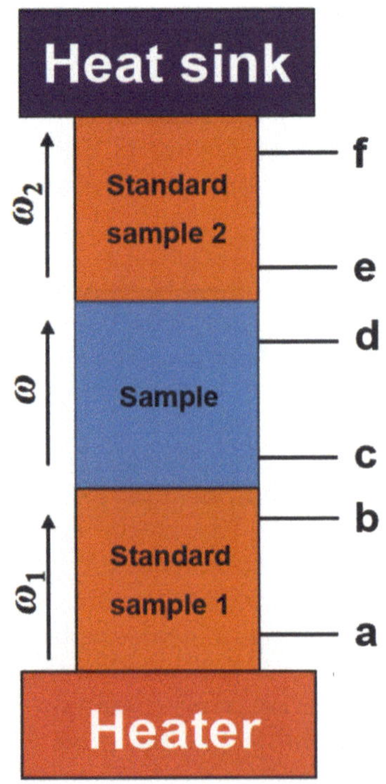


FIGURE 3.4 Schematic diagram of comparative steady-state method for thermal conductivity measurement. The points **a** ~ **f** are the measuring points of temperatures, $\omega$, $\omega_1$, $\omega_2$ are the flux through the measuring sample and reference samples, respectively.

interfaces would become obvious. Therefore appropriate size and regular shape of the measuring sample should be adopted for reliable measurement. For bismuth telluride thermoelectric material, Goldsmid [11] recommended the appropriate size of 10 $mm^2$ in cross-sectional area and millimeters in length.

#### 3.2.3.2 Nonsteady-state method

The nonsteady-state method has been developed to deal with the drawback in the steady-state method, such as long measurement period, difficulty to satisfy the boundary conditions mainly caused by the surface heat losses, and thermal contact resistance between the specimen and its associated heat source and sinks. According to the characteristic of the applied heat source, the nonsteady-state method mainly includes two types: periodic heat flow method and transient heat flow method. Basically, a periodic heat flow or transient (pulse) heat flow is applied on the sample and then the temperature variation is detected. The thermal conductivity is then calculated using the detected temperature changes and sample dimension.

The laser flash method is a typical transient heat flow method developed in the 1960s [12] and has become one of the most common and mature techniques for characterizing thermal diffusivity and thermal conductivity at high temperatures. Considering a flat sample, when a laser pulse is flashed on one side and then absorbed in the front surface of a thermally insulated sample, the temperature of the rear surface will increase gradually by heat conduction. For a sample with a thickness of $L$, taking the initial temperature distribution as $T(x,0)$ under thermal insulation, the temperature distribution $T(x,t)$ at any later time $t$ after flashing the pulse laser is expressed as [12]

$$T(x,t) = \frac{1}{L}\int_0^L T(x,0)dx + \frac{2}{L}\sum_{n=1}^{\infty} \exp\left(\frac{-n^2\pi^2\alpha t}{L^2}\right) \times \cos\frac{n\pi x}{L}\int_0^L T(x,0)\cos\frac{n\pi x}{L}dx \tag{3.9}$$

where $\alpha$ is the thermal diffusion coefficient ($cm^2/s$). If a pulsed radiant energy $Q$ ($cal/cm^2$) is instantaneously and uniformly absorbed in the small depth $g$ at the front surface $x = 0$, the temperature distribution inside the sample at that instant ($t = 0$) is given by

$$T(x,0) = \frac{Q}{DC_p g} \quad (0 < x < g)$$

$$T(x,0) = T_0 \quad (g < x < L)$$

With this initial condition, Eq. (3.9) can be rewritten as

$$T(x,t) = \frac{Q}{DC_pL}\left[1 - 2\sum_{n=1}^{\infty}\cos\frac{n\pi x}{L}\frac{\sin(n\pi g/L)}{(n\pi g/L)} \times \exp\left(\frac{-n^2\pi^2}{L^2}\alpha t\right)\right] \quad (3.10)$$

where $D$ is the density ($g/cm^3$) and $C_p$ is the heat capacity ($cal/(g \cdot K)$) of the sample. Because $g$ approaches zero for opaque materials, it follows that $\sin(n\pi g/L) \approx n\pi g/L$. On the other hand, the temperature variation (temperature history as time) on the rear surface ($x = L$) can be deduced as

$$T(L,t) = \frac{Q}{DC_pL}\left[1 + 2\sum_{n=1}^{\infty}(-1)^n\exp\left(\frac{-n^2\pi^2}{L^2}\alpha t\right)\right] \quad (3.11)$$

Then, it is convenient to define two dimensionless parameters

$$V(L,t) = \frac{T(L,t)}{T_{\max}} \quad (3.12)$$

$$\varpi = \frac{\pi^2\alpha t}{L^2} \quad (3.13)$$

Here, $T_{\max}$ is the maximum temperature rise at the rear surface after laser irradiation (corresponding to infinite time $t = \infty$, on the assumption that there is no heat exchange between the measuring sample and surrounding medium after laser irradiation). The combination of Eqs. (3.11)–(3.13) yields the following relationship:

$$V = 1 + 2\sum_{n=1}^{\infty}(-1)^n\exp(-n^2\varpi) \quad (3.14)$$

By taking $V$ equal to 0.5 (i.e., the temperature rise value just reaches half of the maximum temperature rise), one can easily obtain the value of $\varpi = 1.38$ by solving the Eq. (3.14). Therefore the thermal diffusivity $\alpha$ can be determined as

$$\alpha = \frac{1.38L^2}{(\pi^2 t_{1/2})} \quad (3.15)$$

where $t_{1/2}$ is the time required for the rear surface to reach half of the maximum temperature rise as shown in Fig. 3.5B.

It can be seen from the above derivation that it is not necessary to measure the total amount of heat absorbed by the front surface, but only need to detect the temperature change of the back surface of the sample. The thermal diffusion coefficient is obtained by determining $t_{1/}$

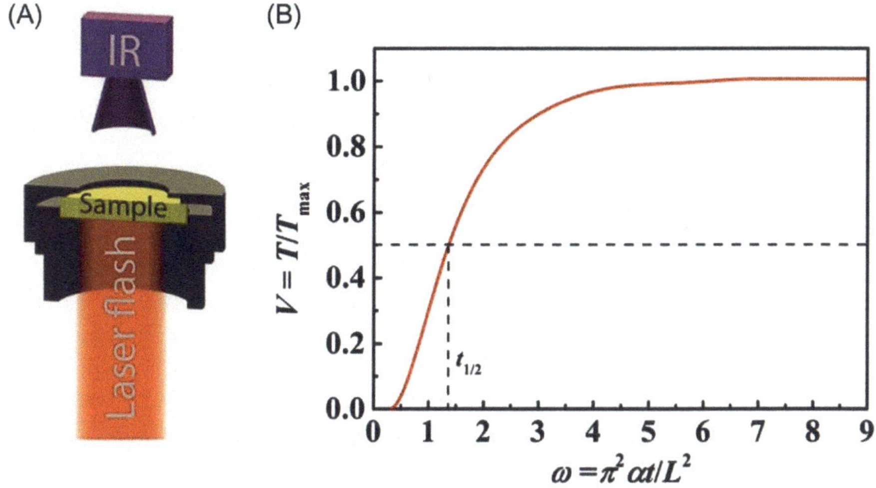


FIGURE 3.5 (A) The schematic diagram of laser flash method for thermal diffusivity measurement; (B) the temperature variation versus time at the rear face of laser-flashed sample.

$_2$, and then the thermal conductivity of the sample is calculated using the heat capacity ($C_P$) and density based ($D$) on

$$\kappa = \alpha D C_P$$

Fig. 3.5A shows a schematic diagram of laser flash method for thermal diffusivity measurement. Here, the sample is placed horizontally on a supporter, and the laser pulse is irradiated vertically. The temperature of the up surface of the sample is detected by the infrared detector. The measuring chamber is equipped with heating or cooling elements to enable the measurement at different temperatures. This method requires the laser heat absorbed by the sample to be transferred unidirectionally along the thickness direction. Therefore the sample size and the contact condition between sample and supporter (holder) have great impact on the measurement reliability. Needle-like supporter is preferred to eliminate the heat transfer between the measuring sample and holder. The sample thickness and diameter should be chosen matching with the number of diffusivities. Generally, the lower the thermal diffusivity, the thinner the sample thickness is needed. Large thickness ($L$) will enlarge the ratio of side-transferred heat and suppressed the maximum temperature rise and therefore induce larger measurement error, typically for most of thermoelectric materials that have low thermal conductivity. To eliminate the measurement deviation of laser flash method, a number of correction procedures have been proposed. The temperature distribution [Eq. (3.9)] expression is deduced and the value of $\varpi$ parameter is produced under the ideal conditions (sample homogeneity, one-dimensional conduction, impulse input, adiabatic boundary). In practice, the adiabatic boundary is very difficult to be maintained, which becomes the major cause of the measurement error. Especially, as radiation loss increases, the $\varpi$ constant decreases in a nonlinear fashion.

If the irradiation and convection heat loss from the sample surfaces is not ignorable, the temperature rise at the rear surface will not saturate at the maximum value but decrease after reaching the maximum temperature rise. Cowan [13] considered the problem of heat loss by looking at the behavior of the cooling part of the temperature history caused from heat loss and presented a dimensionless solution for varying heat losses. By Cowan's modification, the real value of $\alpha t_{1/2}/L^2$ (or $\varpi$) is determined by comparing the cooling part of the measured temperature history with the theoretical behavior without heat loss. Clark and Taylor [14] proposed further correction procedure by considering the influence of all available radiation heat losses, typically not ignoring the exchange loss from the edge of sample (radial radiation) and the emissivity of the sample surface, by combining theoretical analysis and experiment data. These two correction procedures have been adopted in most of the commercial laser flash measurement instruments. Furthermore, the duration of the laser pulse and the uniformity of the pulse energy distribution have also influence on the measurement accuracy. Taylor gave a detailed discussion on these factors and work out a calibration scheme to suppress the measurement error to a very small level of less than 1%.

## 3.3 Measurement for thin films

Obviously, the measurement accuracy for all thermoelectric parameters, typically for thermal conductivity and electrical conductivity, is seriously affected by the dimension factors (shape and size) of the sample as discussed above. Therefore it is difficult to apply the methods and techniques described in Section 3.2 directly to the measurement of thermoelectric properties of low-dimensional materials such as thin films and nanowires. In the past two decades, a number of new methods for measuring thermoelectric transport properties of low dimensional materials have been developed. In this section, some of newly developed measurement techniques for low dimensional materials are highlighted.

### 3.3.1 Measurement of thermal conductivity of thin films

The $3\omega$ method is a common way to measure the thermal conductivity of thin films, which was firstly introduced by Cahill and Pohl [15].

For a given thin film on which a metal line (AB) with certain width and loop-like electrodes (A, B, C, and D) is deposited, when an alternative current with a frequency $\omega$ is passed through metal line AB, a thermal wave with a frequency $2\omega$ ($T_{2\omega}$) is caused by the Joule effect and spreads across the thin films (Fig. 3.6). Accordingly, the electric resistance of the metal line also changes alternatively with a frequency $2\omega$ ($R_{2\omega}$), because the electric resistivity of the metal line has a linear correlation with temperature in a narrow temperature range. It is theoretically deduced that the interaction of the alternating current ($I_\omega$) with frequency $\omega$ and the resistance ($R_{2\omega}$) with frequency $2\omega$ will produce an alternating voltage signal with a frequency of $3\omega$ ($V_{3\omega}$), and the amplitude of this alternating voltage ($V_{3\omega}$) is proportional to the amplitude of the thermal wave ($T_{2\omega}$),

$$V_{3\omega} = \frac{1}{2} I_0 R_0 \varepsilon T_{2\omega} \tag{3.16}$$

where $I_0$ is the current amplitude passing through the metal line AB, $R_0$ is the electric resistance of metal line AB at room temperature, and $\varepsilon$ is the temperature coefficient of resistance ($dR/dT$).

To note that the temperature wave $T_{2\omega}$ is related not only to the heating power but also to the thermal conductivity of the thin film, which provides the possibility to correlate the $3\omega$ voltage signal with the thermal conductivity of the thin film. Since the metal line and the thin film are closely bonded, it is reasonable to neglect the heat transfer from the metal line to the surrounding air and then suppose that the thermal power generated by the metal line is completely absorbed by the thin film. The absorbed heat power is then transferred to the thin film in a semicylindrical conduction manner. By simplifying the metal line heater as a one-dimensional heat source, the amplitude of the thermal wave $T_{2\omega}$ of the heating metal line is inversely

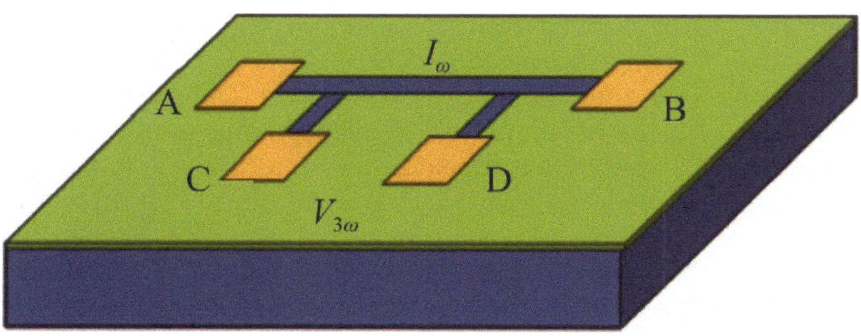


**FIGURE 3.6** Schematic diagram of the $3\omega$ method for the measurement of thermal conductivity of thin film.

proportional to the thermal conductivity of the thin film and logarithm of the angular frequency $\omega^{-1}$

$$T_{2\omega} = \frac{P}{2\pi l \kappa}(C - \ln \omega) \tag{3.17}$$

where $l$ is the length of the metal line AB, $P$ is the amplitude of the alternating thermal power absorbed by the thin film, and $\kappa$ is the thermal conductivity of the thin film.

Combining Eqs. (3.16) and (3.17), it is easy to obtain the linear relation of alternating voltage ($V_{3\omega}$) with the logarithm of heating frequency as

$$\frac{dV_{3\omega}}{d \ln \omega} = \frac{1}{4} \cdot I_0 \cdot \varepsilon \cdot \frac{P}{\pi l \kappa} = \frac{\varepsilon V_0^3}{4\pi R_0 l} \cdot \frac{1}{\kappa} \tag{3.18}$$

where $V_0$ is the heating voltage of the metal line. According to the above relation, the thermal conductivity can be obtained from the slope of $V_{3\omega}$–$\ln \omega$ diagram by measuring a series of alternating voltages ($V_{3\omega}$) under different heating frequencies $\omega$.

The measurement accuracy of $3\omega$ method is less affected by the radiation loss, because the thermal conductivity is determined by using the changes of voltage (caused by the temperature wave and alternating current) with heating frequency (i.e., the slope of $V_{3\omega}$–$\ln \omega$) but not the absolute values of voltage or temperature. Therefore it validates for the measurement of thermal conductivity at wide temperature range. However, for $3\omega$ method, the thickness of thermal penetration across the thin film is required smaller than that of measuring thin film to keep a uniform semicylindrical heat conduction localized within the thin film. Because the thermal penetration thickness is inversely proportional to the square root of the heating frequency $\omega$, high heating frequency is preferred, typically when the thickness is very small.

Fiege et al. developed a microscopic $3\omega$ method to locally measure or map thermal conductivity of thin film at a resolution of submicron or nanometer scale by combining the macroscopic $3\omega$ method with the scanning thermal microscopy [16]. The basic principle of micro $3\omega$ method is similar to the macroscopic $3\omega$ method, and the most modification is based on the development of microthermal probes. A curved line microprobe is used as both heater and detector, which is made of thermal resistive metal and shaped with a certain radius of curvature. When the probe contacts onto the thin film, part of the heating power flows into the thin film through the contact area. The absorbed heat transfers in the film at the same way with the case in macroscopic measurement system. Although the amount of the absorbed heat ($P_s$) and the dimensional factor ($l$) are difficult to be determined, the correlation among the absorbed heat, frequency of alternating current, and sample's thermal conductivity presented as Eq. (3.18) shall be basically tenable in the microscopic measurement system. Therefore it is reasonable to establish the relation between the voltage and the sample's thermal conductivity

$$\frac{dV_{3\omega}}{d\ln\omega} \propto \frac{P_s(\kappa)}{\kappa} \tag{3.19}$$

Here, $P_s(\kappa)$ is the heat power absorbed by the thin film sample, which is dependent on the sample's thermal conductivity ($\kappa$), or exactly to say, the ratio of the sample's thermal conductivity to the radiation conductivity. As a typical example, giving a fixed heating power on the microprobe, the relation between $P_s(\kappa)$ (here, it is convenient to use the ratio of the absorbed heat to the total heating power, $\Delta P_{sample}/\Delta P_{tip}$, instead of the absolute heating power $P_s$) and $\kappa$ is estimated by the finite element analysis as shown in Fig. 3.7. Apparently, $P_s(\kappa)$-$\kappa$ is a nonlinear relation. In order to quantitatively estimate the thermal conductivity, the $P_s(\kappa)$-$\kappa$ curve can be divided into three parts on the basis of different thermal conductivities: low thermal conductivity [1–10 W/(mK)], medium thermal conductivity [10–100 W/(mK)], and high thermal conductivity [>100 W/(mK)]. For each part, the $P_s(\kappa)$-$\kappa$ curve can be fitted approximately as simple formula correspondingly. And then, Eq. (3.19)

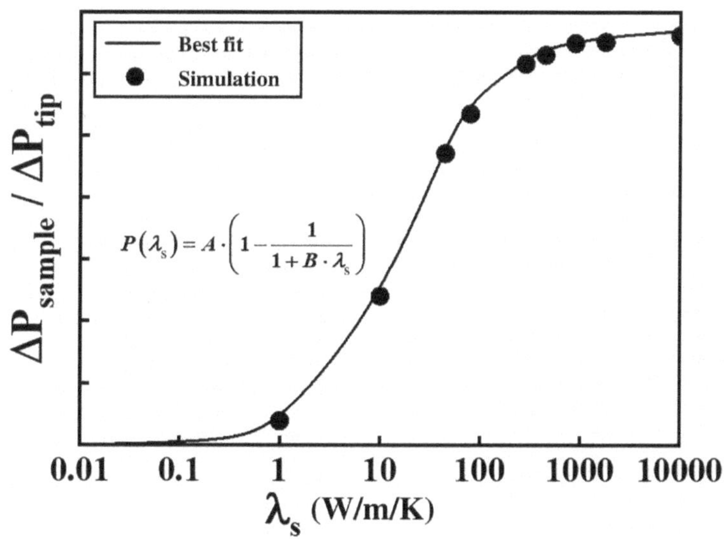


FIGURE 3.7 Relation between heating power absorbed by the sample and its thermal conductivity.

is rewritten to (3.20)–(3.22) for the three different temperature regions, respectively.

$$\frac{dV_{3\omega}}{d\ln\omega} \propto \frac{P_s(\kappa)}{\kappa} \propto k_1\kappa + k_2 \tag{3.20}$$

$$\frac{dV_{3\omega}}{d\ln\omega} \propto \frac{P_s(\kappa)}{\kappa} \propto \frac{k_1}{\kappa} + k_2 \tag{3.21}$$

$$\frac{dV_{3\omega}}{d\ln\omega} \propto \frac{P_s(\kappa)}{\kappa} \propto \frac{k_1}{\kappa} \tag{3.22}$$

where $k_1$ and $k_2$ are environment (i.e., probe, surface condition, contact status) related parameters, which can be calibrated experimentally by using standard samples under the same measurement condition. The thermal conductivity can be obtained from the slope of $V_{3\omega}$–ln $\omega$ diagram by collecting a series of alternating voltages ($V_{3\omega}$) under different heating frequencies $\omega$.

The $3\omega$ scanning thermal microscope can be used not only to provide the high-resolution map of the distribution of thermal conductivity in thin films, but also can conduct the in situ characterization of microarea thermal conductivity. Fig. 3.8 shows an example of thermal image accompanied with the morphological image of $SiO_2$/Si-nanowire composite film [17]. Such microscopic thermal image is usable for characterizing the local thermal transport behavior of thin films. Fig. 3.9 shows the micro $3\omega$ measurement results of Bi-Sb-Te thermoelectric thin film. From the $V_{3\omega}$–ln $\omega$ plot and based on the calibration using reference sample (glass), the thermal conductivity of Bi-Sb-Te film at microarea is calculated as 1.67 W/(mK), which is close to the thermal conductivity of the corresponding single crystal [1.70 W/(mK)] [18]. Furthermore, Zhang et al. proposed a modified microarea conduction model in the

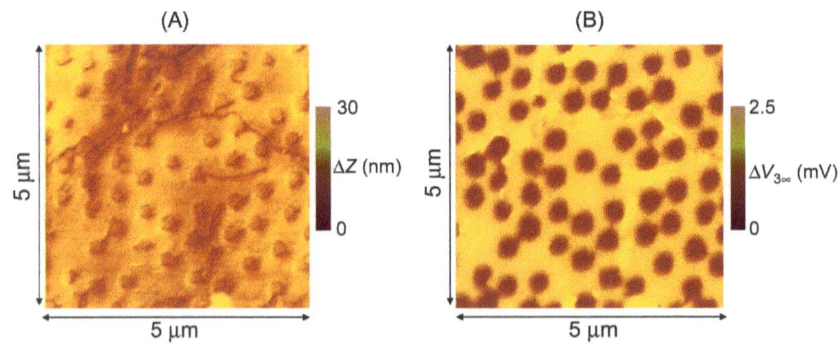


FIGURE 3.8 Morphological image (A) and thermal image (B) of $SiO_2$/Si nanowire composite.

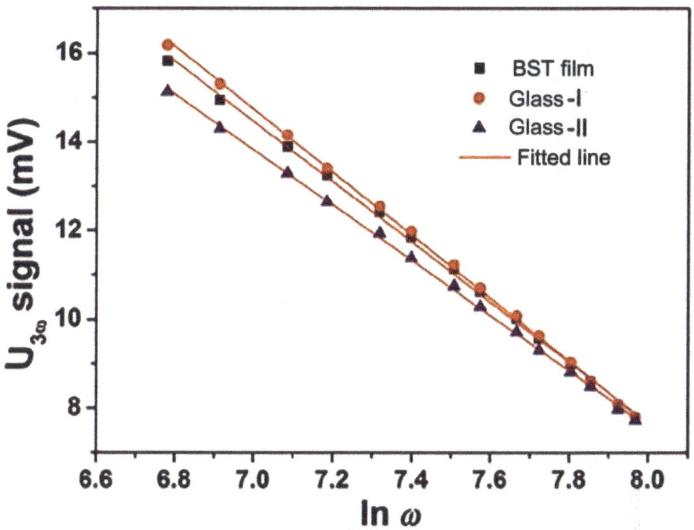


**FIGURE 3.9** $V_{3\omega}$–ln $\omega$ plot of Bi-Sb-Te thermoelectric thin film measured by micro $3\omega$ method.

measurement system consisting of microprobe and thin film [19]. By using this simplified model, microarea thermal conductivity can be obtained only by collecting the temperature rise ($\Delta T$) of and the heat power ($P$) supplied to the microprobe. Fig. 3.10 shows the measured $\Delta T - P$ plot of $Bi_2Te_3$ and $Bi_2Se_3$ nanograin thermoelectric thin films, from which the values of thermal conductivity of $Bi_2Te_3$ and $Bi_2Se_3$ nanograin thin films are 0.36 and 0.52 W/(mK), respectively [19].

## 3.3.2 Measurement of electrical resistivity of thin films

Four-probe method and van der Pauw method are commonly used for measuring the electrical resistivity of thin films. The procedure of four-probe method used for thin films is similar to that for bulk materials. As shown in Fig. 3.11, when a current ($I$) is passed from point 1 to point 4, the resistance ($R$) can be obtained using formula $R = V/I$ by collecting the voltage ($V$) between points 2 and 3. Then electrical resistivity is easily calculated using the dimensional factors (distance $L$ between point 2 and 3, width $D$, and thickness $d$ of the thin film) by $\rho = RdD/L$. All the possible issues affecting the measurement reliability for the bulk four-probe method discussed in Section 3.1 are also valid for the thin film measurement. In addition, the measurement accuracy of

FIGURE 3.10 $\Delta T - P$ plot of $Bi_2Te_3$ and $Bi_2Se_3$ nanograin thermoelectric thin films measured by modified $3\omega$ method.

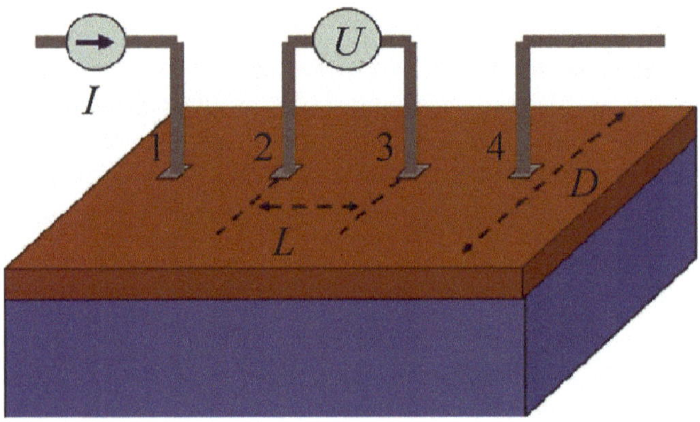


FIGURE 3.11 Schematic diagram of four-probe method for measuring electrical resistivity of thin films.

the thickness becomes most sensitive to the measurement accuracy for thin films.

Fig. 3.12 briefly shows the procedure of van der Pauw method for measuring the electrical resistivity of thin films. Four electrodes are connected onto the four corners of the sample. Let current pass between

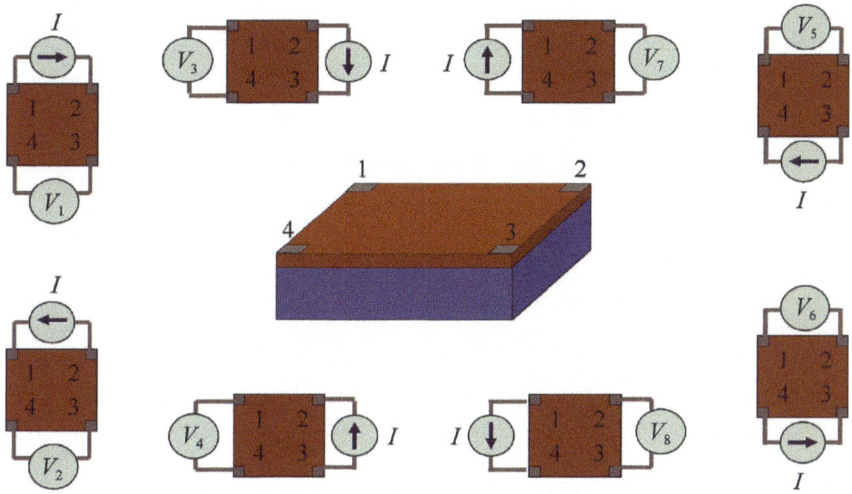


**FIGURE 3.12** Schematic diagrams of van der Pauw method to measure electrical resistivity of thin films.

two electrodes on the same side and measure the voltage between other two electrodes. As shown in Fig. 3.12, eight voltages $V_1-V_8$ are measured for each sample, and then two average resistances in the two directions, vertical resistance $R_{vert}$ and horizontal resistance $R_{hori}$, are defined as

$$R_{vert} = \frac{(V_1 + V_2 + V_5 + V_6)}{4I} \tag{3.23}$$

$$R_{hori} = \frac{(V_3 + V_4 + V_7 + V_8)}{4I} \tag{3.24}$$

Using $R_{vert}$ and $R_{hori}$ one can further calculate the square resistance of the thin film $R_s$ ($\Omega$/sq)

$$e^{-\pi R_{hori}/R_s} + e^{-\pi R_{vert}/R_s} = 1 \tag{3.25}$$

The square resistance $R_S$, resistivity $\rho$ and thickness $t$ of the thin film satisfy the following equation:

$$R_s = \frac{\rho}{t} \tag{3.26}$$

Principally, the van der Pauw method can be applied for the measurement of thin films with any shape. However, in practice, two dimensionally regular samples (i.e., rectangle or square) are usually used. In case of the two dimensionally regular thin films, the resistivity is simply calculated by using the average resistance ($R = (R_{vert} + R_{hori})/2$) of the two directions

$$\rho = \frac{\pi R}{t \ln 2} \tag{3.27}$$

Comparing with the four-probe method, influence of the Seebeck voltage produced by temperature difference on the measurement accuracy is greatly eliminated in the van der Pauw method. The electrodes should be located at as closer as to the sample edge and contacted tightly with sample. Point contacts and large sample area are preferred. Good homogeneity of the sample is of course required.

### 3.3.3 Measurement of Seebeck coefficient of thin films

The procedure to measure Seebeck coefficient of thin films is similar to that of bulk materials. As shown in Fig. 3.13, a temperature gradient is imparted upon the thin films by using a resistance heater or other ways, and the temperatures of points 1 and 2 are measured together with the thermovoltage $\Delta V$ in between. Seebeck coefficient can be obtained from the $\Delta V - \Delta T$ slope by varying the temperature gradient. All the measures to eliminate the measurement errors for bulk materials discussed in Section 3.2 should be also considered for the measurement of thin films.

The increased attention and investigation on nanostructured thermoelectric materials have greatly promoted the development of technology

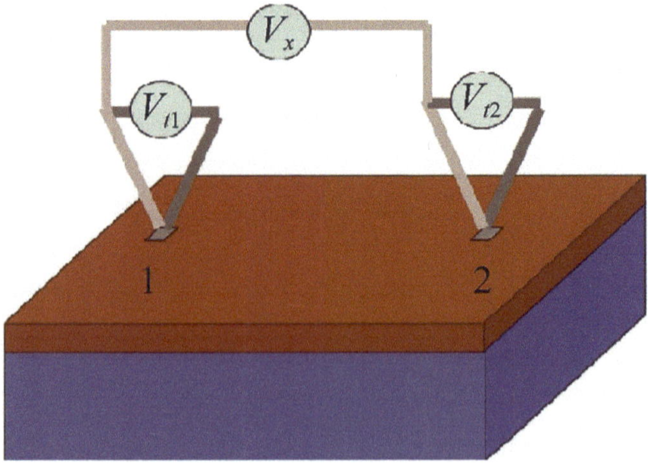


**FIGURE 3.13** Schematic of Seebeck coefficient measurement for thin films.

for Seebeck coefficient measurement at micrometer scale or even nanoscale. Seebeck microscopy has been gradually developed on the basis of scanning tunneling microscopy (STM), scanning thermal microscopy, and conductive atomic force microscopy (CAFM). For example, the scanning thermal microscopy can be simply retrofitted into a microarea Seebeck coefficient detector by combing the probe's heating function and the in situ electrical/thermal measuring function. Fig. 3.14A shows the schematic diagram for thermal probe method of in situ characterizing Seebeck coefficient [19]. In this scheme, when a high-frequency alternating current is applied on the probe, the thin film is locally heated at the contacting area by the Joule heat produced in the probe, and then a temperature gradient between contacting area and basal thin film is produced. Similar to the bulk measurement, the Seebeck coefficient at the microarea can be obtained by collecting thermovoltages and the corresponded temperature drops between the probe and the basal body of thin film. Because the measured temperatures of probe shall not reflect the exact temperature of the contacting point in thin film, calibration is required by using reference samples. Fig. 3.14B shows the Seebeck voltages of $Bi_2Te_3$ and $Bi_2Se_3$ thin films under different temperature gradients measured by thermal probe method. The calculated Seebeck coefficients are 106 and 1.9 μV/K for $Bi_2Te_3$ and $Bi_2Se_3$ thin films, respectively. The spatial resolution of thermal probe method when measuring Seebeck coefficient is majorly determined by the dimension of the probe and is currently in submicron scale.

CAFM is further modified in order to realize measurement of Seebeck coefficient at an extremely high spatial resolution under ambient condition by using a microcantilever heater and nanoprobe [20]. As shown in Fig. 3.15, a micro heater is set at one end of the cantilever beam and a probe with a curvature of nanometer is set at the other end.

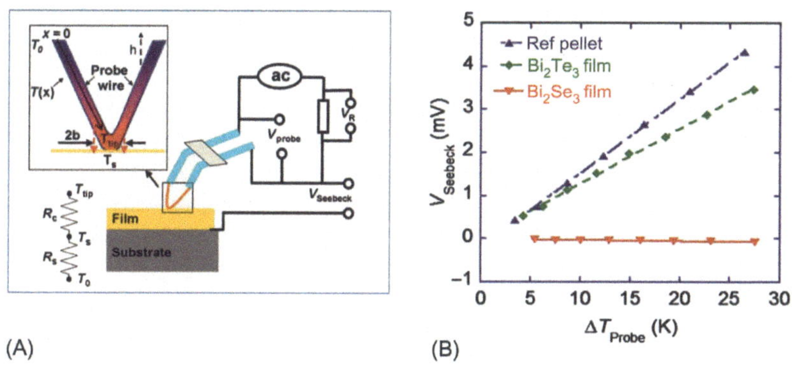


**FIGURE 3.14** (A) Thermal probe method for characterizing microarea Seebeck coefficient; (B) characterization of microarea Seebeck coefficient of $Bi_2Te_3$ and $Bi_2Se_3$ thin films.

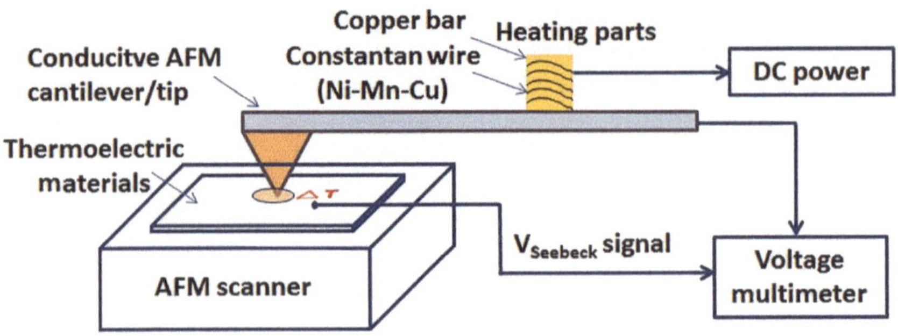


FIGURE 3.15 Schematic diagram of modified conductive atomic force microscopy for nanoscale measurement of Seebeck coefficient.

The usage of nanoprobe enables the improvement of spatial resolution of Seebeck coefficient up to as high as 15 nm (Fig. 3.16).

STM has also been used to characterize thermoelectric behavior in nanoscale [21–27]. Fig. 3.17 shows the schematic of modified STM for the characterization of thermoelectric effect in nanoscale under ultra-high vacuum. By introducing a heating wire sticking underneath the sample, a temperature gradient up to 30K can be implemented between the two points on the thin film contacting to STM probe tip and the heating wire. Using scanning tunneling thermoelectric microscopy (STeM), thermopower can be detected in a high spatial resolution as 2 nm. For example, the discontinuous change of thermoelectric behavior across the n-p junction caused by different types of carriers can be clearly observed by STeM measurement (Fig. 3.17B).

### 3.3.4 Measurements of electrical conductivity and Seebeck coefficient of nanowires

The basic principle of measuring electrical conductivity and Seebeck coefficient of nanowires is the same as those for bulk material. Because of the limitation of materials dimension, the most challenge is to highly integrate multiple techniques, such as electric transport measurement, microfabrication, and microstructure characterization, into a comprehensive platform. The measurement of single nanowire's electrical conductivity is usually done on the microcircuit made by lithography. The resistance can be characterized using four-probe method or two-probe method ($R = V/I$) by detecting the voltage ($V$) induced by passing a current ($I$). The section area ($A$) of the nanowire and the distance ($L$) between the two measurement points are obtained by scanning

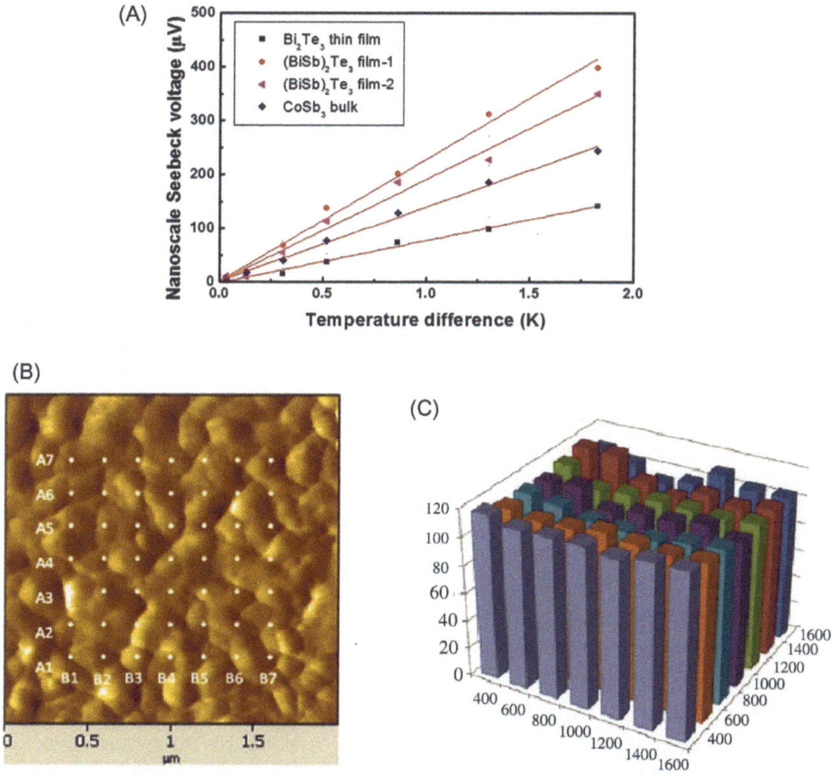


FIGURE 3.16 (A) Measurement results of nanoscale Seebeck coefficient for typical thermoelectric materials; (B) AFM morphology of $(BiSb)_2Te_3$ thin films; (C) cartogram of nanoscale Seebeck coefficient of the $7 \times 7$ matrix presented in (B). The distribution of Seebeck coefficient reveals the inhomogeneity of the thermoelectric properties in nanoscale.

microscope or tunneling microscope, then the electrical conductivity can be obtained by $\sigma = L/(RA)$.

Measuring the Seebeck coefficient of a single nanowire is more challenging. A nanoheater and a thermocouple or an electric resistance thermometer are integrated on a microcircuit to detect the temperature gradient between two sides of the nanowire. The thermal power of nanoheater is supplied in several to dozens of microwatt by altering current to produce small temperature gradients ($\Delta T < 10\text{K}$). It is skilled to control the heating current appropriately according to the material composition, structure of the microcircuit, and the hot-cold distance. Based on the structure design of microcircuit, two typical techniques, preset circuit method and postcircuit method, have been developed for

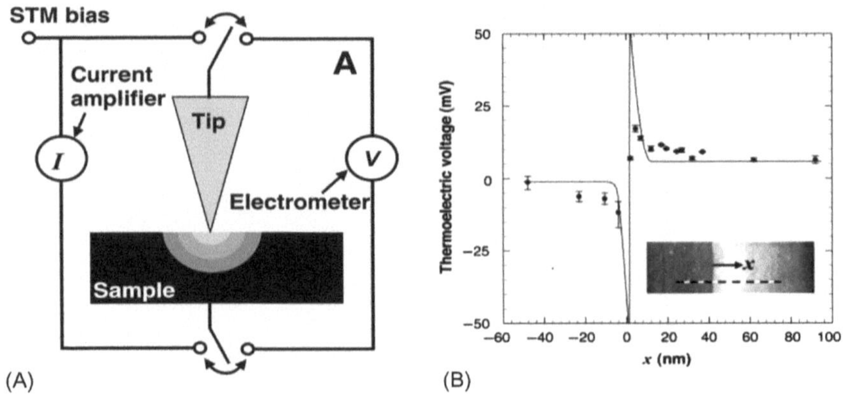


FIGURE 3.17 (A) Schematic of scanning tunneling thermoelectric microscopy (STeM) developed on STM under ultrahigh vacuum; (B) nanoscale characterization of thermoelectric behavior across the $p-n$ junction by STeM.

the measurement of electrical conductivity and Seebeck coefficient of nanowires.

In preset circuit method, microcircuit is processed in advance on a substrate, and then nanowires either are directly grown on the preset microcircuit or are transferred to the microcircuit by a nanomanipulator. This method is suitable for measurement of nanowires that are easy to be separated and transferred or are easy to be grown on substrate with controllable orientation. Shin et al. [28] used this method to measure the Seebeck coefficient and conductivity of $Bi_2Te_3$ nanowires. Fig. 3.18 shows the schematic diagram of a prefabricated microcircuit. The two white zigzag lines on the top and bottom are Pt nanoheaters. The four Pt lines are printed between the up and bottom heaters, of which two act as electrodes and two as resistance thermometers. These nanosized probes and thermometers are connected to measurement instruments (including power supply and nanovoltmeter) through a series of micrometer-sized film electrodes made of Au layer (155 nm in thickness) and Ni layer (45 nm in thickness). The $Bi_2Te_3$ nanowire is adsorbed and moved onto the board by a micromanipulator-controlled tungsten probe. Then Ni/Al composite layer is deposited onto the connecting points of nanowire and microcircuit by radio-frequency -sputtering as welding agent to enhance the adhesion between nanowire and probes. The measurement procedure in such a microcircuit scheme follows that in bulk measurement, while the resistance of Pt wire is used to determine the temperature because of the linear temperature dependency of Pt resistance between 200K and 400K.

In the postcircuit technique, nanowires are deposited in advance on a substrate and then the measuring circuit is directly fabricated on the

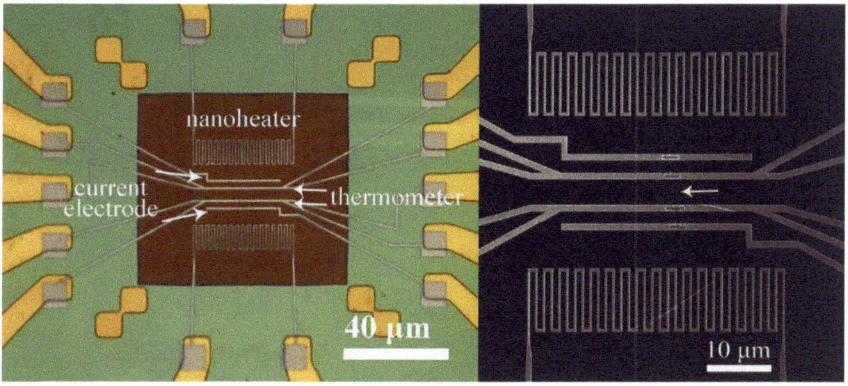


**FIGURE 3.18** Schematic diagrams of preset circuit method for measuring transport properties of nanowires.

substrate by using lithographic technique or others. Before measurement, the redundant nanowires are removed remaining only one appropriately bridged on the circuit. Kim et al. [29] fabricated Si nanowire arrays with 50 nm of diameter on boron-doped silicon-on-insulatorsubstrate by using thermal oxidization and hydrofluoric acid corrosion process. They further deposited Cr lines (0.2 nm in width) and Pt lines (2.3 nm in width) by masking magnetron sputtering forming a microcircuit. Then the measurement of electrical conductivity and Seebeck coefficient can be conducted similarly with the preset circuit method. A similar approach was used by Yee et al. to measure the conductivity and Seebeck coefficient of one-dimensional nanoarrays of Te/PEDOT:PSS composites as shown in Fig. 3.19, in which the nanoarrays were synthesized by prefabricated photoresist and acetone dissolving process [30].

In the measurement of electrical conductivity of nanowires, it is vital to get accurate dimensions by scanning microscope, tunneling microscope, or atomic force microscope in order to obtain reliable results. Seebeck coefficient is not affected by dimension factor but it is more challenging to accurately determine the temperature gradient on the nanowire. In actual measurement, usually the calibration of $\Delta T$ based on both experiment and simulation calculation is needed to guarantee the measurement accuracy.

## 3.3.5 Measurement of thermal conductivity of nanowires

Basically, similar processing with the microcircuit fabrication for the measurement of Seebeck coefficient of nanowires is also needed for

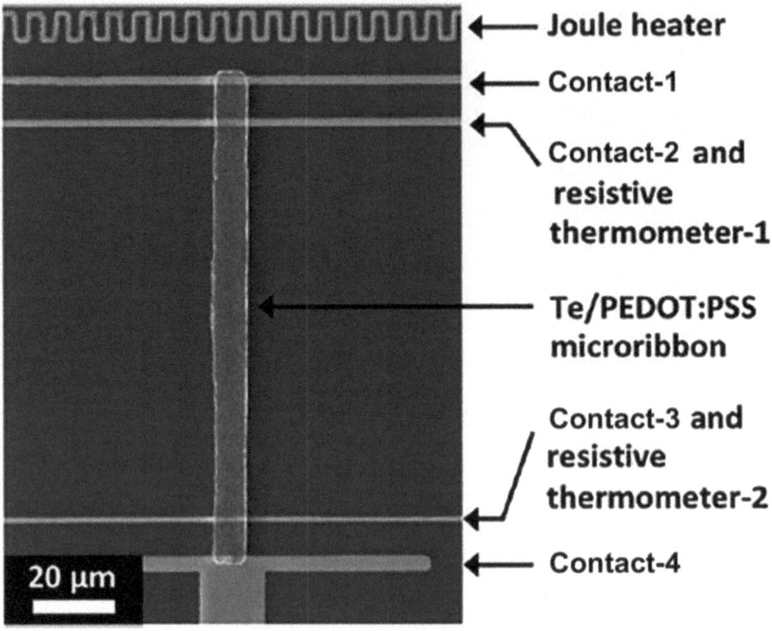


FIGURE 3.19 Measuring thermoelectric properties of Te/PEDOT:PSS nanoarrays.

thermal conductivity measurement of nanowires. Two probe statistic technique as shown in Fig. 3.20 can be used to measure thermal conductivity, in which nanowire is bridged on the microcircuit and then temperature gradient is measured when one end is heated. Actually, it is able to measure all three thermoelectric parameters, electric conductivity, Seebeck coefficient, and thermal conductivity, by using this microcircuit on the same nanowire.

The nanowire is suspended on the microcircuit by dispensing or micromanipulator, so that the heating point, nanowire, and resistance thermometer are adiabatic with the environment. Meanwhile, Pt/C composite is deposited on the connecting area of nanowire and heater by focused electron beam to enhance the thermal connection. Heat is transferred through the nanowire from one end (heating point) to the other end (receiving point) when electric current passes through the Pt circuit. By measuring the change of the resistances $R_h$ and $R_s$ on the heating point and receiving point, respectively, the heating power can be determined using $P = I^2(R_h + R_I/2)$ ($R_I$ is the resistance of the conducting wire connecting the microcircuit). The temperature changes at the heating point and receiving point, $\Delta T_h$ and $\Delta T_s$, can also be estimated. Therefore

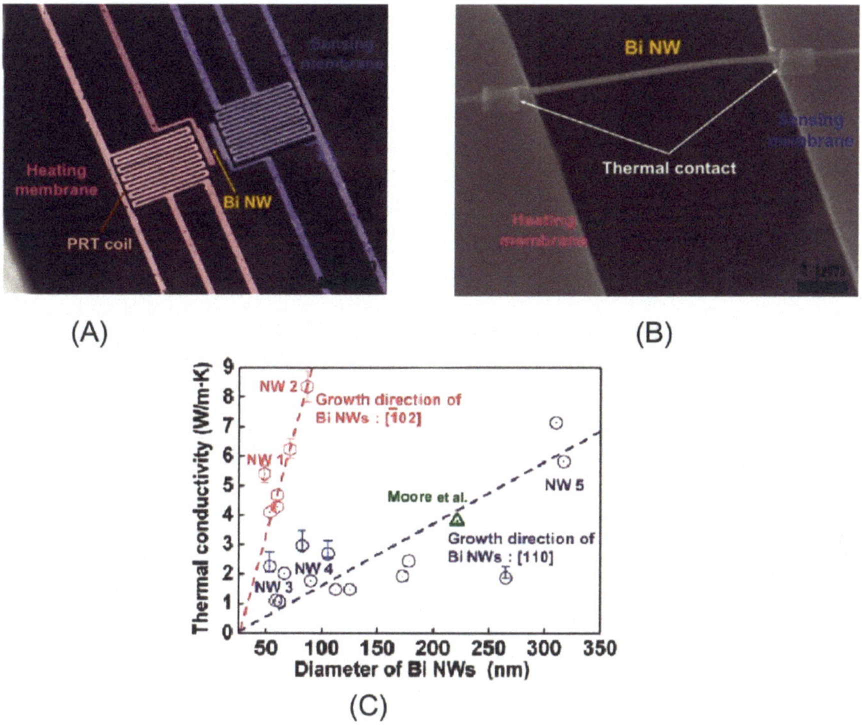


**FIGURE 3.20** Measuring thermal conductivity of single Bi nanowire using suspending method. (A) Schematic diagram of the microcircuit of two-point contact; (B) ensure Bi nanowire being well connected with electrodes at both cold and hot ends; (C) the relation of the diameters of the Bi nanowires and their thermal conductivity measured by suspending method.

the thermal conductance $G_w$ of the nanowire can be determined in Eq. (3.28)

$$G_w = \frac{P}{\Delta T_h - \Delta T_s}\left(\frac{\Delta T_s}{\Delta T_h + \Delta T_s}\right) \tag{3.28}$$

According to the equation $\kappa = G_w l/A$ (where $l$ is the distance between the heating point and receiving point; $A$ is the cross-sectional area of the nanowire), the thermal conductivity of the nanowire can be calculated. Roh et al. measured the conductivity of many Bi nanowires with different growing orientations by using this method [31]. They found that the thermal conductivity of nanowire decreases greatly when its diameter decreases, and Bi nanowires with different growing orientations have different thermal conductivity. For example, the difference of thermal conductivity between two nanowires (with the same diameter of 100 nm) growing along two different directions, (102) and (111), is as large as 7 W/(mK).

Compared with electrical property measurement of nanowires, measuring thermal conductivity is more difficult and the measuring accuracy is more sensitive to the environment. Firstly, accurately measuring temperatures is essential for both the thermal conductivity and Seebeck coefficient measurements. Secondly, high vacuum is typically required for thermal measurement to prevent heat losses caused by convection. Furthermore, point connection (as small area as possible) between nanowire and microcircuit is needed. In order to ensure the measurement accuracy, the calibration for the measurement system can be done using standard samples. Amorphous $SiO_2$ nanowire is a good candidate as the standard sample, because it possesses the minimum thermal conductivity independent of its morphology and dimension.

## 3.4 Conclusion

The dimensionless figure of merit $ZT$ ($=S^2\sigma T/\kappa$) characterizes the thermoelectric performance of a material. It is a composite parameter calculated from several intrinsic physical parameters including Seebeck coefficient, electrical conductivity, and thermal conductivity. Currently, there is no established technique to measure the above three physical parameters simultaneously on one piece of specimen. Therefore the measurement errors of each parameter add together and enlarge the deviation of calculated $ZT$. Furthermore, thermoelectric properties are usually dependent on its crystal orientation and microstructures. The possible variation of crystalline orientation (or texture) and the fluctuation of chemical composition in different samples will inevitably enlarge the deviation of calculated $ZT$. In Chapter 7, Design and Fabrication of Thermoelectric Devices, a method of directly estimating $ZT$ using the refrigerating ability of device is described, but it is still an approximate way to characterize $ZT$. It is really a big challenge to develop a new technique to characterize the electrical conductivity, Seebeck coefficient, and thermal conductivity all in one experiment with high accuracies. In addition, comparing to the characterization of bulk materials, the uncertainty of measurement for low-dimension and nanoscale materials becomes much more serious, because the thermal and electric resistances of the interfaces and the thermal losses contribute more sensitively to the weak signals detected on nanomaterials. Meanwhile, the accurate measurement and control of small power heat flow are more difficult. These factors together increase the difficulty to measure precisely the thermoelectric properties of low-dimensional and nanoscale materials.

## References


[1] H.H. Wieder, Laboratory Notes on Electrical and Galvanomagnetic Measurements, Elsevier Scientific Publishing Company, Amsterdam, 1979.

[2] D.J. Ryden, Techniques for the Measurement of the Semiconductor Properties of Thermoelectric Materials, UKAEA, Harwell, 1973.

[3] R.B. Roberts, The absolute scale of thermoelectricity, Philos. Mag. 36 (1) (1977) 91–107.

[4] R.B. Roberts, The absolute scale of thermoelectricity II, Philos. Mag. B 43 (6) (1981) 1125–1135.

[5] R.B. Roberts, F. Righini, R.C. Compton, Absolute scale of thermoelectricity III, Philos. Mag. B 52 (6) (1985) 1147–1163.

[6] J.R. Drabble, H.J. Goldsmid, Thermal Conduction in Semiconductors, Pergamon Press, London, 1961.

[7] R.P. Tye, Thermal Conductivity, Academic Press, London, 1969.

[8] C.M. Bhandari, D.M. Rowe, Thermal Conduction in Semiconductors, Wiley Eastern Ltd., New Delhi, 1988.

[9] H.D. Young, Statistical Treatment of Experimental Data, McGraw-Hill, New York, 1962.

[10] T.M. Tritt, Thermal Conductivity: Theory, Properties, and Applications, Springer, New York, 2004.

[11] H.J. Goldsmid, The electrical conductivity and thermoelectric power of bismuth telluride, Proc. Phys. Soc. 71 (4) (1958) 633–646.

[12] W.J. Parker, R.J. Jenkins, C.P. Butler, et al., Flash method of determining thermal diffusivity, heat capacity, and thermal conductivity, J. Appl. Phys. 32 (9) (1961) 1679–1684.

[13] R.D. Cowan, Pulse method of measuring thermal diffusivity at high temperatures, J. Appl. Phys. 34 (4) (1963) 926–927.

[14] L.M. Clark III, R.E. Taylor, Radiation loss in the flash method for thermal diffusivity, J. Appl. Phys. 46 (2) (1975) 714–719.

[15] D.G. Cahill, R.O. Pohl, Thermal conductivity of amorphous solids above the plateau, Phys. Rev. B 35 (8) (1987) 4067–4073.

[16] G.B.M. Fiege, A. Altes, R. Heiderhoff, et al., Quantitative thermal conductivity measurements with nanometre resolution, J. Phys. D Appl. Phys. 32 (5) (1999) 13.

[17] E. Puyoo, S. Grauby, J.M. Rampnoux, et al., Scanning thermal microscopy of individual silicon nanowires, J. Appl. Phys. 109 (2) (2011) 024302.

[18] K.Y. Zhao, H.R. Zeng, K.Q. Xu, et al., Scanning thermoelectric microscopy of local thermoelectric behaviors in $(Bi, Sb)_2Te_3$ films, Phys. B: Condens. Matter 457 (2015) 156–159.

[19] Y.L. Zhang, C.L. Hapenciuc, E.E. Castillo, et al., A microprobe technique for simultaneously measuring thermal conductivity and Seebeck coefficient of thin films, Appl. Phys. Lett. 96 (6) (2010) 062107.

[20] K.Q. Xu, H.R. Zeng, H.Z. Yu, et al., Ultrahigh resolution characterizing nanoscale Seebeck coefficient via the heated conductive AFM probe, Appl. Phys. A 118 (1) (2014) 57–61.

[21] D. Hoffmann, J.Y. Grand, R. Möller, et al., Thermovoltage across a vacuum barrier investigated by scanning tunneling microscopy: imaging of standing electron waves, Phys. Rev. B 52 (19) (1995) 13796–13798.

[22] D. Hoffmann, J. Seifritz, B. Weyers, et al., Thermovoltage in scanning tunneling microscopy, J. Electron. Spectrosc. Relat. Phenom. 109 (2000) 117–125.

[23] A. Rettenberger, C. Baur, K. Läuger, et al., Variation of the thermovoltage across a vacuum tunneling barrier: copper islands on Ag(111), Appl. Phys. Lett. 67 (9) (1995) 1217–1219.

[24] J. Seifritz, T. Wagner, B. Weyers, et al., Analysis of a $SeCl_4$-graphite intercalate surface by thermovoltage scanning tunneling microscopy, Appl. Phys. Lett. 94 (11) (2009) 113112.
[25] H.K. Lyeo, A.A. Khajetoorians, L. Shi, et al., Profiling the thermoelectric power of semiconductor junctions with nanometer resolution, Science 303 (2004) 816–818.
[26] C. Evangeli, K. Gillemot, E. Leary, et al., Engineering the thermopower of $C_{60}$ molecular junctions, Nano Lett. 13 (5) (2013) 2141–2145.
[27] J. Park, G. He, R.M. Feenstra, et al., Atomic-scale mapping of thermoelectric power on graphene: role of defects and boundaries, Nano Lett. 13 (7) (2013) 3269–3273.
[28] H.S. Shin, J.S. Lee, S.G. Jeon, et al., Thermopower detection of single nanowire using a MEMS device, Measurement 51 (2014) 470–475.
[29] J. Kim, Y. Hyun, Y. Park, et al., Seebeck coefficient characterization of highly doped n- and p-type silicon nanowires for thermoelectric device applications fabricated with top-down approach, J. Nanosci. Nanotechnol. 13 (9) (2013) 6416.
[30] S.K. Yee, N.E. Coates, A. Majumdar, et al., Thermoelectric power factor optimization in PEDOT: PSS tellurium nanowire hybrid composites, Phys. Chem. Chem. Phys. 15 (11) (2013) 4024–4032.
[31] J.W. Roh, K. Hippalgaonkar, J.H. Ham, et al., Observation of anisotropy in thermal conductivity of individual single-crystalline bismuth nanowires, ACS Nano 5 (5) (2011) 3954–3960.

CHAPTER

# 4

# Review of inorganic thermoelectric materials

## 4.1 Introduction

Although the thermoelectric effects were discovered in the early 1820s, thermoelectric materials had not attracted world-wide attention until 1950s. In 1950s and 1960s, as the establishment of solid state physics and semiconductor physics, especially the electron and phonon transport models in solids, Ioffe et al. [1] opened up a systematic study on thermoelectric performance in narrow bandgap semiconductors, which led an epoch-making discovery of a series of high-performance thermoelectric materials such as $Bi_2Te_3$ and its solid solutions, PbTe, and Si-Ge alloys. In the late 1990s, G. Slack proposed the concept of "phonon-glass electron-crystal" (PGEC) [2] to screen and discover ideal thermoelectric materials with glass-like low thermal conductivity and good electrical properties as those in crystals. Inspired by the PGEC concept, many new compounds with high thermoelectric performance have been discovered successively with typical examples in special caged compounds, such as filled skutterudites and clathrates. After entering the 21st century, with continuous efforts on comprehensive understanding on the electronic and thermal transports, thermoelectrics came into a new era with an explosive discovery of new TE materials and a leap forward in materials' $ZT$. Fig. 4.1 summarizes the timeline of the dimensionless figure of merit ($ZT$) in the past six decades putting emphasis on the recent progress since 2000. This chapter gives an overview on the presentative inorganic thermoelectric materials focusing on expounding their electronic and thermal transport behavior and the optimization of thermoelectric performance from the perspective of structure manipulation.

*Thermoelectric Materials and Devices*
DOI: https://doi.org/10.1016/B978-0-12-818413-4.00004-1

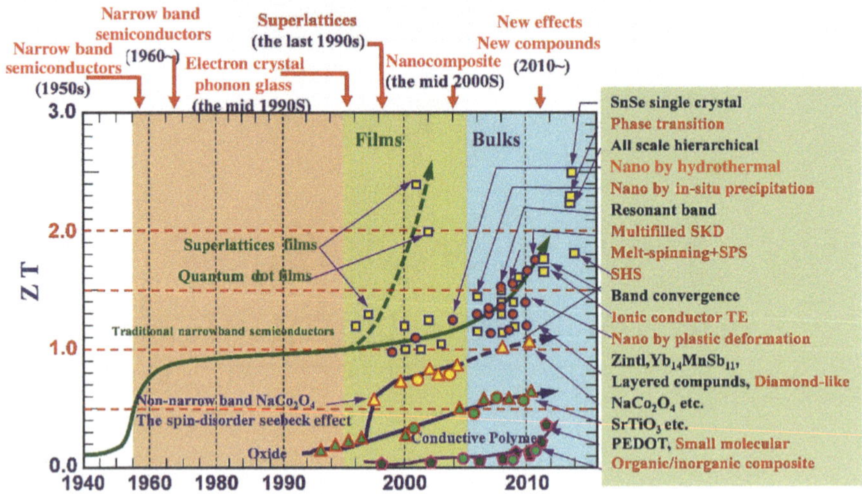


FIGURE 4.1 Timeline of dimensionless figure of merit *ZT*.

## 4.2 Bismuth telluride and its solid solutions

Bismuth telluride ($Bi_2Te_3$)–based thermoelectric materials, which were discovered in 1950s, possess excellent thermoelectric properties and are widely used in both refrigeration and power generation near room temperature. $Bi_2Te_3$ is a typical narrow bandgap semiconductor with a band gap about 0.15 eV. It crystallizes in rhombohedral structure with the space group of $R\bar{3}m$. $Bi_2Te_3$ has a layered structure which is arranged by Bi atom layer and Te atom layer following the sequence: $Te^1$-Bi-$Te^2$-Bi-$Te^1$, as shown in Fig. 4.2A. Covalent bonds exist between Bi atoms and Te atoms, while the two adjacent layers of Te atoms are weakly bonded with van der Waals force, so the bismuth telluride and its solid solutions are easily cleaved between the layers. The band structure of $Bi_2Te_3$ is quite complex. The crystal symmetry gives rise to six anisotropic carrier pockets in both the conduction and valence bands as shown in Fig. 4.2B, which is conducive to improving the state density and the effective mass [3]. At the same time, the small difference in electronegativity between Bi atoms and Te atoms can help to obtain high carrier mobility. For example, the electron mobility of $Bi_2Te_3$ along the in-plane direction can reach 1200 $cm^2/(Vs)$ at 293K, so $Bi_2Te_3$ presents good electrical transport property. In addition, large atomic masses of Bi and Te atoms and the low melting point of $Bi_2Te_3$ contribute to its low lattice thermal conductivity. Due to the good electrical transport performance and the low lattice thermal conductivity, $Bi_2Te_3$-based materials become the best thermoelectric materials around room temperature.

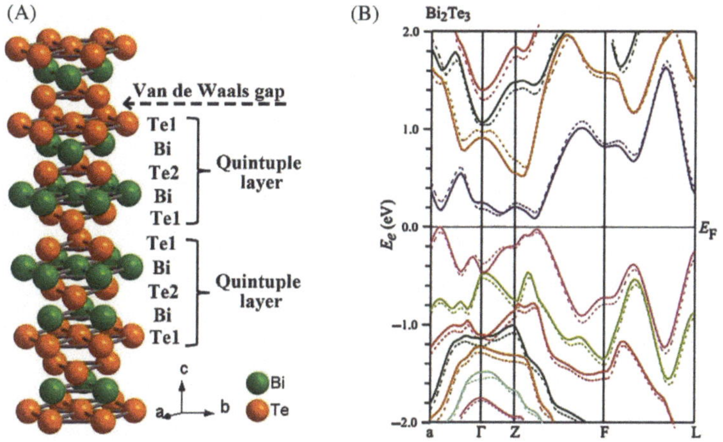


FIGURE 4.2 (A) Crystal structure of $Bi_2Te_3$ and (B) electronic band structure of $Bi_2Te_3$ along the high-symmetry lines with spin-orbit coupling. The solid and dashed lines are for the results with and without the $p^{1/2}$ corrections included, respectively.

In $Bi_2Te_3$, due to the similar physical properties between Bi atoms and Te atoms, Bi atoms can easily occupy the position of Te atoms to form antisite defects $Bi_{Te}^{'}$ to act as acceptor impurities, leading to p-type conduction [4]. This is confirmed by the experiment in which binary $Bi_2Te_3$ prepared by zone-melting (ZM) method usually is of p-type conduction. However, polycrystalline $Bi_2Te_3$ materials synthesized by powder processing and sintering technique tend to be of n-type conduction. In the powder process, atoms slip near the grain surface caused by mechanical grinding may induce the formation of Te vacancies ($V_{Te}^{\bullet\bullet}$), acting as donor impurities [5].

In practical applications, alloying $Bi_2Te_3$ with $Sb_2Te_3$ is usually used to prepare p-type materials with the optimal composition of about $Bi_{0.5}Sb_{1.5}Te_3$ [6]. Due to the similar crystal structures between $Sb_2Te_3$ and $Bi_2Te_3$, continuous solid solutions can be achieved as shown in the Fig. 4.3A. Compared with Bi atoms, the differences in atomic size and electronegativity between Sb and Te atoms are small. Thus, the formation energy of antisite defects $Sb_{Te}^{'}$ is lower than $Bi_{Te}^{'}$ to display stronger p-type conduction. In addition to tuning hole concentrations, dissolving $Sb_2Te_3$ into $Bi_2Te_3$ also introduces point defects to significantly reduce lattice thermal conductivity. Similarly, $Bi_2Se_3$ has the same crystal structure with $Bi_2Te_3$. As shown in the pseudo-binary phase diagram of $Bi_2Se_3$-$Bi_2Te_3$ (Fig. 4.3B), $Bi_2Se_3$-$Bi_2Te_3$ solid solutions can form at high temperature [7]. Due to the larger differences in size and electronegativity between Bi and Se atoms than those between Bi and Te, the formation of antisite defects $Bi_{Se}^{'}$ in $Bi_2Se_3$ is suppressed. Furthermore, the

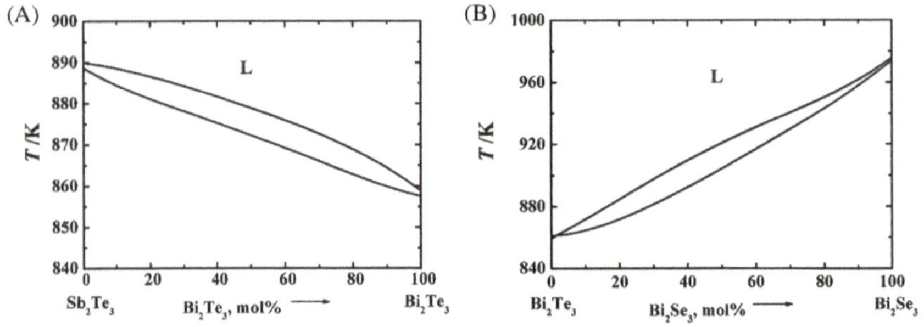


FIGURE 4.3 Pseudo-binary phase diagrams of (A) $Bi_2Te_3$-$Sb_2Te_3$ and (B) $Bi_2Te_3$-$Bi_2Se_3$.

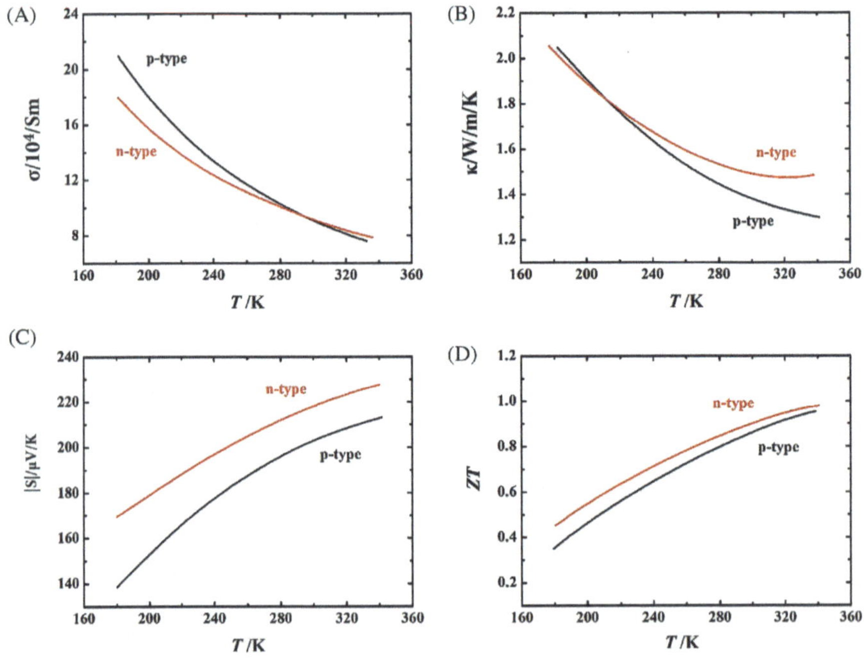


FIGURE 4.4 Temperature dependence of (A) electrical conductivity, (B) thermal conductivity, (C) Seebeck coefficient, and (D) *ZT* for p-type and n-type $Bi_2Te_3$-based materials.

larger vapor pressure of Se makes it easy to form Se vacancy which is compensated by electrons. Therefore, n-type solid solutions can be formed between $Bi_2Te_3$ and $Bi_2Se_3$, and the carrier concentrations can be tuned by varying $Bi_2Se_3$ content. The optimized composition is generally located at $Bi_2Te_{2.7}Se_{0.3}$. The p-type and n-type $Bi_2Te_3$-based materials are conventionally fabricated by directional growth process and their thermoelectric properties are shown in Fig. 4.4.

The layered crystal structure of bismuth telluride endows the material with significant anisotropic thermal and electronic transport properties. The $Bi_2Te_3$-based materials exhibit larger electrical conductivity and thermal conductivity in the direction parallel to the cleavage planes than those in the direction perpendicular to them. The electrical conductivity and the thermal conductivity in the direction parallel to the layers are usually 3–7 [8] and 2–2.5 times [9] higher than those in the perpendicular direction, respectively. Therefore, improving the grain orientation can obtain better performance along the in-plane direction. In the large-scale production, ZM directional growth process is widely adopted, and the in-plane direction is generally chosen as the direction of heat/current flow in device.

The easy cleavage of the ZM grown $Bi_2Te_3$-based materials bring about great difficulty for device fabrication and may damage the device reliability. In order to improve the mechanical strength of $Bi_2Te_3$-based materials, vast effort has been made on developing powder sintering technology to prepare dense polycrystalline $Bi_2Te_3$-based materials. The sintered polycrystalline materials also show anisotropic character in both mechanical properties and thermoelectric properties to some extent. The degree of the anisotropy is dependent on the sintering process and raw powder characters such as the grain size and shape.

Thermal deformation process was applied to improve the grain orientation of $Bi_2Te_3$-based polycrystalline materials. In this process, the polycrystalline bulk is re-sintered at a certain high temperature under pressure to cause its thermoplastic deformation. The deformation process is usually accompanied by microstructure evolution such as atom slip along the in-plane direction, grain reversal and grain growth. These microstructure evolution processes are beneficial to improving the grain orientation (re-arranging) as shown in Fig. 4.5 [10]. Furthermore, the lattice distortion, dislocation, and nanocrystals can be also introduced into the materials by the hot deformation treatment resulting in significant reduction of lattice thermal conductivity. Fig. 4.6 shows the enhancement of thermoelectric properties of $Bi_2Te_3$-based polycrystalline materials by hot deformation treatment [5,11].

In addition to the control of grain orientation, nanostructuring and nanocomposite have also been widely used to enhance the thermoelectric properties of $Bi_2Te_3$-based materials [12–14]. For example, by using melt-spinning process, the molten liquid $(Bi, Sb)_2Te_3$ can be ultra-rapidly solidified into amorphous and/or nano-grained flake-like powder. The inert atmosphere protected high-energy ball milling is also used to fabricate $Bi_2Te_3$-based solid solution nanoparticles. Then nanostructured bulk materials can be obtained by densifying the nano-powders by spark plasma sintering (SPS) or hot press, in which numbers of nano-sized grains (10–20 nm) are embedded. The lattice thermal conductivity

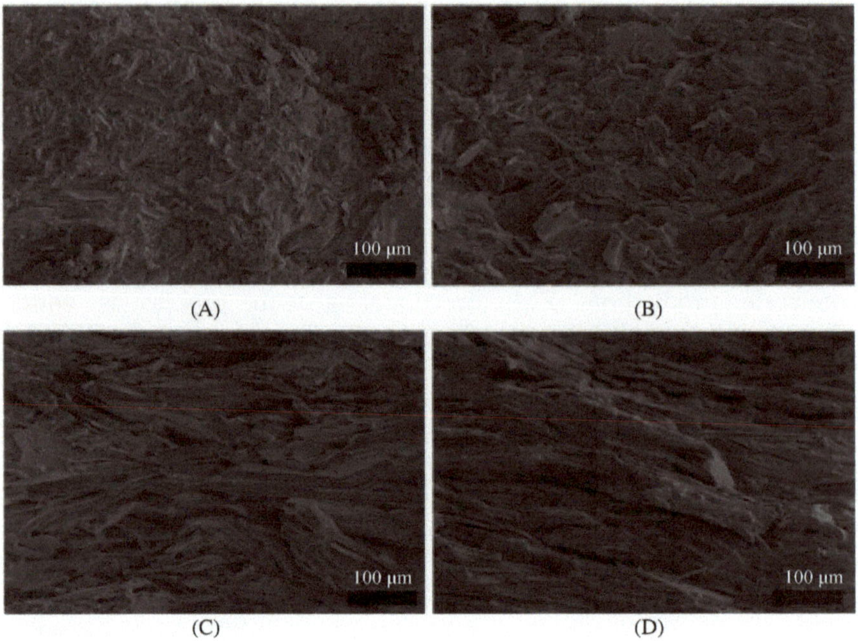


**FIGURE 4.5** Scanning electron microscopy (SEM) of the cross-sections parallel to the pressing direction for the bulk $Bi_2Te_3$-based materials: (A) before hot deformation and (B) after hot deformation for once, (C) after twice and (D) three times of hot deformation [10].

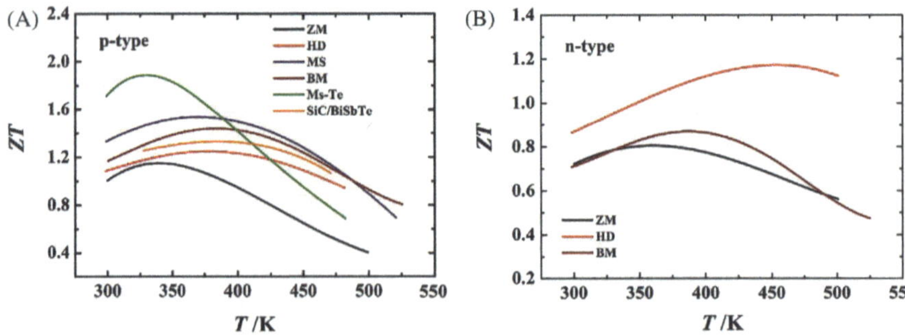


**FIGURE 4.6** Temperature dependence of *ZT* for (A) p-type and (B) n-type $Bi_2Te_3$-based materials prepared by different processes. *BM*: Ball-milling; *HD*: hot deformation; *MS*: melting-spin; *ZM*: zone-melting.

is obviously reduced and then the *ZT* value is significantly increased in the nanostructured bulks [13,14]. For example, the p-type bismuth telluride materials with nanostructure exhibit the *ZT* value as high as 1.4 at 373K [15]. In addition, dispersing nanoparticles of second phase can also

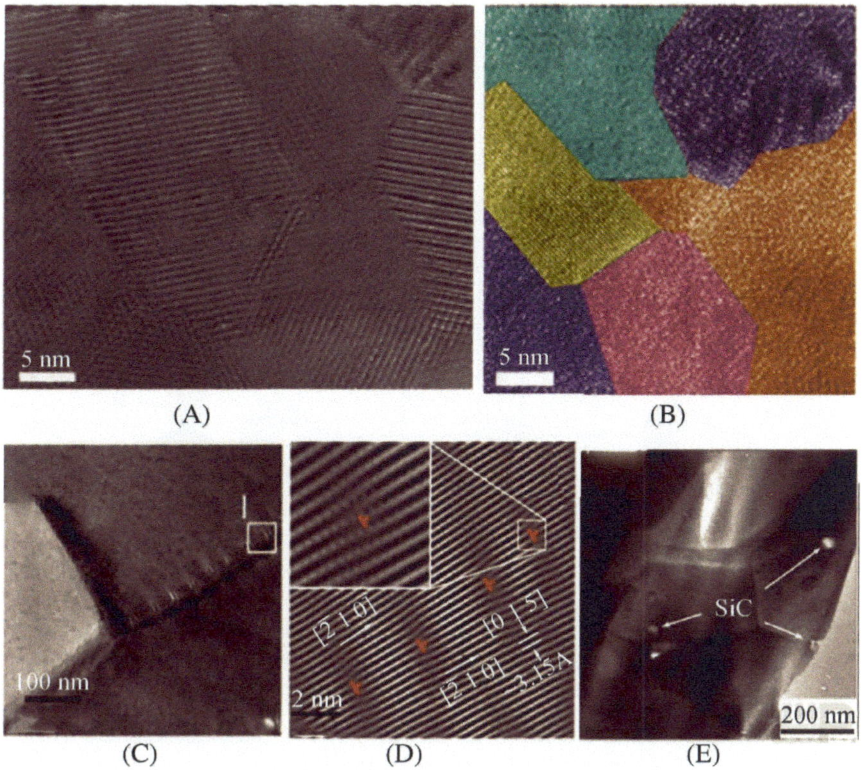


FIGURE 4.7 (A) Microstructure of (Bi, Sb)$_2$Te$_3$ prepared by ball milling; (B) Coherent boundaries in nano-crystalline (Bi, Sb)$_2$Te$_3$ prepared by melt-spinning and SPS; (C) Grain boundary in nano-crystalline (Bi, Sb)$_2$Te$_3$ containing excessive Te prepared by melt-spinning and SPS; (D) Dense dislocation arrays are observed in the selected area of grain boundary in (C); (E) Microstructure of (Bi, Sb)$_2$Te$_3$/nano-SiC composite.

significantly reduce the lattice thermal conductivity. For example, the *ZT* value of bulk (Bi, Sb)$_2$Te$_3$ reaches 1.33 at 373K (Fig. 4.7) when embedding a small amount of SiC nanoparticles [16].

## 4.3 Lead telluride–based compounds: PbX (X = S, Se, and Te)

PbX (X = S, Se, and Te) compounds are one of the earliest discovered thermoelectric materials. Although Seebeck found high thermal power of PbS as early in 1822, the thermoelectric properties of PbX compounds had not been systematically investigated until the Second World War. PbX compounds belong to the family of IV-VI compounds with face-centered cubic NaCl crystal structure and the space group of Fm$\bar{3}$m.

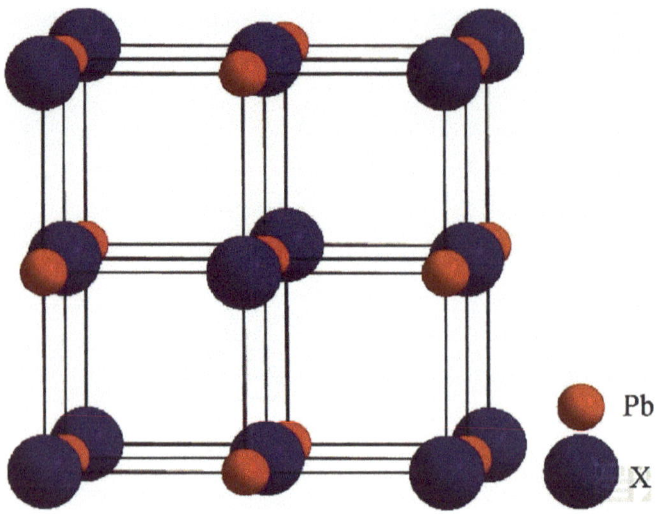


**FIGURE 4.8** Crystal structure of PbX (X = S, Se, and Te) compounds.

**TABLE 4.1** Space group (SG), lattice parameters ($a$/Å), band gap ($E_g$/eV), Debye temperature ($\Theta_D$/K), melting point ($T_M$/K), density [$\rho$/(g/cm$^3$)], Grüneisen's parameter ($\gamma$), the transverse speeds of sound [$\nu_l$/(m/s)], and longitudinal speeds of sound [$\nu_t$/(m/s)] of PbX compounds [17,18].

| PbX | SG | $a$/Å | $E_g$/eV | $\Theta_D$/K | $T_M$/K | $\rho$/(g/cm$^3$) | $\gamma$ | $\nu_l$/(m/s) | $\nu_t$/(m/s) |
|---|---|---|---|---|---|---|---|---|---|
| PbS | Fm$\bar{3}$m | 5.94 | 0.42 | 210 | 1387 | 7.6 | 2 | 1910 | 3460 |
| PbSe | | 6.13 | 0.29 | 191 | 1338 | 8.3 | 1.65 | 1760 | 3220 |
| PbTe | | 6.46 | 0.31 | 163 | 1190 | 8.2 | 1.45 | 1600 | 2900 |

The coordination number of each atom is 6 in these compounds, as shown in Fig. 4.8. PbX (X = S, Se, and Te) compounds are opaque crystals with metallic luster. High brittleness and easy cleavage at low temperatures along (100) plane are also the peculiarity of PbX compounds. However, as temperature increasing, the cleavage gradually becomes insignificant or even disappears (no cleavage above 700°C for PbS, 350°C for PbSe, and 300°C for PbTe). The related physical properties are listed in Table 4.1.

PbX (X = S, Se, and Te) compounds are polar semiconductor with coexistence of ionic and covalent bonds. Early time, they were considered as typical semiconductors dominated by ionic bonds, because their crystal structures present typical ionic bonding characteristics in which the distances between the nearest atoms were close to the sum of ionic radius rather than that of the covalent radius, and their high-frequency dielectric constants differed greatly from the static dielectric constants. Later, it was revealed that charge carriers in PbX are scattered by both optical phonons and acoustic phonons, while being dominated by the acoustic phonon scattering. This confirmed covalent bonding characteristics was dominant in PbX because ionic bonds would principally provide optical phonon scattering as major. However, it is difficult to quantitatively estimate the ratio of ionic bonds to covalent bonds. Nevertheless, the coexistence of ionic and covalent bonds in PbX compounds is undoubted.

Generally, it is easy to achieve n-type or p-type conduction in PbX either by tuning the stoichiometry or by doping with donor or acceptor impurities. Varying the stoichiometric ratio by producing excessive Pb or X, or forming Pb or X vacancies can directly change carrier concentrations. The solubility of excessive Pb/Te and the saturated concentration of Pb/Te vacancies in PbTe is quite low (less than 0.1%), and therefore the highest carrier concentration by varying stoichiometric ratio is usually less than $10^{19}/cm^3$, which is lower than the optimal carrier concentration in PbX compounds. Doping acceptors (Li, Na, K, Rb, Cs, Tl, etc.) can achieve p-type conduction to widely tune carrier concentrations. For example, maximum doping level (molarity) of Na reaches 2%, 0.9%, and 0.5% for PbS, PbSe, and PbTe, respectively [19]. On the other hand, n-type conduction can be achieved by doping Ga, In, La, Sb, Al, and Bi on Pb sites or Cl, Br, and I on X sites. For example, the *ZT* value of Al-doped PbSe reaches 1.3 at 850K [20]. LaLonde et al. replaced Te atoms with I atoms to obtain n-type PbTe with *ZT* value up to 1.4 (at 700K ~ 850K). Fig. 4.9 shows the temperature dependence of thermoelectric properties for typical doped PbX compounds [21].

Solid solution approach is widely used to regulate band structure and optimize thermoelectric performance of PbX compounds, including binary (PbS-PbSe, PbS-PbTe, and PbSe-PbTe) and ternary (PbS-PbSe-PbTe) solutions [22–25]. Continuous solid solution can be formed between PbS and PbSe, while the solid solubility for PbS-PbTe system is quite small at low temperature. PbSe-PbTe forms a limited solid solution below 300°C, but a continuous solid solution above 300°C. As shown in Fig. 4.10, for PbTe materials, there is an $L$ band at $L$ point in the Brillouin zone with the valley degeneracy of 4. In addition, another $\Sigma$ band with a valley degeneracy of 12 locates at the position below the $L$ band with a gap of 0.2 eV at room temperature. By changing the Se content in $PbTe_{1-x}Se_x$ solid solutions, Pei et al. successfully overlapped the $L$ and $\Sigma$ bands at high temperature to realize high valley degeneracy, leading to the optimized electrical transport properties. Meanwhile, lattice thermal conductivity is also suppressed by the alloy scattering. The combined contribution resulted in a *ZT* value as 1.8 at 800K [24].

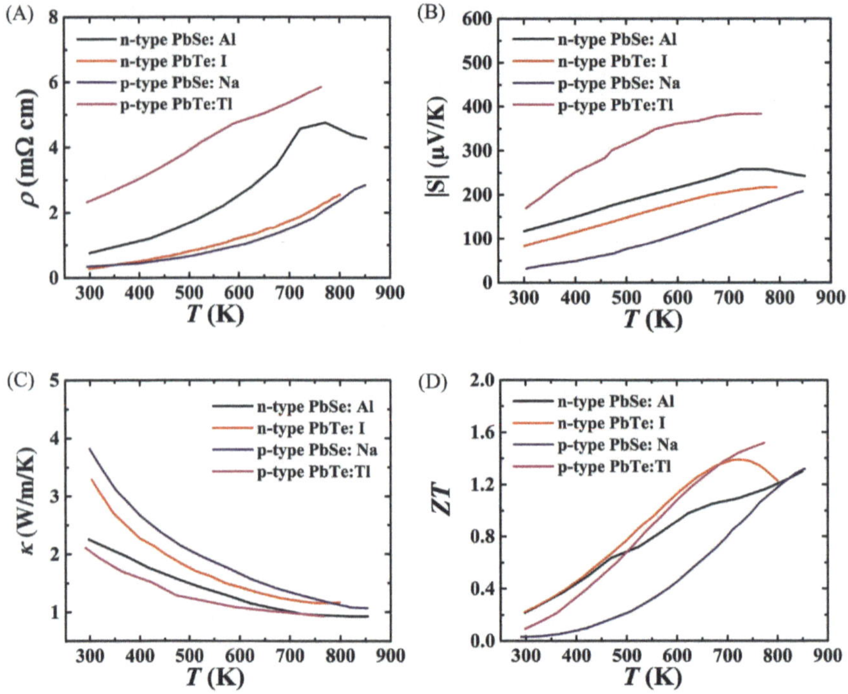


**FIGURE 4.9** Temperature dependence of thermoelectric properties of doped PbX compounds.

By a similar approach of solid solution, Rachel et al. further improved the *ZT* to 2.0 at 800K for $(PbTe)_{1-2x}(PbSe)_x(PbS)_x$ solid solutions with $x = 0.07$ and Na [25] content of 2%.

Nanostructure approach has also been widely used for modulating thermoelectric transport properties of PbX compounds. Especially, because the solid solution solubility in PbX usually varies with temperature, in situ formation of secondary phase at nano-scale (nano-precipitates) becomes practical by changing the synthesis process such as the cooling process. Three typical types of nano-inclusions are observed in PbX-based materials, that is, "coherent," "semicoherent," and "incoherent" according to the structural characteristics of the interface between the PbX matrix and nano-precipitates [26], as shown in Fig. 4.11. The "coherent" nano-inclusions are featured with less mismatch in interface structure and therefore without visual boundary between PbX matrix and nano-precipitates, while the "semicoherent" and "incoherent" inclusions show a clear boundary. All these nano-inclusions can significantly enhance the phonon scattering and reduce the lattice thermal conductivity. It is generally considered that the coherent nano-inclusions act as

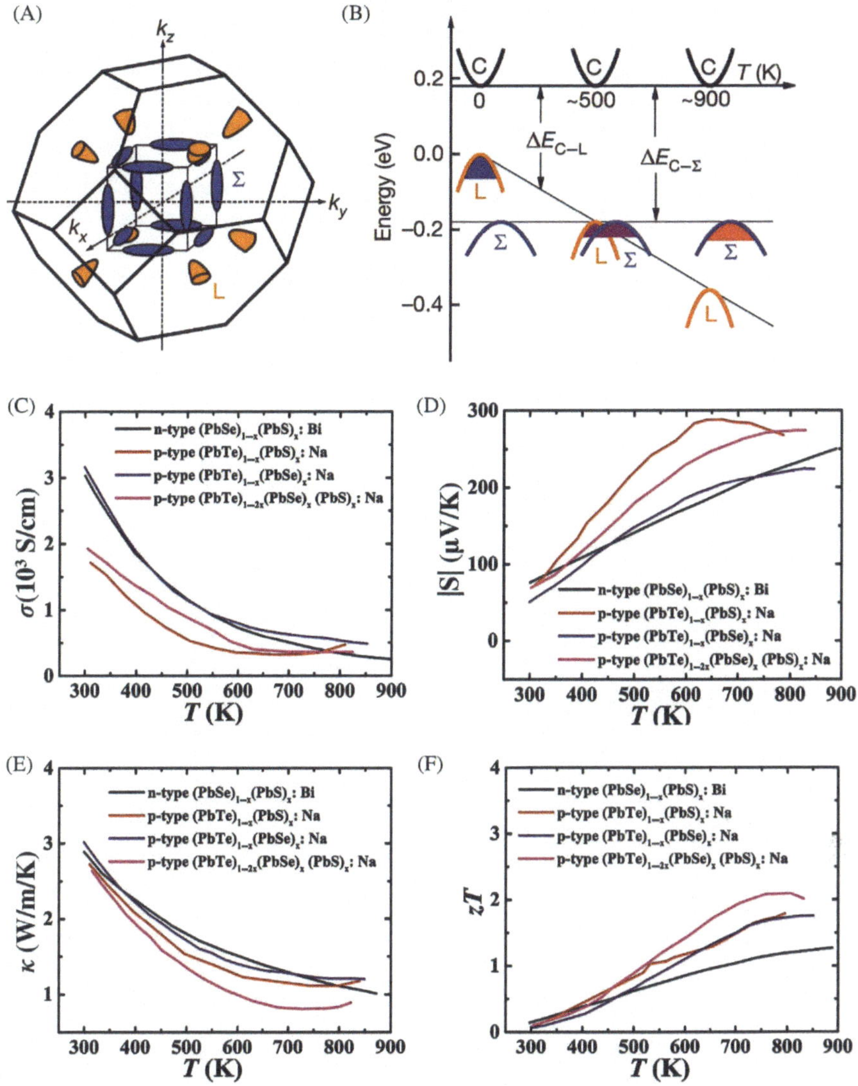


FIGURE 4.10 Schematic map of electronic band structure for PbTe (A and B) and temperature dependence of thermoelectric properties of PbX (X = S, Se, and Te) solid solutions (C–F).

point defects to scatter short wavelength phonons, while "semicoherent" and "incoherent" nanostructures may scatter mid- to long-wavelength phonons. Nevertheless, they also have a slight effect on the electrical transport. Wu et al. prepared 3%Na-doped $(PbTe)_{1-x}(PbS)_x$ by

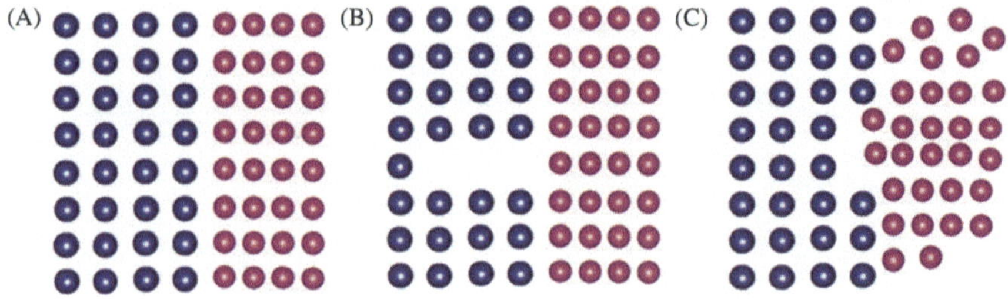


FIGURE 4.11 Schematic illustration of (A) coherent nanostructure, (B) semicoherent nanostructure, and (C) incoherent nanostructure.

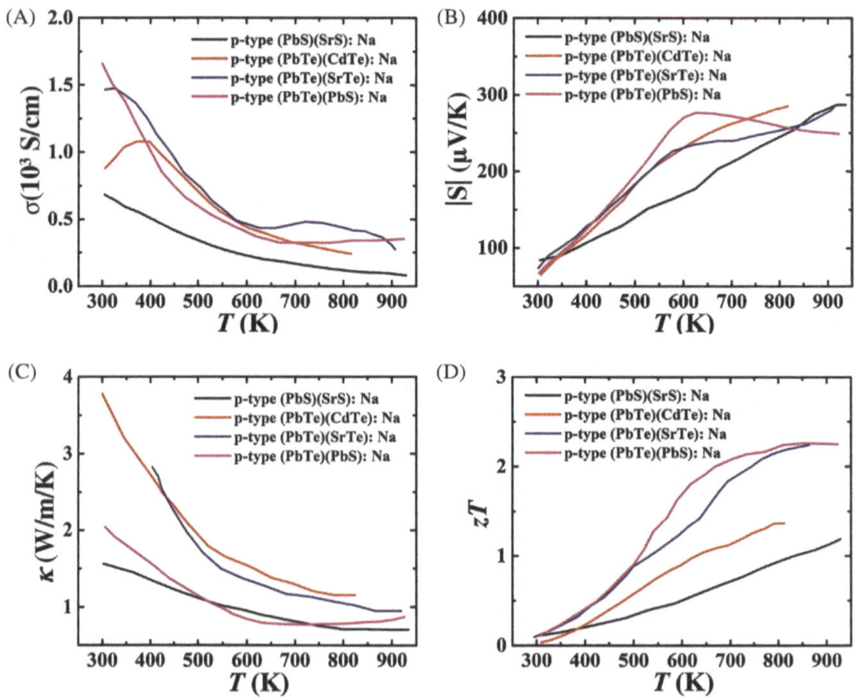


FIGURE 4.12 Temperature dependence of thermoelectric properties of nanostructured PbX (X = S, Se, and Te).

SPS and found that the ultra-low lattice thermal conductivity is majorly attributed to the in situ formed sulfur-rich nano-precipitates. Combined with carrier concentration modulation by the excessive Na, a maximum *ZT* value of 2.3 was realized at 923K for $(PbTe)_{0.8}(PbS)_{0.2}$ [27] (Fig. 4.12).

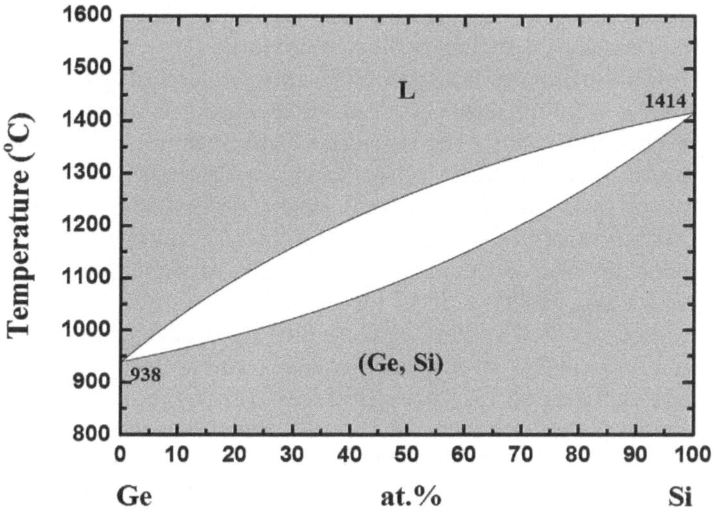


FIGURE 4.13 Phase diagram of Si-Ge binary system.

## 4.4 Silicon-based thermoelectric materials

### 4.4.1 Si-Ge alloys

Element silicon and germanium in group VI crystallize in diamond structure. Although they possess large power factors, both Si and Ge are not good thermoelectric materials due to their high thermal conductivity. As early as 1960s, it was found that alloying Si with Ge can greatly depress the lattice thermal conductivity while maintaining relatively good electron mobility. Subsequently, silicon-germanium alloys became one of the star members of thermoelectric materials especially for the power generation applications at high temperatures due to its large band gap and excellent stability at high temperatures.

Si and Ge form continuous solid solutions as revealed by the Si-Ge phase diagram shown in Fig. 4.13. The physical properties of Si-Ge alloys, such as density, lattice parameters, melting point, Debye temperature, and band gap, vary monotonically with change of the chemical composition, which provides an effective way to adjust their physical properties by simply changing their composition. Practically, Si-rich Si-Ge alloys are generally selected as the thermoelectric material, mainly because of the alloys with high Si content have high melting point, large band gap, low density, and strong antioxidant capacity. In addition, they are also cheaper than Ge-rich alloys. Moreover, Si-Ge alloys with high Si content possess low lattice thermal conductivity. For Si-rich alloys, element doping limits are also high, which provides a large space to modulate electrical properties. Among Si-Ge alloys, $Si_{80}Ge_{20}$ is the most commonly used thermoelectric materials.

The elements of group V, such as P and As, are frequently used as donor dopants, while B and Ga are used as acceptor dopants to modulate carrier concentrations of Si-Ge alloys. The optimal carrier concentration corresponding to the best power factor is about $10^{21}/cm^3$ and $10^{20}/cm^3$ for n-type and p-type Si-Ge materials, respectively. The most successful application of Si-Ge alloy is in the long-life radioisotope thermoelectric generator (RTG) developed by NASA, in which the *ZT* values are 0.5 and 0.93 at 900°C~950°C for p- and n-type alloys, respectively. In later studies, the *ZT* values were further increased to 0.65 for p-type and 1.0 for n-type by optimizing carrier concentrations [28–31]. In 1990s, Dresselhaus et al. [32] proposed low-dimensional and nanostructuring approach to enhance materials' thermoelectric performance, which was also applied to Si-Ge alloys. Joshi et al. made the first breakthrough in thermoelectric properties of Si-Ge bulks by nanostructuring [33]. They fabricated Si-Ge alloy nanoparticles by ball milling (particle size ~20 nm), and then obtained nano-grained polycrystalline Si-Ge bulk by hot press. As compared with the melted ingots or large grain samples, the thermal conductivity is decreased by ~50%, while the electrical properties are not significantly deteriorated. As the result, the maximum *ZT* values of p-type and n-type Si-Ge alloys are increased from 0.5 to 0.95 and from 0.93 to 1.3, respectively. This is contributed mainly by the enhanced phonon scattering through induced dense grain boundaries. In 2012, Bathula et al. prepared n-type nanostructured Si-Ge alloys by the similar mechanical alloying method, and realized its record *ZT* value of 1.5 at 900°C [34]. In 2015, the same group further created the record *ZT* of 1.2 for p-type Si-Ge alloys [35] (Fig. 4.14).

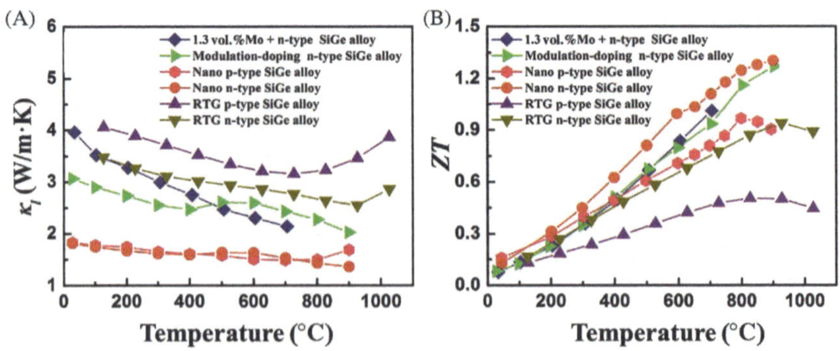


**FIGURE 4.14** Temperature dependence of (A) lattice thermal conductivity and (B) *ZT* value for Si-Ge alloys.

Nanocomposites are also effective to improve the thermoelectric properties of bulk Si-Ge. Mingo et al. [36] introduced silicide or germanide nanoparticles as the second phase into Si-Ge alloys to improve the thermoelectric properties in the whole temperature range. $WSi_2$ and other 17 compound nanoparticles are proposed as possible second phase candidates to effectively reduce lattice thermal conductivity without significantly deteriorating electrical properties. Favier et al. homogeneously dispersed small amount of Mo particles into $Si_{91.3}Ge_{8.0}P_{0.7}$ matrix via ball milling and then converted the Mo particles into $MoSi_2$ nanoparticles by means of solid reaction between Mo and Si-Ge matrix during SPS. The *ZT* value of the obtained nanocomposites is increased by ~43% as compared with the matrix, reaching 1.0 at 700°C [37]. Yu et al. prepared $(Si_{95}Ge_5)_{0.65}(Si_{70}Ge_{30}P_3)_{0.35}$ nanocomposite by using P-doped $Si_{70}Ge_{30}P_3$ nanoparticles and undoped $Si_{95}Ge_5$ particles as raw powders. In this unique "nanocomposite," the charge carriers excited in the doped grains migrate through the undoped matrix with less impurity scattering. And therefore, the carrier mobility and electrical conductivity were increased compared with the homogeneous sample having the same composition. Furthermore, the low lattice thermal conductivity of the $(Si_{95}Ge_5)_{0.65}(Si_{70}Ge_{30}P_3)_{0.35}$ sample proved the effectiveness of phonon scattering by the interfaces between doped and undoped nano-grains [38].

### 4.4.2 $Mg_2X$ (X = Si, Ge, and Sn)

for power generation in middle temperatures. $Mg_2X$ (X = Si, Ge, and Sn) compounds crystalize in the same structure with $Mg_2Si$ and receive general interests as a promising family of thermoelectric materials for power generation in middle temperatures. The constituent elements of $Mg_2X$ compounds are abundant in the earth, low cost, nontoxic, and environmentally friendly. They crystallize in the antifluorite crystal structure (space group $Fm\overline{3}m$) with elements X in the face-centered cubic positions and Mg in the tetrahedral sites as shown in Fig. 4.15. There are three atoms per primitive unit cell. The band gap is 0.75 ~ 0.8 eV for $Mg_2Si$, 0.7 ~ 0.75 eV for $Mg_2Ge$, and 0.3 ~ 0.4 eV for $Mg_2Sn$ [39], respectively, which allows them to be used in the medium-temperature region. The basic physical properties of $Mg_2X$ are shown in Table 4.2. For $Mg_2Si$ and $Mg_2Ge$, the electron effective masses are less than the hole effective mass, and thus the electron mobility is much higher than the hole mobility. $Mg_2Sn$ has the similar electron effective mass and mobility to the hole. Among them, n-type $Mg_2Si$ has a high $\beta$ factor and is one of the most promising thermoelectric materials.

$Mg_2X$ (X = Si, Ge, and Sn) is the only compound in the phase diagram of Mg-X binary system. Because of the same crystal structure and

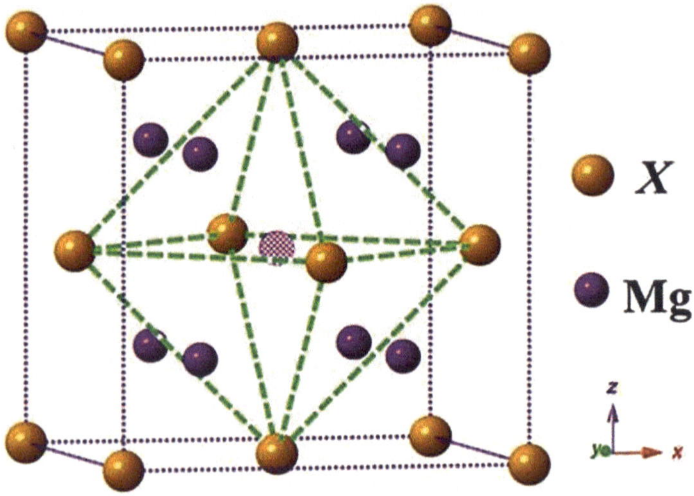


FIGURE 4.15 The crystal structure of $Mg_2X$ (X = Si, Ge, and Sn). The interstitial site is the octahedral center.

TABLE 4.2 Physical properties of $Mg_2X$ (X = Si, Ge, and Sn).

| Compounds | Melting point/K | $a$/ Å | Density / (g/cm$^3$) | $E_g$/ eV | $\mu_n$/[cm$^2$/ (V s)] | $\mu_p$/ [cm$^2$/ (V s)] | $m_n$/ $m_0$ | $m_p$/ $m_0$ |
|---|---|---|---|---|---|---|---|---|
| $Mg_2Si$ | 1375 | 6.34 | 1.88 | 0.77 | 405 | 65 | 0.50 | 0.90 |
| $Mg_2Ge$ | 1388 | 6.38 | 3.08 | 0.74 | 530 | 110 | 0.18 | 0.31 |
| $Mg_2Sn$ | 1051 | 6.76 | 3.57 | 0.35 | 320 | 260 | 1.20 | 1.03 |

similar lattice parameters of the $Mg_2X$ binary compounds, solid solutions can be formed between each two pairs of $Mg_2X$. $Mg_2Si$ and $Mg_2Ge$ can form continuous solid solution, while the solid solution regions in $Mg_2Si$-$Mg_2Sn$ and $Mg_2Ge$-$Mg_2Sn$ phase diagrams are discontinuous. It is generally known that the content range of two-phase regions are about 0.4～0.6 for $Mg_2Si$-$Mg_2Sn$ and about 0.3～0.5 for $Mg_2Ge$-$Mg_2Sn$, though there are still some controversial descriptions of these two quasibinary systems. Fig. 4.16 shows the quasibinary phase diagrams of $Mg_2Si$-$Mg_2Sn$ system.

The easy formation of solid solutions between $Mg_2X$ compounds also provides a favorable space to tune their electrical and thermal properties. $Mg_2X$ compounds are indirect bandgap semiconductors with similar band structures. As an example, Fig. 4.17A shows the band structure

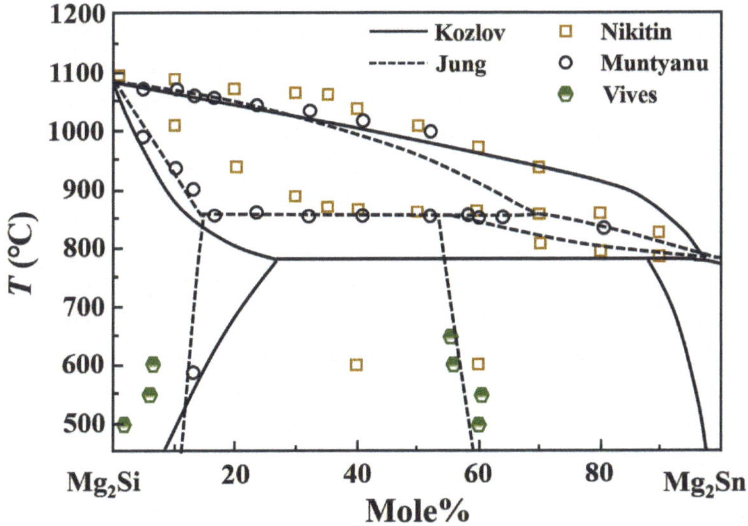


FIGURE 4.16 The quasibinary phase diagram of $Mg_2Si$-$Mg_2Sn$.

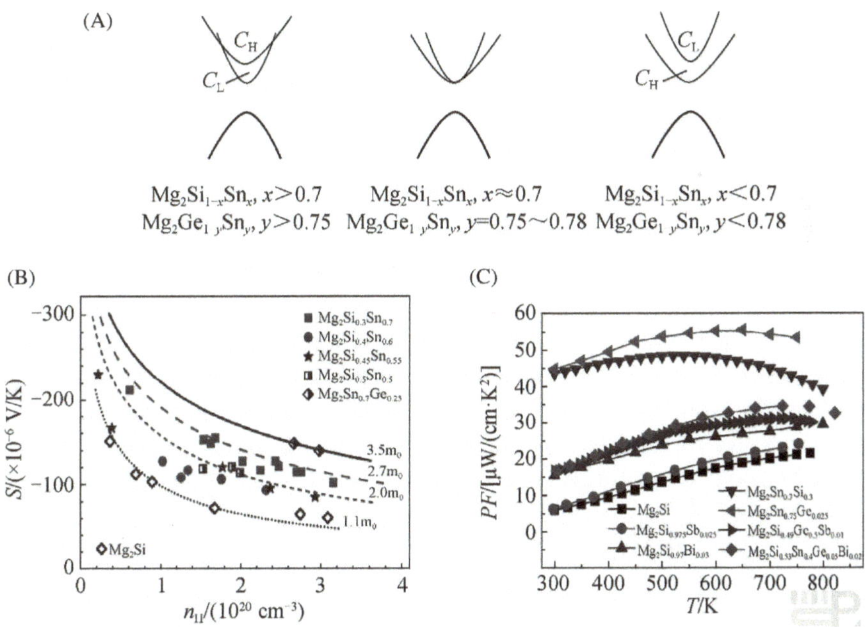


FIGURE 4.17 (A) Schematic diagram of band convergence, (B) the Seebeck coefficient as the function of the carrier concentration, and (C) temperature dependence of the power factor for $Mg_2X$ solid solutions.

of $Mg_2Si$. The valence band maximum is at point $\Gamma$ and the conduction band minimum is at point $X$. Another conduction band is 0.4 eV below the conduction band minimum. The two bands have different effective masses and the heavy band is below the light band. $Mg_2Ge$ has a similar band structure, and the energy gap between these two bands is 0.58 eV. For $Mg_2Sn$, the light band is below the heavy band, and the energy gap between the light band and the heavy band is 0.6 eV. As shown in Fig. 4.17B, in $Mg_2Si_{1-x}Sn_x$ solid solutions, the position of the two bands at the conduction band minimum can be changed by varying Sn content ($x$). The heavy band moves down gradually, and the light band falls after rising firstly, then the two bands overlap near $x = 0.7$, where the two bands will participate in the electron transport producing band convergence as shown in Fig. 4.17A. Because of the converged bands, the effective mass of $Mg_2Si_{1-x}Sn_x$ ($m^* = 2.0 \sim 2.7\, m_0$) becomes higher than those of $Mg_2Si$ ($m^* = 1.0\, m_0$) and $Mg_2Sn$ ($m^* = 1.3\, m_0$). The effective mass of $Mg_2Ge_{1-y}Sn_y$ ($m^* = 3.5\, m_0$) is also higher than those of $Mg_2Sn$ and $Mg_2Ge$. The higher effective mass of electrons contributes significant enhancement of Seebeck coefficient. Liu et al. verified the band convergence in $Mg_2Si_{1-x}Sn_x$ solid solutions through combining the experiment and first principles calculations [40]. Due to the improvement of the power factor, the $ZT$ value of $Mg_2Si_{1-x}Sn_x$ reaches 1.3 at 750K. The alloying induced band degenerate effect is also proven in $Mg_2Ge_{1-y}Sn_y$ solid solutions by Ren et al., in which the heavy and light bands overlap when Sn content ($y$) is at $0.78 \sim 0.75$ and then the maximum $ZT$ value of 1.4 is obtained at 723K [41].

### 4.4.3 High manganese silicide

Mn and Si are very abundant elements in the earth. According to the Mn-Si binary phase diagram (see Fig. 4.18), Mn and Si can form multiple compounds with different stoichiometries. Among them, $MnSi_{1.75-x}$ compositions (Si and Mn atomic ratios change between 1.72 and 1.75) exhibit typically high power factors, and thus receive great attention as potential thermoelectric materials in the medium-temperature region. In this series of compounds, Mn element is at its highest valence state, so they are called high manganese silicon (HMS) compounds. There are four well-known compounds reported with similar composition and structure including $Mn_4Si_7$, $Mn_{11}Si_{19}$, $Mn_{15}Si_{26}$, and $Mn_{27}Si_{47}$. The schematic diagram of the crystal structure of $Mn_4Si_7$ is given in Fig. 4.19. The unit cell consists of a tetragonal sublattice occupied by Mn atoms resembling a "chimney" and a "ladder" sublattice of pairs of Si atoms forming two coupled helices inside the "chimney." If it is regarded as a primitive, other compounds can be considered as its repeated

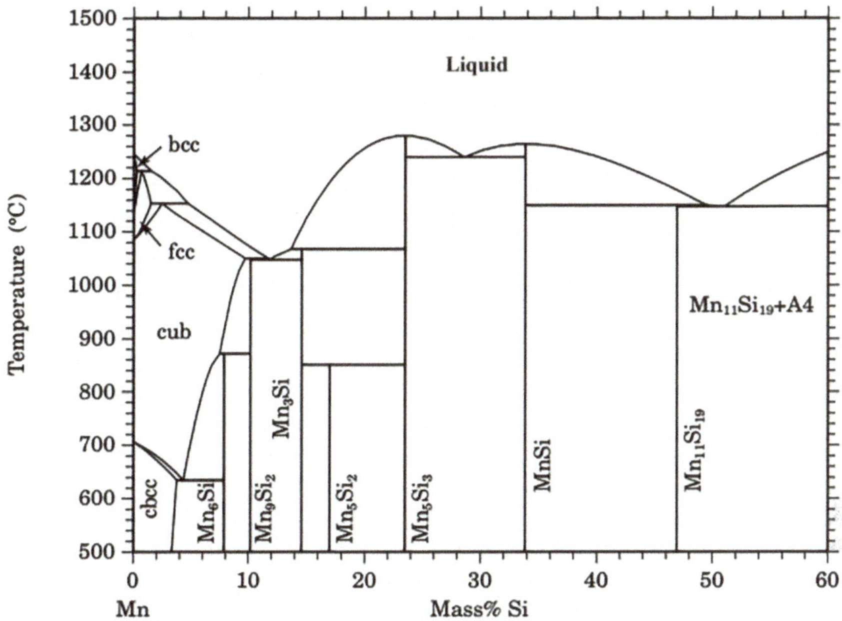


FIGURE 4.18 The Mn-Si binary phase diagram.

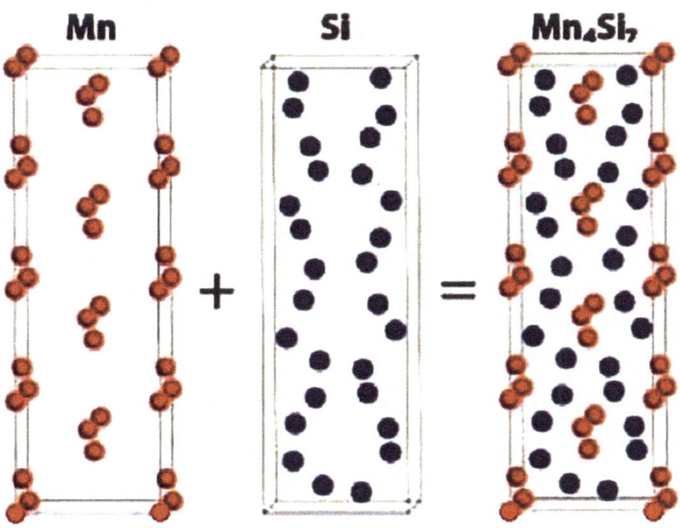


FIGURE 4.19 Crystal structure of $Mn_4Si_7$.

arrangement, which is also called Nowotny chimney-ladder (NCL) structure [42]. According to the phase diagram, a peritectic reaction shall occur in the melting-cooling process, MnSi and Si are often embodied as impurity phases in the HMS materials [43].

For all compounds with NCL structure, the conducting characters are determined mainly by the average valence electron count (VEC) per metal atom. When VEC = 14, it behaves as an intrinsic semiconductor; when VEC < 14, it presents a p-type semiconductor [44]. The VEC of $Mn_4Si_7$ is 14, while other three HMS compounds are all p-type semiconductors. Because the structure and composition of the four HMS compounds are very close to each other, the experimentally synthesized HMS materials are always mixed phases and show p-type conduction.

The easy precipitation of MnSi metallic phase during crystal growth has obsessed researcher attempting to synthesize HMS single crystals. The periodically precipitated thin MnSi layers tend to orient perpendicularly to the c axis of the HMS matrix, which gives adverse impact on thermoelectric performance. Polycrystalline HMS materials without peritectic precipitates can be fabricated by powder process. Umemoto et al. prepared polycrystalline HMS materials by mechanical alloying and SPS, and the *ZT* value reached 0.5. Rapid cooling processes, such as levitation melting, arc melting, etc., are also used to fabricate HMS polycrystalline with fine-grained or nano-grained structure. Zhao et al. [45] and Shi et al. [46] successfully reduced the content of MnSi in HMS materials and improved the phase purity of the HMS through these rapid cooling methods.

Doping approach in HMS is conducted mainly through substituting Mn or Si to form solid solutions, or simultaneously substituting two sites with multiple atoms. It is easy to use Ge to replace Si in HMS, because Ge and Si belong to the same group IVA and can dissolve in any proportion. She et al. [47] prepared $Mn(Si_{1-x}Ge_x)_{1.75}$ by ultra-fast thermal explosion and a rapid plasma–activated sintering (PAS) technique at high temperature and found that the power factor significantly increases when a slight amount of Ge replaces Si. The maximum *ZT* value reaches 0.62 at 850K when Ge doping content at 0.015. Chen et al. [48] found both Al and Ge replacement on Si site can increase the carrier concentration.

### 4.4.4 $\beta$-$FeSi_2$

$\beta$-$FeSi_2$ is a kind of thermoelectric materials which can realize thermoelectric conversion in a wide temperature range from room temperature to 900°C due to its advantages of high-temperature oxidation resistance, nontoxicity, and low cost. Nevertheless, its *ZT* value is relatively low compared with other silicon-based thermoelectric material

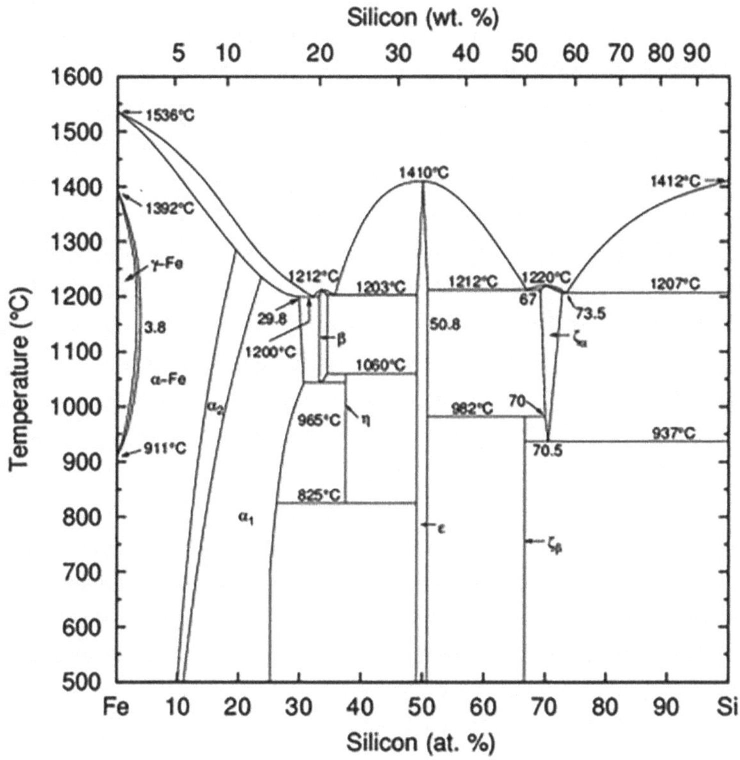


FIGURE 4.20 Phase diagram of Fe-Si binary system.

mainly due to its intrinsically poor electronic conductivity and high thermal conductivity.

Fe and Si are the most abundant elements in the earth and can form a series of silicides with different Fe/Si ratios, such as $\alpha$-$Fe_2Si_5$, $\beta$-$FeSi_2$, $\varepsilon$-FeSi, and $Fe_3Si$ and so on, as shown by the Fe-Si binary phase diagram in Fig. 4.20. Among these binary compounds, $\alpha$-$Fe_2Si_5$, $\varepsilon$-FeSi, and $Fe_3Si$ are metallic or semimetallic phases; only $\beta$-$FeSi_2$ possesses semiconducting behavior. The thermoelectric properties of $\beta$-$FeSi_2$ were firstly studied by Ware and McNeill in 1964 [49]. $\beta$-$FeSi_2$ crystallizes in orthogonal structure (a = 0.9683 nm, b = 0.7791 nm, and c = 0.7833 nm) [50] with the space group of Cmca (Fig. 4.21A) at low temperature. There are 48 atoms in the unit cell. Fe and Si have two kinds of occupation, with slightly different distance of the neighboring atoms. As shown in Fig. 4.21B, $\beta$-$FeSi_2$ is an indirect bandgap semiconductor. The valence band maximum is at point $Y$ and the conduction band minimum is between point $\Gamma$ and $Z$ (about $0.6 \times \Gamma$-$Z$) with the indirect band gap

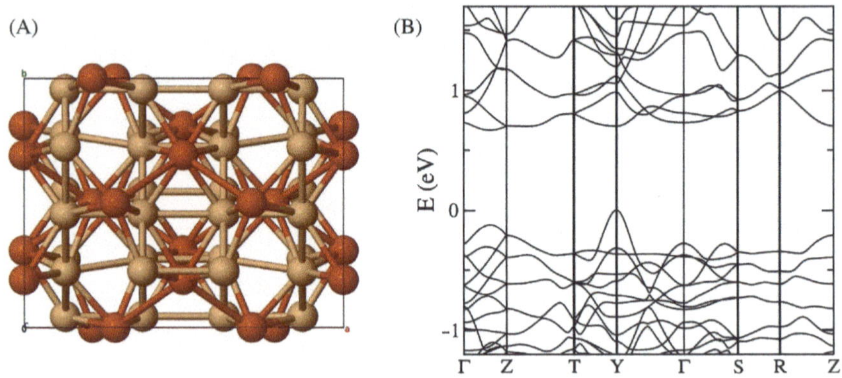


**FIGURE 4.21** (A) The crystal structure and (B) band structure of $\beta$-$FeSi_2$.

about 0.85 eV. The large band gap leads to a very low level of carrier concentration about $10^{17}/cm^3$. Both the conduction band and valence band of $\beta$-$FeSi_2$ are relatively flat, resulting in a large effective mass and low mobility (about $0.3 \sim 0.4\ cm^2/V\ s$) for the band edge carriers [51].

According to Fe-Si phase diagram, the formation of $\beta$-$FeSi_2$ is a complex process and it often takes long term to prepare pure phase in practical. When the Fe-Si melt is cooled down, $\alpha$-$Fe_2Si_5$ and $\varepsilon$-FeSi precipitate from the melt firstly and then $\beta$-$FeSi_2$ is generated at the boundary of $\alpha$ and $\varepsilon$ phases through peritectoid reaction. The peritectoid reaction and the growth of $\beta$ phase are dominated by the Fe/Si diffusion through $\beta$ phase; therefore the growth rate of $\beta$ phase quickly slows down as the reaction progressing. Meanwhile, high dense stack faults easily form in the fresh $\beta$-$FeSi_2$ layer which also hinders the phase transition. Furthermore, the elimination of impurity is also difficult because of the strict chemical stoichiometry of $\beta$ phase. The slow condensation by melting processes generates coarse $\alpha$ and $\varepsilon$ grains, and therefore the transformation of $\alpha/\varepsilon$ to $\beta$ phase is often incomplete remaining large amount of metallic phase in the $\beta$ matrix even after a long time of annealing. Yamauchi et al. [52,53] found that adding small amount of Cu to Fe-Si raw material could significantly accelerate the peritectoid reaction by suppressing the formation of stacking fault in $\beta$ phase, and then shorten the annealing time in several orders. Rapid solidification is effective to produce a fine grain microstructure and accelerate the formation of $\beta$-$FeSi_2$. Chen et al. [54] prepared thin ribbon-like powder of $FeSi_2$ alloy by melt spinning. The product powder consists of $\alpha$ and $\varepsilon$ phases of about 100 nm in grain size. By hot pressing the melt-spun powder, bulk samples of pure $\beta$-$FeSi_2$ with average grain size of about 2 μm are fabricated.

Undoped $\beta$-$FeSi_2$ is an intrinsic semiconductor with poor thermoelectric performance. Generally, element doping is conducted in Fe sites. For example, Mn [55], Al [56], Cr [57], and V [58] are often used as p-type dopants, while Co [59], Ni [58], and Pt [60] are as n-type dopants. Ware and McMeill et al. [49] studied the influence of Al and Co doping on the thermoelectric performance of $\beta$-$FeSi_2$, and realized the relatively large Z ($>2\times10^{-4}$/K) over a wide temperature range (150°C–650°C). Zhao et al. [56] prepared $FeAl_xSi_2$ by melt spinning followed by nitriding treatment. In the Al-containing system, an Al-rich liquid phase tends to be formed at the grain boundary because of low melting point of Al, which would accelerate the peritectoid reaction. Then, nitriding treatment could significantly improve the Seebeck coefficient for Al-doped $\beta$-$FeSi_2$. A maximum *ZT* value of 0.12 is obtained for $FeAl_{0.05}Si_2$ at 723K. Tani et al. [55] doped Mn into $\beta$-$FeSi_2$ and obtained a maximum *ZT* of 0.17 at 873K. They also doped Pt [60] and Co [59] into $\beta$-$FeSi_2$ and turn it into n-type conduction. The maximum *ZT* values for Pt- and Co-doped $FeSi_2$ are 0.14 and 0.25, respectively. Du et al. [61] chose the Co-doped $\beta$-$FeSi_2$ ($Fe_{0.94}Co_{0.06}Si_2$) as the matrix and then alloyed the element Ru at Fe sites to further suppress the lattice thermal conductivity; a maximum *ZT* value of 0.33 at 900K is obtained for $Fe_{0.89}Ru_{0.05}Co_{0.06}Si_2$.

Nanocomposite approach has also been used to optimize $\beta$-$FeSi_2$ through reducing lattice thermal conductivity. Ito et al. [62] mixed a small amount of oxides (such as $ZrO_2$, $Y_2O_3$, $Nd_2O_3$, $Sm_2O_3$, and $Gd_2O_3$) into $Fe_{0.98}Co_{0.02}Si_2$ using mechanical alloying. The uniformly dispersed submicrometer oxide particles significantly reduce thermal conductivity through the enhanced grain boundary scattering to phonons. Morikawa et al. [63] found that the thermal conductivity could be reduced by about 50% via introducing appropriate $Ta_2O_5$ particles into $\beta$-$FeSi_2$. Qu et al. [64] successfully coated ZnO nano-layer on the surface of $\beta$-$FeSi_2$ particles by sol-gel process. Fig. 4.22 shows the representative progress of $\beta$-$FeSi_2$-based thermoelectric materials [54,56,57,60,61,63–72].

## 4.5 Skutterudites and clathrates

In the 1990s, Slack proposed the concept "phonon-glass electron-crystals" (PGEC) for designing thermoelectric compounds, showing that a good thermoelectric material should possess good electrical transport like crystals but low thermal conductivity like glasses. Inspired by this conceptual strategy, cage-structured compounds came into researchers' view as potential thermoelectric materials, because there have framework structure and intrinsic voids in the crystal to provide independent optimization of electrical and thermal transports. Caged clathrates and

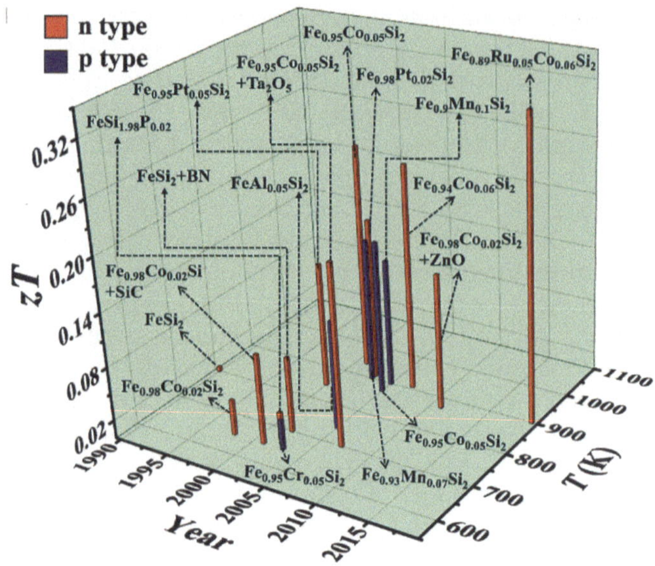


FIGURE 4.22 The representative progress of $\beta$-$FeSi_2$-based thermoelectric materials.

skutterudites exhibit typical PGEG characteristics. Especially, filled skutterudites have been recognized as one of the most important thermoelectric materials for power generation at medium temperatures.

## 4.5.1 Filled skutterudites

The name of skutterudite was firstly given to the mineral $CoAs_3$ originally reposited in Skutterud, a small town in Norway. Afterwards it has been extended to represent a family of compounds having the similar structure with $CoAs_3$. Skutterudites crystallize in body-centered cubic structure with space group of $Im\overline{3}$ as shown in Fig. 4.23A. The general formula is written as $MX_3$, where M can be Co, Rh, or Ir, while X can be P, As, or Sb. There are 32 atoms in each unit cell containing 8 $MX_3$. 8 M atoms occupy the 8c position, and 24 X atoms occupy the 24 g position, while the two icosahedral voids surrounded by 12 X atoms occupy 2a position. M is at the center of regular octahedrons that are composed of 6 X atoms and connected by sharing vertex. The most striking feature of skutterudites is that 4 X atoms form a nearly four atomic ring $[X_4]^{4-}$ at the center of a small cube composed of 8 M atoms. These planar four atomic rings $[X_4]^{4-}$ are orthogonal to each other and parallel to the cubic axis of the crystal. In addition, the icosahedral voids composed of 12 X atoms can be filled with guest atoms, forming filled skutterudite, as shown in Fig. 4.23B. In the skutterudite structure, each M atom and each

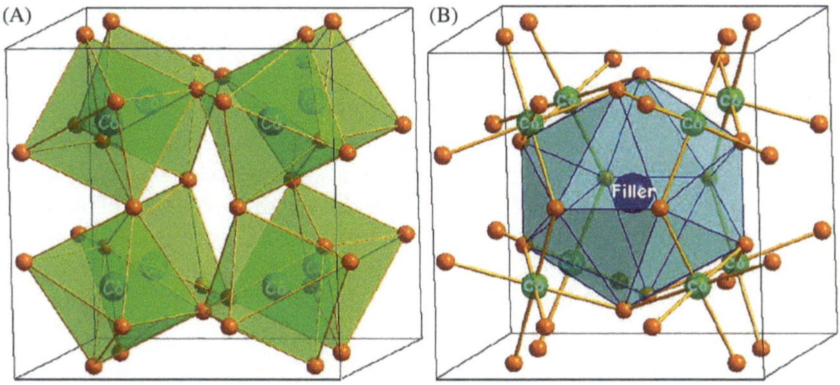


FIGURE 4.23 Crystal structure of (A) binary $CoSb_3$ and (B) filled skutterudites.

TABLE 4.3 Lattice constants ($a$), density ($\rho$), void radius ($r$), melting point ($T_M$), and band gap ($E_g$) for typical binary skutterudites.

| Compounds | $a$/Å | $\rho$/(g/cm$^3$) | $r$/Å | $T_M$/°C | $E_g$/eV |
|---|---|---|---|---|---|
| $CoP_3$ | 7.7073 | 4.41 | 1.763 | >1000 | 0.43 |
| $CoAs_3$ | 8.2043 | 6.82 | 1.825 | 960 | 0.69 |
| $CoSb_3$ | 9.0385 | 7.64 | 1.892 | 873 | 0.23 |
| $RhP_3$ | 7.9951 | 5.05 | 1.909 | >1200 | – |
| $RhAs_3$ | 8.4427 | 7.21 | 1.934 | 1000 | >0.85 |
| $RhSb_3$ | 9.2322 | 7.90 | 2.024 | 900 | 0.8 |
| $IrP_3$ | 8.0151 | 7.36 | 1.906 | >1200 | – |
| $IrAs_3$ | 8.4673 | 9.12 | 1.931 | >1200 | – |
| $IrSb_3$ | 9.2533 | 9.35 | 2.040 | 1141 | 1.18 |
| $NiP_3$ | 7.819 | – | – | >850 | metal |
| $PdP_3$ | 7.705 | – | – | >650 | metal |

X atom contributes nine electrons and three electrons, respectively, to form covalent bonds, and therefore each $M_4X_{12}$ has 72 electrons in total. Basically, skutterudites are semiconductors. The lattice constants, density, void radius, melting point, and band gap for typical binary skutterudites are listed in Table 4.3 [73,74]. Besides, some transition metals (Ni and Pd) and P can also form skutterudite structures, such as $PdP_3$. However, because Ni and Pd have one more electron than M atoms (Co, Rh, or Ir), $NiP_3$ and $PdP_3$ exhibit metallic conduction character.

Among various skutterudites, $CoSb_3$ has received widest attention as thermoelectric material due to its environmentally friendly composition, suitable band gap (~0.2 eV) and high carrier mobility [75]. However, the relatively high lattice thermal conductivity (larger than 10 W/mK at room temperature) had become the biggest obstacle to enhance its thermoelectric performance. In 1996, Sales et al. found that the icosahedral voids in $CoSb_3$ can be partially filled by rare-earth elements if Co is partially substituted by Fe forming filled skutterudite $La(Fe, Co)_4Sb_{12}$, and the lattice thermal conductivity of the filled skutterudite was remarkably decreased as compared with unfilled skutterudite $CoSb_3$. Because substituting Fe provides less electrons than Co, the filled skutterudite $La(Fe, Co)_4Sb_{12}$ exhibits p-type conduction, and the $ZT$ value was dramatically enhanced from 0.5 for $CoSb_3$ to 0.9 for $La(Fe, Co)_4Sb_{12}$ [76]. By measuring the atomic displacement parameters (ADP) of the constituent elements, Sales et al. concluded that the rattling behavior of the filled (guest) atoms contributes the great suppression of lattice thermal conductivity. Due to large radius of the icosahedral voids, the bonds between guest atoms and host atoms (Sb) are much weaker than those between the framework atoms (Co-Sb and Sb-Sb) and therefore the guest atoms generate highly localized vibrations within the lattice voids, which is called "rattlers" by Sales [76]. The rattling mode produces significant scatter to the low-frequency phonons and depresses lattice thermal conductivity.

As for $CoSb_3$-based skutterudites, p-type materials can be obtained by partially substituting Co with Fe, and the carrier and lattice thermal conductivity can be tuned by the filling content of rare-earth element and the Fe substitution content. The charge compensation generated by partial substitution of Fe for Co makes high-valence rare-earth elements easy to be filled into the cage, and the filler amount can be adjusted in a wide range. At the early stage, n-type skutterudites are obtained by partially substituting Co with Ni and/or partially substituting Sb by Te. The filling fraction of rare-earth elements is very low in the $CoSb_3$ binary structure without charge compensation, which makes the space of modulating thermoelectric properties of n-type filled skutterudites very small. The improvement of $ZT$ value of n-type filled skutterudites had lagged behind than p-type skutterudites for a long time. It is not until 2001, when Nolas et al. [77] and Chen et al. [78] discovered Yb- and Ba-filled skutterudites, respectively, that the research on n-type skutterudites was greatly activated. The filling fractions of Yb and Ba in the Sb-icosahedral voids in $CoSb_3$ are much larger than the previously reported rare-earth elements. The large filling fraction provides large space to tune the carrier concentrations and lattice thermal conductivity, and enables dramatic enhancement of $ZT$ of n-type filled skutterudites up to 1.0–1.2.

In 2005, Shi et al. theoretically established the criterion—electronegativity principle—to judge the stability of filled skutterudites. According to this principle, guest atom I can be filled into the Sb-icosahedral voids to form stable filled skutterudites when the difference in electronegativity between the filler atom I and Sb atom is greater than 0.8 ($\chi_{Sb} - \chi_I > 0.8$). They further proposed a method to predict the filling fraction limit (FFL) of alkali metals, alkaline-earth metals and rare-earth elements in skutterudites [79]. These attainments encouraged exploration of various new filled skutterudites such as $Na_xCo_4Sb_{12}$ and $K_xCo_4Sb_{12}$.

The phonon scattering behavior of guest atoms filled in the crystal voids is related to their vibrations. Principally, the vibration branch frequencies introduced by the guest atoms in the phonon spectrum of filled skutterudites are sensitive to the atomic radius, mass, and electronegativity of the filler atoms. Yang et al. calculated the vibration acoustic branch position of the guest atoms (I) in $I_xCo_4Sb_{12}$ based on the resonant model, and found that the acoustic branch frequency decreases for the fillers from alkali metals to alkaline-earth metals, and then to rare-earth metals [80]. In other words, the vibration frequency of alkali metals is the highest, and rare-earth metals are the lowest (see Table 4.4). On this basis, Shi et al. [81,82] proposed the multiple filling approach to strengthen the resonant vibration-induced phonon scattering, that is, introducing different kinds of filler atoms with different resonant vibration frequencies into skutterudites to form multiple-filled skutterudites to achieve broad-frequency phonon scattering. This is experimentally proven in several ternary-filled skutterudites. For example, the lattice thermal conductivity of (Ba, La, Yb)$_x$$Co_4Sb_{12}$ is significantly reduced than those of single- or double-filled skutterudites as shown in Fig. 4.24.

**TABLE 4.4** Elastic constant $k$ and resonance frequency $\omega_0$ in the [111] and [100] directions of $R_{0.125}Co_4Sb_{12}$, where R = La, Ce, Eu, Yb, Ba, Sr, Na, and K [80].

| | | [111] | | [100] | |
|---|---|---|---|---|---|
| R | Mass/$10^{-26}$kg | $k$/(N/m) | $\omega_0$/cm$^{-1}$ | $k$/(N/m) | $\omega_0$/cm$^{-1}$ |
| La | 23.07 | 36.10 | 66 | 37.42 | 68 |
| Ce | 23.27 | 23.72 | 54 | 25.18 | 55 |
| Eu | 25.34 | 30.16 | 58 | 31.37 | 59 |
| Yb | 28.74 | 18.04 | 42 | 18.88 | 43 |
| Ba | 22.81 | 69.60 | 93 | 70.85 | 94 |
| Sr | 14.55 | 41.62 | 90 | 42.56 | 91 |
| Na | 3.819 | 16.87 | 112 | 17.18 | 113 |
| K | 6.495 | 46.04 | 141 | 46.70 | 142 |

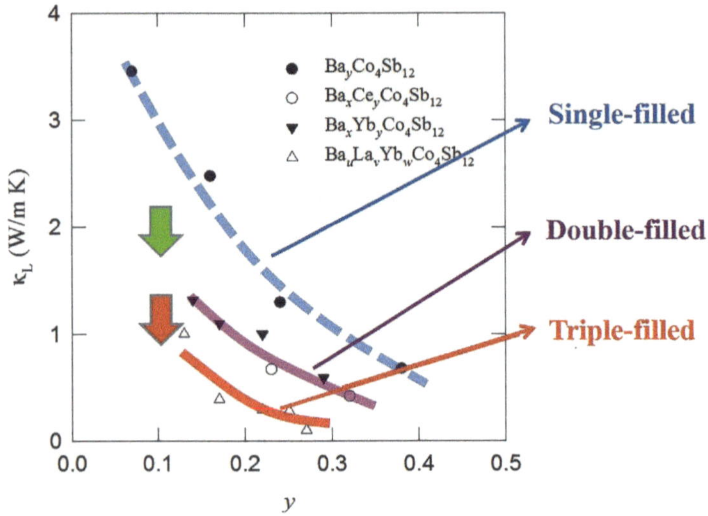


**FIGURE 4.24** Lattice thermal conductivity of single-, double-, and multiple-filled skutterudites.

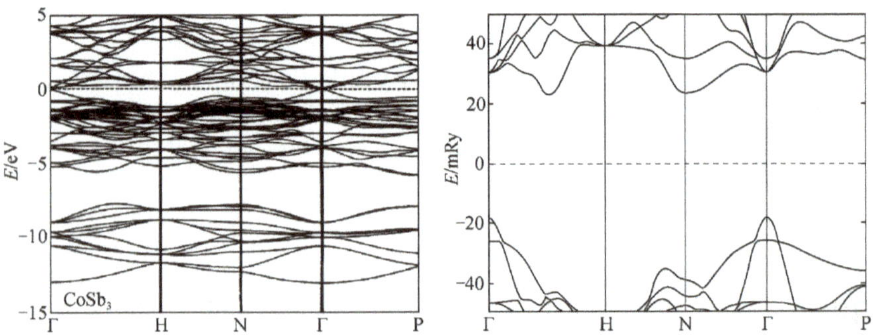


**FIGURE 4.25** The band structure of $CoSb_3$ (left) and $LnFe_3CoSb_{12}$ (right).

The electrical transport properties of $CoSb_3$ and its filled compounds are mainly determined by the framework structure of $CoSb_3$, while less affected by fillers. Fig. 4.25 is the band structure of $CoSb_3$. Due to the antibonding characteristics of Sb *p* electrons, the *p* electrons form a very discrete single band at the valence band maximum, which has a very close linear dispersion relation near point *Γ*, resulting in a very small effective mass. The conduction band minimum is a triple degenerate band formed by the Sb 5p electron band and the Co 3d electron band, resulting in a relatively large effective mass. So the electrical transport performance in n-type binary $CoSb_3$ can be optimized by modifying the carrier concentrations. In filled skutterudites, it is nearly ionic bond

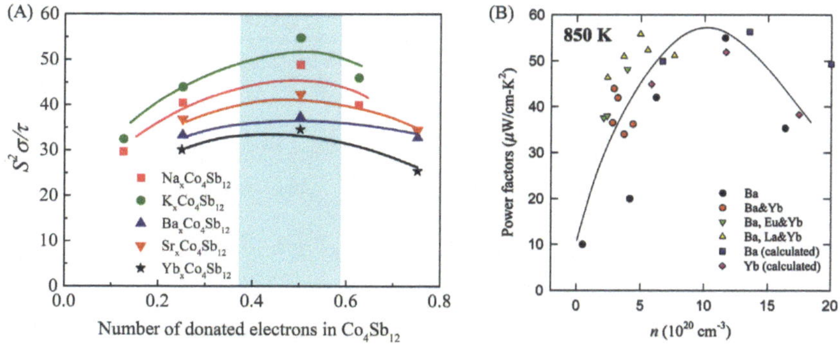


**FIGURE 4.26** Electrical properties of n-type filled skutterudites: (A) calculated electrical properties of single-filled skutterudites and (B) power factors at 850K as a function of carrier concentration for single-, double-, and multiple-filled skutterudites. The solid line shows a trend for both the calculated and measured data.

between the guest and Sb atoms, and their valence electrons are almost completely donated to the frame atoms, which has little effect on the band structure. Therefore, the relationship between Seebeck coefficient and carrier concentration, as well as the relationship between mobility and carrier concentration follow the same law in all n-type unfilled and filled skutterudites. The optimal carrier concentration corresponding to the maximum power factor is about $6 \times 10^{20}/cm^3$, at which each unit cell will obtain 0.4–0.6 electrons from guest atoms, as shown in Fig. 4.26. This gives the guidance of optimal total filling fraction to obtain the maximum power factor. Therefore, combining the multiple filling approach to minimum thermal conductivity, the optimization of thermal and electrical properties can be independently realized in n-type multiple-filled skutterudites, by selecting the kinds of filler elements and total filling fraction, respectively. The optimally designed composition was $Ba_{0.08}Yb_{0.04}La_{0.05}Co_4Sb_{12}$ for n-type filled skutterudites with the maximum *ZT* of 1.7 at 850K [81]. The situation of p-type filled skutterudites is more complex than n-type ones, because p-type filled skutterudites are usually doped by Fe at Co sites. A 3d heavy band at the valence band maximum introduced by Fe atoms can effectively increase the density of the valence band maximum and contribute to the optimization of electrical properties. By simultaneously adjusting doping and filling, the optimal composition is obtained in $LnFe_3CoSb_{12}$ or $LnFe_{3.5}(Co/Ni)_{0.5}Sb_{12}$, with maximum *ZT* values of about 1.1. The multiple filling does also work in p-type filled skutterudites but not as significant as that in n-type ones [83], because the coeffects of filling and Fe substitution already bring their lattice thermal conductivity to a very low level and the strengthened electron scattering by Fe substitution makes it more difficult to further manipulate electrical properties (Fig. 4.27).

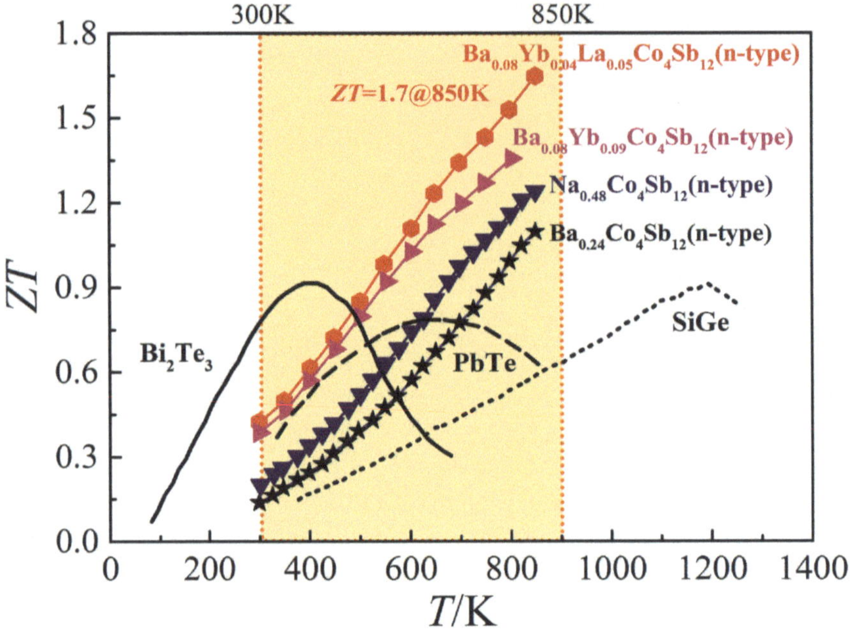


**FIGURE 4.27** Temperature dependence of $ZT$ for single-, double-, and multiple-filled n-type skutterudites.

## 4.5.2 Clathrates

"Clathrate" is firstly used by Powell after the Latin word "Clathratus" as the name of a kind of compounds with a caged structure. These compounds have the following characteristics in crystalline structure in addition to having a large void [84]: (1) large unit cell with high symmetry; (2) guest atoms or molecules are physically trapped in the large voids of lattice, and also stabilize the frame structure; and (3) guest atoms or molecular lattices may be nonstoichiometric, and the interaction between the main lattice frame and guest atoms or molecules will affect the performance of clathrates.

Similar to skutterudites, inorganic clathrates have a three-dimensionally periodic open frame structure. For example, one or multiple host atoms of Al, Si, Ga, Ge, and Sn can form polyhedral voids by tetrahedral bonding, and appropriate guest atoms can be filled in the voids. The numbers of guest atoms and host atoms need to satisfy the Zintl–Klemm valence electron rule. According to the cage number and shape, inorganic clathrates are usually divided into different types, as shown in Table 4.5 [85]. Among them, the structure and physical properties of type I and type II clathrates are widely studied.

**TABLE 4.5** Structure characteristics of inorganic clathrates.

| Type | Formula | Polyhedron | Space group | Example |
|---|---|---|---|---|
| I | $A_2B_6E_{46}$ | $[5^{12}6^2]_6[5^{12}]_2$ | $Pm\bar{3}m$ | $K_8Ge_{46-x}$; $Ba_8Al_{16}Ge_{30}$ |
| II | $A_{16}B_8E_{136}$ | $[5^{12}6^4]_8[5^{12}]_{16}$ | $Fd\bar{3}m$ | $Na_xSi_{136}$; $Cs_8Na_{16}Ge_{136}$ |
| III | $A_{10}B_{20}E_{172}$ | $[5^{12}6^2]_{16}[5^{12}6^3]_4$ | $P4_2/mnm$ | $Cs_{30}Na_{1.33x-10}Sn_{172-x}$, $x = 9.6$ |
| IV | $A_6B_8E_{80}$ | $[5^{12}6^2]_4[5^{12}6^3]_4$ | $P6/mnm$ | $K_7Ge_{40-x}$, $x = 2$ |
| V | $A_8B_4E_{68}$ | $[5^{12}6^4]_4[5^{12}]_8$ | $P6_3/mnm$ | – |
| VI | $B_{16}E_{156}$ | $[4^35^96^27^3]_{16}[4^45^4]_{12}$ | $I\bar{4}3m$ | – |
| VII | $B_2E_{12}$ | $[4^66^8]_2$ | $Im\bar{3}m$ | – |
| VIII | $A_8E_{46}$ | $[3^34^35^9]_8$ | $I\bar{4}3m$ | $Ba_8Ga_{16}Sn_{30}$; $Eu_8Ga_{16}Ge_{30}$ |
| IX | $A_xB_8E_{100}$ | $[5^{12}]_2 + \cdots$ | $P4_132$ | $K_8Sn_{25}$; $Ba_6In_4Ge_{21}$ |

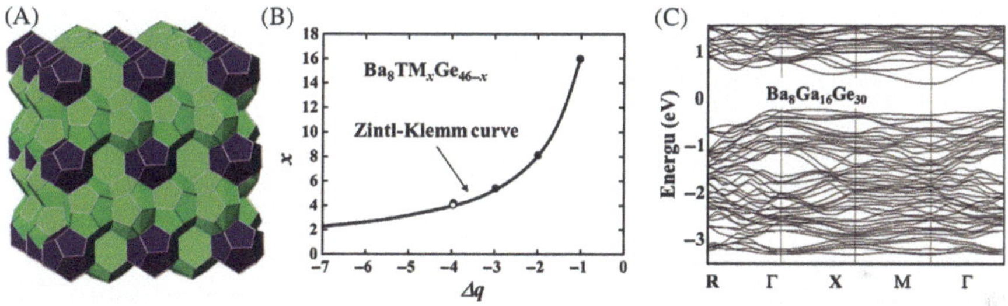


**FIGURE 4.28** (A) Crystal structure of type I clathrate; (B) relationship between the composition ($x$) and the nominal charge $\Delta q$ for type I clathrates $Ba_8TM_xGe_{46-x}$ (TM: 3d transition metal; $\Delta q$: difference of the valence electrons between TM and Ge); and (C) band structure of type I clathrate $Ba_8Al_{16}Ge_{30}$. The solid line in (B) presents the Zintl–Klemm rule $x = 16/\Delta q$.

As shown in Fig. 4.28A, type I clathrates with primitive cubic structure (space group $Pm\bar{3}m$) have the general formula $A_2B_6E_{46}$, where the subscript is the number of atoms. E is the host atoms, which usually form the main frame by tetrahedral coordination. A and B present alkali metals, alkaline-earth metals, or rare-earth atoms, which locate as guest atoms at two different polyhedral frames $E_{20}$ and $E_{24}$. There are 2 dodecahedrons and 6 tetrakaidecahedrons in each $A_2B_6E_{46}$ unit cell. Dodecahedron is enclosed by 12 pentagons, containing 20 host atomic positions. Tetrakaidecahedron is enclosed by 12 pentagons and 2 regular hexagons, containing 24 host atomic positions. The dodecahedron is coplanar with the tetrahedron. 8 polyhedrons can hold guest atoms A and B. The atoms of the main-group III can partially replace atoms of the main-group IV forming compounds with a general formula $A_2B_6D_{16}E_{30}$, in which there are three types of crystallographically atomic positions in the framework: 6c, 16i, and 24k. For the eight guest atoms, six atoms occupy the center of the tetrahedron 6d, while two atoms occupy the center of the dodecahedron 2a.

Clathrates exhibit intrinsically low lattice thermal conductivity because of the violent vibration of the guest atoms scatters phonons significantly similar to the filled skutterudites. Among the inorganic clathrates, Si-, Ge-, and Sn-based clathrates ($A_xB_yE_z$, E = Si, Ge, Sn) have attracted extensive interest as thermoelectric materials. The *ZT* value of single crystal $Ba_8Ga_{16}Ge_{30}$ synthesized by Czochralski method reaches 1.35 at 900K [86]. Single crystal type I clathrate $Ba_8Au_{5.3}Ge_{40.7}$ prepared by the Bridgman method has a low lattice thermal conductivity and its *ZT* value reaches 0.9 at 680K [87]. Gaining high power factor ($S^2\sigma$) has been the most priority in clathrate thermoelectrics. Shi et al. pointed out that the substitution of host atoms needed to satisfy the Zintl–Klemm rule, and found that introducing ionized impurities into the framework could distort the local electrical potential field and enhance thermal power. And then they fabricated polycrystalline bulk of Ni-doped type I clathrates $Ba_8Ni_yGa_zGe_{46-z}$ with a high *ZT* up to 1.2 at 1000K [88].

## 4.6 Superionic conductor thermoelectric materials

Most of the state-of-the-art thermoelectric materials are crystalline semiconductors, and the reduction of lattice thermal conductivity is the common approaches used to enhance the thermoelectric performance. According to the phonon transport theory, the reduction of lattice thermal conductivity in a solid matter is limited due to the long-range order of the crystal structure. There exists a minimum lattice thermal conductivity in a solid, which is equal to the value in its completely disordered glass state. The lattice thermal conductivity is in proportion to both the phonon meaning free path and volumetric heat capacity ($C_v$). Recent study on the thermal transport of ionic conductors revealed that the lattice thermal conductivity can be reduced not only by decreasing phonon meaning free path but also by diminishing the lattice volumetric heat capacity [89]. It was firstly discovered in $Cu_2Se$ ionic conductor [90]. The high-temperature β-phase $Cu_2Se$ crystalizes in antifluorite structure and is a superionic conductor. The Se atoms in β-$Cu_2Se$ form a relatively rigid face-centered cubic sublattice, providing a crystalline pathway for electrons (or more precisely holes), while the Cu ions are highly disordered around the Se sublattice and can easily migrate at high temperatures. The ion migration activation energy is about 0.14 eV. The disorderedly

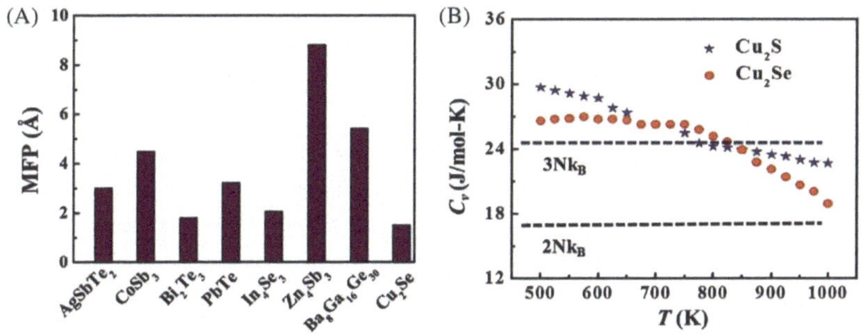


FIGURE 4.29 (A) Minimum phonon mean free paths of $Cu_2Se$ and several typical thermoelectric materials and (B) temperature dependence of heat capacity of $Cu_2Se$ and $Cu_2S$.

distributed Cu ions strongly scatter phonons to diminish the phonon mean free path. As shown in Fig. 4.29A, the phonon mean free path in β-$Cu_2Se$ is estimated about only 1.2 Å, which is much smaller than the values in other conventional thermoelectric materials [90]. In particular, the heat capacity of β-$Cu_2Se$ (both $C_p$ and $C_v$) shows abnormal decrease with increasing temperature, which disobeys the Dulong–Petit law for a solid material as shown in Fig. 4.29B. This unusual behavior is explained by the "liquid-like" behavior of copper ions. In the Dulong–Petit law, the volumetric heat capacity ($C_V$) of a solid material is equal to $3Nk_B$. However, the limit value of $C_V$ is $2Nk_B$ by Trachenko. This is because the shear wave vibration modes shall not contribute to the total state density of phonons when $\omega$ is less than a certain characteristic frequency $\omega_0$ in liquids [91]. In $Cu_2X$ (X = S, Se, Te) ionic conductors, due to the easy migration of Cu ions, some shear vibrational modes shall be weakened or even eliminated, showing a damping effect for shear waves. Therefore, the $C_V$ shall be less than the value in solids ($3Nk_B$) at high temperatures. Instead, it will approach to the limit value in liquids ($2Nk_B$), that is, its values are between $2Nk_B$ and $3Nk_B$. Structurally, there are two sublattices in $Cu_2Se$. One is the Se-rigid sublattice to maintain a good solid and another is liquid-like Cu ion sublattice to scatter phonons and show abnormally low $C_V$. Furthermore, the electronic band structures near the Fermi level are mainly contributed by the Se sublattice, which is less affected by Cu ions. Such synergetic characters are used to independently (or partially independently) tune the thermal and electrical transports of ionic conductors to realize extremely low thermal conductivity and high power factors. Accordingly, a novel concept of "phonon-liquid and electron-crystal" (PLEC) is used to describe such abnormal and interesting properties in ionic conductors, which can be regarded as the good extension of "phonon-glass electron-crystal" (PGEC) concept proposed by Slack. Nevertheless, the new category of PLEC thermoelectric materials are frequently called "liquid-like" thermoelectric materials. Various synthesis techniques, such as traditional long-time melting and annealing method, self-propagating high-temperature synthesis (SHS), melt-rapid cooling crystallization, and ball milling, have been used to fabricate $Cu_{2-x}X$ (X = S, Se, Te) chalcogenides and their thermoelectric performance has been successively improved with recorded $ZT$ values ranging from 1.5 to 2.1 [90,92–96], which are among the highest values of existing thermoelectric materials. Especially, the low cost and environmental friendliness make them receive much attentions for power generation applications.

A unique phenomenon was observed in $Cu_2Se$ during its second-order phase transition from $\alpha$ phase to $\beta$ phase near 400K [97]. During the phase transition, the cubic sublattice formed by Se atoms changes very small, while the arrangement of Cu ions embedded in the Se sublattice gradually changes with temperature, approaching to the high-temperature sites. Accordingly, the crystal structure of $Cu_2Se$ gradually changes from layered monoclinic structure to cubic structure as temperature increases. This process differs from the rapid structure changes in the first-order phase transitions such as the one in $Ag_2Se$ [98]. Such gradual structure change in second-order phase transition in $Cu_2Se$ produces extremely large fluctuations in the density and crystallographic structure, and thus leads to extremely remarkable scattering to both electrons and phonons around the critical temperature. Such critical scattering results in dramatic decrease of charge mobility, distinguished increase of Seebeck coefficient, and significant reduction of lattice thermal conductivity. For example, the Seebeck coefficient of $Cu_2Se$ near the critical temperature is about 170 μV/K, which is twice of that at room temperature, while the thermal conductivity near the critical temperature is reduced to 0.2 W/(mK). The thermoelectric performance of $Cu_2Se$ during the phase transition is therefore greatly improved, although the electrical conductivity is also much decreased.

The thermoelectric properties of the Cu-based binary chalcogenides $Cu_{2-x}X$ (X = S, Se, Te) are also optimized by forming solid solutions of two or three of the chalcogenides. For example, $Cu_2S$-$Cu_2Te$ solid solutions show extremely low lattice thermal conductivity and a large $ZT$ of 2.1 at 1000K [99]. In $Cu_2S$-$Cu_2Te$ solid solutions, element S and Te have much different atomic sizes and masses, and thus generate extremely large lattice distortions. However, all the elements are homogeneously distributed at room temperature in the macro- and micro-scales. As shown in Fig. 4.30A, peculiar mosaic structure is observed in the single-phase polycrystalline $Cu_2(S, Te)$ in the nano-scale. Within a single crystal-like (quasisingle crystal) grain, there are several subgrains or "nano-blocks" with the size of 10~20 nm that have a nearly identical crystal orientation but slightly tilt or rotate to maintain high lattice

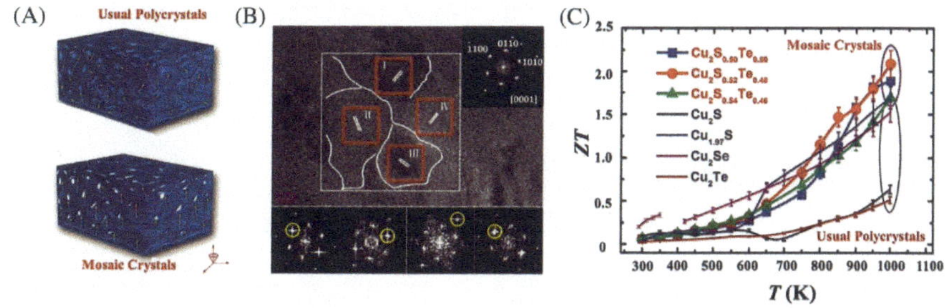


**FIGURE 4.30** (A) Structural characteristics of a nano-mosaic crystal, (B) high-resolution transmission electron microscopy (HRTEM) image, and its fast Fourier transform (FFT) diffractogram, where the nano-crystallites labeled as I-IV show nearly identical diffractograms. Variations in the brightness of different reflections corresponded to a slight misorientation with respect to each other. (C) temperature dependence of *ZT* values for $Cu_2S$-$Cu_2Te$ solid solutions.

coherence. These mosaic grains ensure excellent electronic transport properties, while introduce long-range disorder scattering to thermal phonons. Hence a glass-like thermal conductivity is observed in this nano-mosaic structure. In addition, such mosaic crystalline structure may be an ideal example to display a nanostructure imported into bulk materials to produce strong quantum confinement effects and perhaps even electron barrier filtering, leading to much enhanced electron effective masses and Seebeck coefficient. And therefore, the optimization of electron and phonon transport is expected to be simultaneously promoted to achieve ultrahigh thermoelectric performance.

Similar to $Cu_2X$, a number of ternary superionic conductors, such as $CuCrSe_2$ and CuAgSe, also show PLEC feature with interesting thermoelectric properties. Compared with binary liquid-like materials, ternary superionic materials have more complex chemical compositions and crystal structures, and therefore provide a larger space for the optimization of thermoelectric properties. For example, $CuCrSe_2$ is fabricated and its thermoelectric properties are studied. The Cu ions in $CuCrSe_2$ are disorderedly distributed at high temperatures ($>$365K), leading to a low lattice thermal conductivity of 0.6–0.8 W/(mK) between 300K and 800K. Accordingly, its *ZT* reaches 1.0 at 773 K [100]. CuAgSe is another typical superconductor having the same crystal structure as the $\beta$-$Cu_2Se$ above 470K [101]. P-type CuAgSe is demonstrated having a high *ZT* value of 1.0 at 673K, and is regarded as a promising thermoelectric material at medium-temperatures [102]. $Cu_7PSe_6$ is a kind of argyrodite-type compounds and a typical complex superionic semiconductor. Argyrodite-type compounds have a general formula of $A_{(12-y)/x}{}^{x+}B^{y+}Q_6{}^{2-}$, in which x and y are the valence states of A-cations ($Li^+$, $Cu^+$, and $Ag^+$) and B-cations ($Ga^{3+}$, $Si^{4+}$, $Ge^{4+}$, $Sn^{4+}$, $P^{5+}$, and $As^{5+}$), respectively, and Q = S,

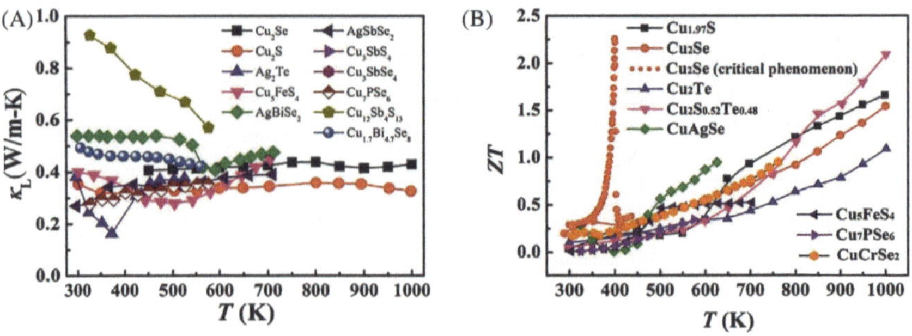


**FIGURE 4.31** The lattice thermal conductivity and *ZT* values of representative superionic conductor thermoelectric materials.

Se, or Te. The crystal structure of $Cu_7PSe_6$ can be understood as what some $Se^{2-}$ anions form a face-centered cubic close packing, and the excess $Se^{2-}$ anions and tetrahedral $[PSe_4]^{3-}$ units occupy the tetrahedral interstices alternately. Cu ions fill in the anionic skeleton structures, and can move along a specific diffusion path in $[PSe_6]^-$ skeletons as temperature increases. The thermal conductivity of $Cu_7PSe_6$ compound is only $0.3\sim0.4$ W/(mK) between 300K and 600K, and its *ZT* value is about 0.35 at 575K [103]. In 2014, Qiu et al. [104] studied the thermoelectric properties of sulfide bornite compound $Cu_5FeS_4$. Like other superionic conductors, it also exhibits an ultra-low lattice thermal conductivity with the value only about $0.3\sim0.4$ W/(mK) and a maximum *ZT* value of 0.3 at 700K. By forming solid solutions between $Cu_5FeS_4$ and $Cu_2S$, the electrical conductivity is effectively improved and the *ZT* reaches 1.2 at 900K. In addition, a number of superionic conductors such as $Ag_8GeTe_6$ [105], $Ag_8SnSe_6$ [106], $TmCuTe_2$ [107], and so on also show excellent thermoelectric performance. Fig. 4.31 illustrates the lattice thermal conductivity and *ZT* values of representative superionic conductor thermoelectric materials.

## 4.7 Oxide thermoelectric materials

It is not until the late 1990s, the oxides receive increasing interest as thermoelectric materials. Especially, the exited report on a high *ZT* of above 1.0 in $Na_xCoO_{2-\delta}$ monocrystalline (long whisker-like single crystal) in 2001 set off a new upsurge of research on oxide thermoelectrics [108]. Oxide thermoelectric materials have obvious advantages over intermetallic compounds, such as excellent high-temperature thermal stability, high oxidation resistance, low cost, nontoxic, and simple synthesis.

In layered oxide thermoelectric materials, there is an important concept called "block module" [109] to independently tune electrical and

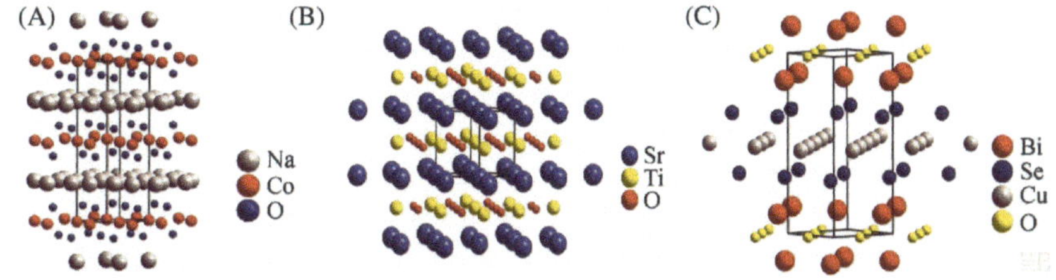


**FIGURE 4.32** Crystal structures of several typical thermoelectric oxides: (A) $Na_xCoO_2$; (B) $SrTiO_3$; and (C) BiCuSeO.

thermal transports. Taking Na-Co-O and Ca-Co-O as examples, they have the $CoO_2$ layers, which are similar to $CdI_2$-typed layered compounds. In $Na_xCoO_2$, there are two different types of layers. One is the incomplete $Na^+$ layer and another is the conductive $CoO_2$ layers. $Na^+$ is randomly distributed in the layers between the $CoO_2$ layers, which is formed by the edge-sharing $CoO_6$ octahedrons, as shown in Fig. 4.32A. This structure leads to low electrical resistivity and moderate Seebeck coefficient, and thus relatively large power factors around 50 $\mu W/(cmK^2)$. The high Seebeck coefficient in $Na_xCoO_2$ can be explained by the spin entropy [110]. In experiment, it is shown that the Seebeck coefficient increases when increasing Na content [111]. The distortion of $Na^+$ layer leads to a relatively low lattice thermal conductivity. Generally, K, Sr, Y, Nd, Sm, Yb, etc., can be doped at the Na sites, while Mn, Ru, Zn, etc., can be doped at the Co sites to tune their thermoelectric properties. $Ca_3Co_4O_9$ is another example having "block" layer structure containing $[Ca_2Co_3]_m[CoO_2]$ $[Ca_2Co_3]_m$ mismatch layers, where $CoO_2$ is conductive and $Ca_2CoO_3$ is insulating [112]. Such complex layer structure produces a large Seebeck coefficient and relatively low resistivity. Because of the high similarity of the layered structure in cobaltate compounds, interlayer composites are prepared by cogrowth of different cobaltates. For example, Liu et al. [113] synthesized $Ca_3Co_4O_9/\gamma$-$Na_{0.66}CoO_2$ composite in which $Ca_3Co_4O_9$ and $\gamma$-$Na_{0.66}CoO_2$ nano-layers are stacked up alternately layer by layer forming an interlayer structure, and found that the power factors of the interlayer composite are enhanced more than 50% than the $Ca_3Co_4O_9$ matrix.

Perovskite oxides, such as $CaMnO_3$ and $SrTiO_3$, have also attracted attention as promising thermoelectric materials [114–117]. $SrTiO_3$ crystalizes in cubic structure (Fig. 4.32B) with a wide bandgap ($E_g \sim 3.2$ eV). Generally, $SrTiO_3$ exhibits n-type conduction due to the existence of intrinsic oxygen vacancy defects. La and/or Dy are often used as dopants. The *ZT* value of La-doped $SrTiO_3$ reaches 0.27 at 1073K [114], while La and Dy Co-doping enhanced the *ZT* up to 0.36 at 1045K [115]. $CaMnO_3$ is orthogonal perovskite structure with the space group of

Pnma. The substitution at Ca site using rare-earth elements (Yb, Tb, Nd, and Ho) with different ionic radius sensitively influences the thermoelectric properties of $CaMnO_3$ ceramics [116]. The maximum *ZT* value of Yb-doped $CaMnO_3$ reaches about 0.2, and that of Nb-doped sample reaches over 0.3 at about 1000K [117].

In addition to the above two families of thermoelectric oxides, BiCuSeO-based oxides are a new but outstanding oxide thermoelectric material. It crystalizes in tetragonal structure (Fig. 4.32C) with the space group of P4/nnm, which is similar to ZrSiCuAs [118]. This structure can be considered as an alternating stack consisting of conductive layers $(Cu_2Se_2)^{2-}$ and insulating layers $(Bi_2O_2)^{2+}$ along the c axis. BiCuSeO possesses a wide band gap (about 0.8 eV). BiCuSeO has low lattice thermal conductivity due to weak chemical bond (Young's modulus $E$ is about 76.5 GPa) and strong inharmonic characteristics (Grüneisen parameter $\gamma$ is about 1.5). Compared with other thermoelectric materials, BiCuSeO possesses moderate Seebeck coefficient but poor electrical conductivity. Hence, for BiCuSeO-based oxides, optimizing electrical transport properties through forming solid solution or doping has been the major action to modulate its thermoelectric properties [119]. Substituting bivalent ions on Bi sites can effectively increase the electrical conductivity and maintain high Seebeck coefficient. For example, when the heavy metal Pb and alkaline-earth metal Ba are Co-doped into BiCuSeO-based oxide, the maximum *ZT* value reaches above 1.1 at high temperature [120] (Fig. 4.33).

## 4.8 Others

### 4.8.1 Half-Heusler (HH) compounds

The general formula for half-Heusler (HH) alloys can be written as XYZ, where X is the most electronegative transition metal or rare-earth metal such as Hf, Zr, Ti, Er, V, or Nb, Y is the least electronegative transition metal such as Fe, Co, or Ni, and Z is a main-group element such as Sn or Sb. Half-Heusler alloys crystallize in the MgAgAs structure with the space group of $F\bar{4}3m$. Their structure can be described as consisting of four interpenetrating face-centered cubic sublattices, where X, Y, and Z atoms occupy the crystallographic positions of 4a (0,0,0), 4c (1/4, 1/4, 1/4), and 4b (1/2, 1/2,1/2), respectively, while the fourth position 4d (3/4, 3/4, 3/4) is vacant. Because one of the four sublattices is unoccupied, these compounds are called half-Heusler alloys with the crystal structure given in Fig. 4.34A. When the fourth position 4d is fully occupied by Y atoms, these compounds are full-Heusler alloys ($XY_2Z$). Most of the full-Heusler alloys are metals, while half-Heusler compounds are half-metals or semiconductors. For half-Heusler compounds, due to the

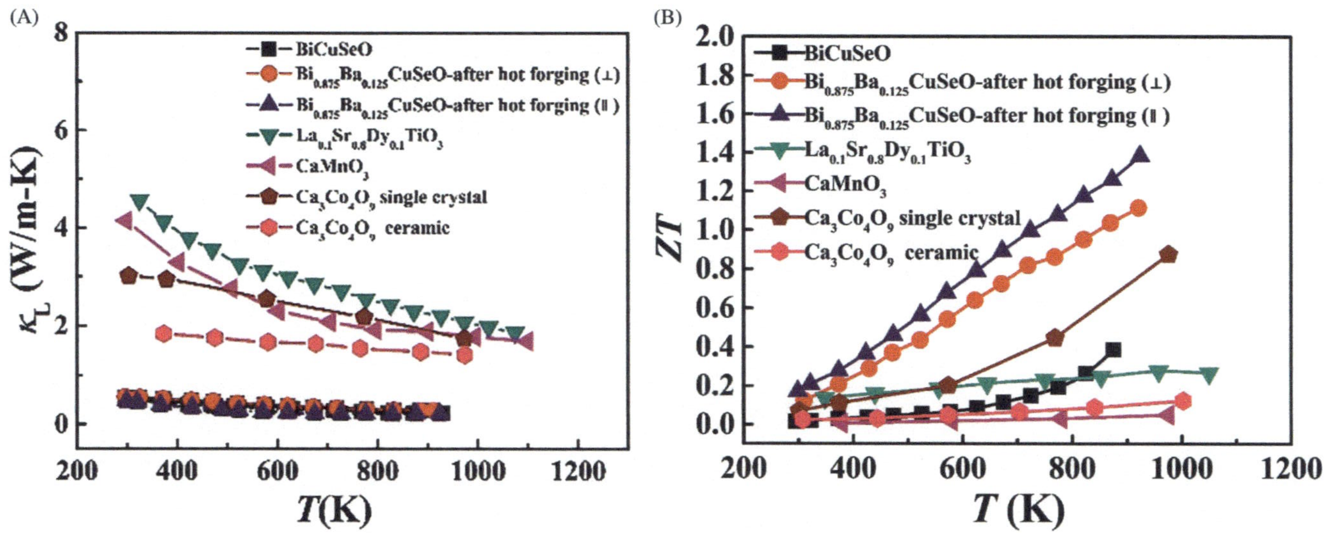


FIGURE 4.33 Temperature dependence of (A) lattice thermal conductivity and (B) *ZT* value for typical oxide thermoelectric materials.

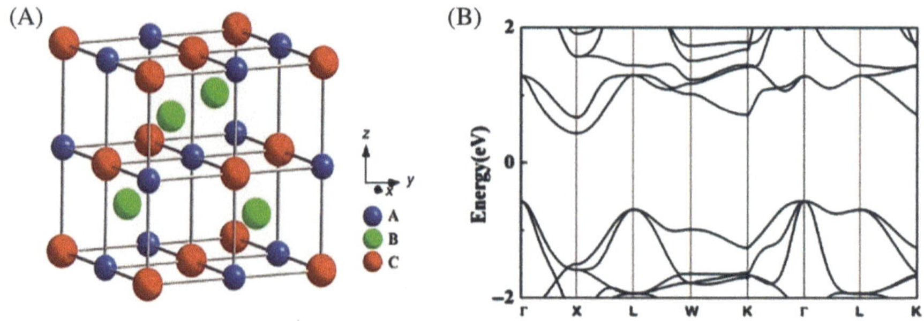


**FIGURE 4.34** (A) Crystal structure of HH alloys and (B) band structure of TiCoSb.

presence of the vacant sites in the lattice, the number of bonds between X and Z atoms is reduced, and the distance between X, Y, and Z atoms is increased, then the orbital overlaps of *d* electrons are weakened and the band gap is formed. That is the reason why half-Heusler compounds are half-metallic materials or semiconductors. In HH compounds, the physical properties and band structure are dictated by the VEC. When VEC = 8 or 18, Fermi level locates in the middle of the band gap, and HH compound shows semiconducting behavior. First principles calculations indicate that HH alloys are stable semiconductor with the band gap of 0~1.1 eV when they have 18 electrons per unit cell [121,122]. Some unique physical features make HH compounds suitable for thermoelectric applications. The highly symmetrical crystal structure makes HH compounds have high valley degeneracies. The density of states near the Fermi level is usually dominated by the *d* electron state of transition metals, which makes the conduction band or valence band have a large effective mass. Thus a high Seebeck coefficient and a moderate electrical conductivity can be obtained. Among HH thermoelectric materials, MNiSn-based, MCoSb-based (M = Ti, Zr, or Hf), and XFeSb-based (X = V, Nb, or Ta) HH alloys have been widely studied. MNiSn-based alloys are the most popular n-type materials, while MCoSb-based and XFeSb-based compounds are commonly used as p-type materials. High power factor about 30 $\mu W/(cm/K^2)$ is obtained by substituting Sb for Sn in (Ti, Zr, Hf)NiSn-based materials [123].

The biggest obstacle for HH alloys as thermoelectric materials is its very high lattice thermal conductivity, which is mainly due to its simple crystal structure. Half-Heusler alloy is composed of three different atoms, and isoelectronic alloying and chemical doping are available at all these three atomic positions. This provides a large room to optimize the thermoelectric performance. For example, the lattice thermal conductivity can be efficiently decreased by isoelectronic alloying. Yu et al. prepared $Hf_{1-x}Zr_xNiSn_{1-y}Sb_y$ alloy by combining levitation melting and SPS [124].

Substituting Sn for Sb is used to optimize carrier concentrations and improve the power factor, while isoelectronically alloying Zr by Hf is effective to reduce the lattice thermal conductivity. Through these synergetic approaches, the *ZT* value of $Hf_{0.6}Zr_{0.4}NiSn_{0.98}Sb_{0.02}$ has been improved up to 1.0 at 1000K [125]. Nanocomposite is also utilized to enhance *ZT* of n-type HH materials by combining the arc-melting, ball-milling, and hot-pressing processes. Considering the high cost of Hf element, Ren et al. tried to replace Hf by more Zr and achieved comparable *ZT* in composition of $Hf_{0.25}Zr_{0.75}NiSn_{0.99}Sb_{0.01}$, which is more acceptable for large-scale applications [126].

Similar with n-type HH alloys, isoelectronic alloying at all three atomic positions can be used to tune the thermoelectric properties of MCoSb-based p-type materials. In earlier work, researchers focused on optimizing the carrier concentrations by substituting Sb with Sn and reducing lattice thermal conductivity by element alloying at Hf site, and moderate value of *ZT* about 0.5 is obtained for $ZrCoSb_{0.9}Sn_{0.1}$ and $Zr_{0.5}Hf_{0.5}CoSb_{0.8}Sn_{0.2}$ at high temperatures [127,128]. Further improvement of *ZT* is realized in more complex isoelectronic alloys and nanostructured bulk materials, such as $ZT = 0.8$ at 973K for nano-grained $Zr_{0.5}Hf_{0.5}Ti_{0.12}CoSb_{0.8}Sn_{0.2}$ [129] prepared by combining ball milling and hot pressing, and $ZT = 1.0$ at 1073K for $Hf_{0.44}Zr_{0.44}Ti_{0.12}CoSb_{0.8}Sn_{0.2}$ [130]. Fig. 4.35 summarizes the temperature-dependent *ZT* for several typical half-Heusler alloys.

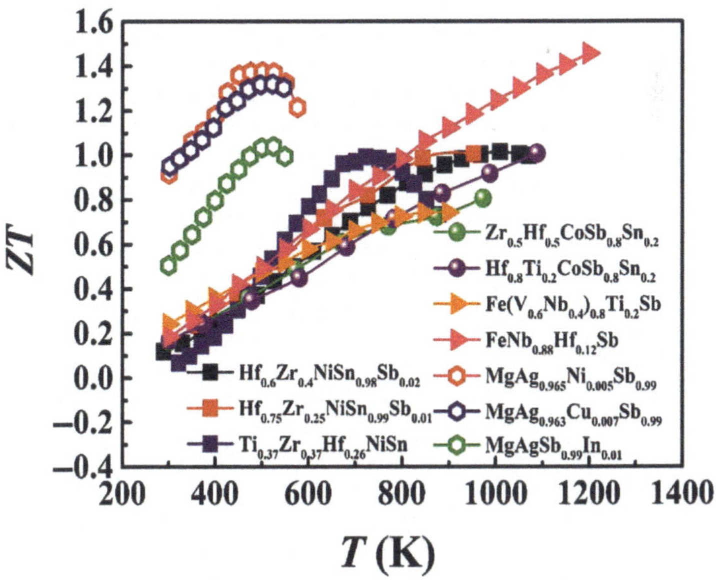


**FIGURE 4.35** Temperature dependence of *ZT* value for typical half-Heusler alloys.

XFeSb-based compounds (X = V, Nb, and Ta) are another kind of promising HH thermoelectric materials [131,132]. Undoped XFeSb is an n-type semiconductor with a narrow band gap of about 0.32 eV, and its power factor reaches 48 μW/(cm/$K^2$) at 300K. However, the thermal conductivity of XFeSb is generally higher than that of MNiSb alloys, which restricts *ZT* value below 0.25. Compared with n-type materials, p-type XFeSb-based alloys possess better thermoelectric performance. The valence band edges of MNiSn-based and MCoSb-based alloys are triple degeneracy at the center of Brillouin region (at the point *Γ*), while the valence band edge of XFeSb-based alloys is double degeneracy which is located at the point *L* in Brillouin region. Since HH alloys have a cubic structure, *L* point has four symmetrical carrier energy valleys, which means that the degeneracy of its valence band is 8 (=4 × 2) in XFeSb-based alloys, which is higher than that of MNiSn-based and MCoSb-based alloys ($N_v = 3$). The hole concentration can be tuned by substituting V/Nb with Ti for p-type $V_{0.6}Nb_{0.4}FeSb$ alloy [133]. In addition, the *ZT* value of nanostructured Hf-free NbFeSb-based HH alloys can reach 1.0 at 973K [134]. Moreover, the *ZT* value of NbFeSb-base HH alloy is further improved greatly by Hf doping with a value of 1.5 at 1200K [135].

### 4.8.2 Diamond-like compounds

Diamond-like compounds are a big family of compounds having the distorted diamond structures. Around 2009, the thermoelectric properties of the quaternary diamond-like compounds, $Cu_2CdSnSe_4$ [136] and $Cu_2ZnSnSe_4$ [137], were firstly studied and moderately *ZT* values of 0.65 at 700K and 0.95 at 850K were reported, respectively. Since then, the diamond-like compounds gradually came under spotlight as a new category of thermoelectric materials, and a vast number of new diamond-like compounds with high *ZT* have been discovered, such as $Cu_2ZnSn_{0.9}In_{0.1}Se_4$ ($ZT = 0.95$ at 850K) [137], $Cu_2Sn_{0.9}In_{0.1}Se_3$ ($ZT = 1.14$ at 850K) [138], $Ag_{0.95}GaTe_2$ ($ZT = 0.77$ at 850K) [139], $CuInTe_2$ ($ZT = 1.18$ at 850K) [140], $CuGaTe_2$ ($ZT = 1.4$ at 900K) [141], and so on. Temperature dependence of the *ZT* values for typical diamond-like compounds is given in Fig. 4.36.

The diamond-like structure provides a huge space to design and discover new compounds with tunable electronic structures and therefore thermoelectric properties. Fig. 4.37 shows the crystal structure of several typical diamond-like compounds and its relationship with the structures of diamond and/or binary sphalerite. The physical properties of typical diamond-like structure compounds are listed in Table 4.6. Using group I atom Cu and group III atoms (Ga, In) to replace two divalent Zn atoms and doubling the cubic unit cell of ZnSe along the c axis, tetragonal

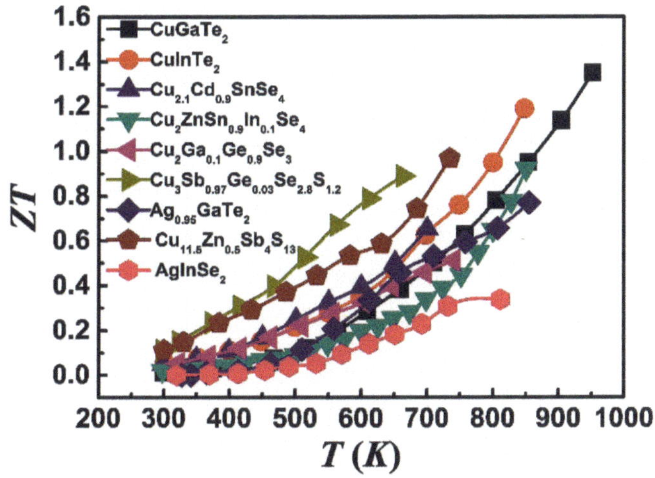


FIGURE 4.36 Temperature dependence of *ZT* values for typical diamond-like compounds.

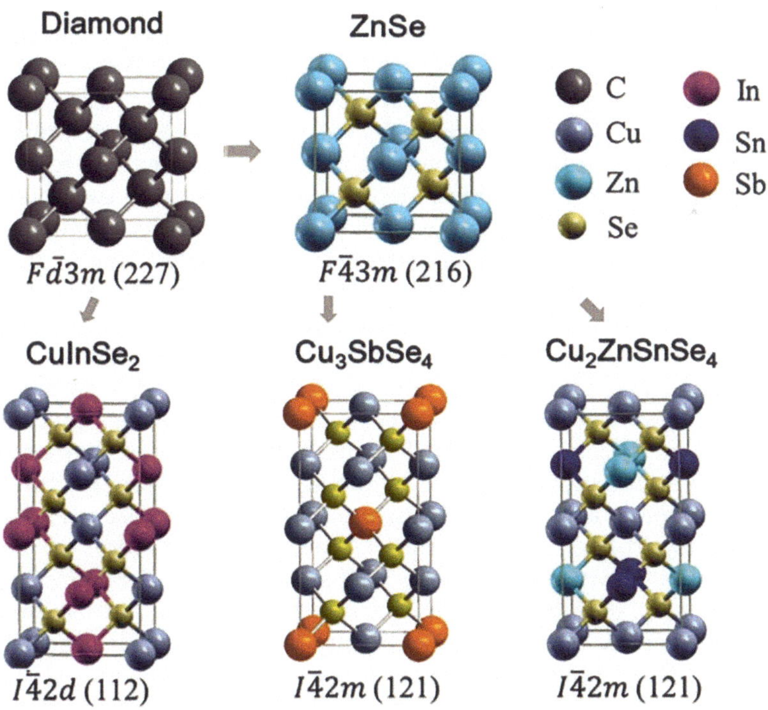


FIGURE 4.37 The crystal structures of diamond and several typical diamond-like compounds.

TABLE 4.6 The physical properties of typical diamond-like structure compounds.

| Compound | $E_g$/eV | $T_M$/K | $\rho/(g/cm^3)$ | 2-c/a | $\kappa$/[W/(mK)] |
|---|---|---|---|---|---|
| ZnSe | 2.8 | 1793 | 5.26 | 0 | 19 |
| $CuInSe_2$ | 1 | 1260 | 5.65 | 0 | 2.9 |
| $CuInTe_2$ | 1.02 | 1050 | 6.02 | 0.0046 | 5.8 |
| $Cu_2SnSe_3$ | 0.7 | 971 | 5.76 | 0.01 | 3.5 |
| $Cu_3SbSe_4$ | 0.3 | 734 | 5.77 | 0.007 | 3.0 |
| $Cu_2ZnSnSe_4$ | 1.43 | 1081 | 5.67 | 0.007 | 3.2 |
| $Cu_2CdSnSe_4$ | 0.9 | 1063 | 5.77 | 0.05 | 1.01 |

chalcopyrite $CuAX_2$ (A = Ga or In; X = Se or Te) is deduced. When three Zn atoms are replaced by two Cu atoms and one tetravalent Sn or Ge atom, binary sphalerite structure mutates exclusively into the $Cu_2BSe_3$ (B = Sn or Ge) configuration which has many possible phases with the cubic sublattice. Similarly, the ternary tetrahedrite compound $Cu_3SbSe_4$ can be obtained by substituting three Cu and one Sb with four Zn. Using two Cu and a tetravalent atom (Sn or Ge) to replace a trivalent atom A (Ga or In) in tetragonal chalcopyrite $CuAX_2$, one can further obtain the quaternary compounds $Cu_2ABSe_4$ (A = Zn or Cd; B = Sn or Ge).

The ternary and quaternary compounds derived from sphalerite structure commonly have a typical tetrahedral coordination configuration, where each anion is surrounded by four cations and each cation is surrounded by four anions. In general, substituting cations shall decline the symmetry of anion atom sublattice. The substituted anions will displace from their ideal center of tetrahedron resulting in elongation along different directions for different structures, for example, elongating along (1/4, 1/4, 1/4) direction in sphalerite structure, ($x$, 1/4, 1/8) in chalcopyrite, ($x$, $x$, $z$) in stannite and ($x$, $y$, $z$) in kesterite. That is to say, the anion-centered tetrahedron structure deviates from the regular tetrahedron with a certain distortion. The distortion degree of diamond-like structure is generally expressed by using the distortion parameter $\eta$ ($\eta = c/2a$, where $a$ and $c$ are the lattice parameters). Apparently, the distorted diamond-like structure shall cause the reduction of lattice thermal conductivity which mostly well satisfies the $T^{-1}$ relation due to the dominant phonon-phonon Umklapp scattering as shown in Fig. 4.38A. Element alloying can further reduce the thermal conductivity of diamond-like compounds. For example, $CuInTe_2$-$CuGaTe_2$ [142] and $Cu_3SbSe_4$-$Cu_3SbS_4$ [143] solid solutions show lower thermal conductivity (see Fig. 4.38B) than those in the matrix compounds. Furthermore, the lattice thermal conductivity can also be effectively reduced by

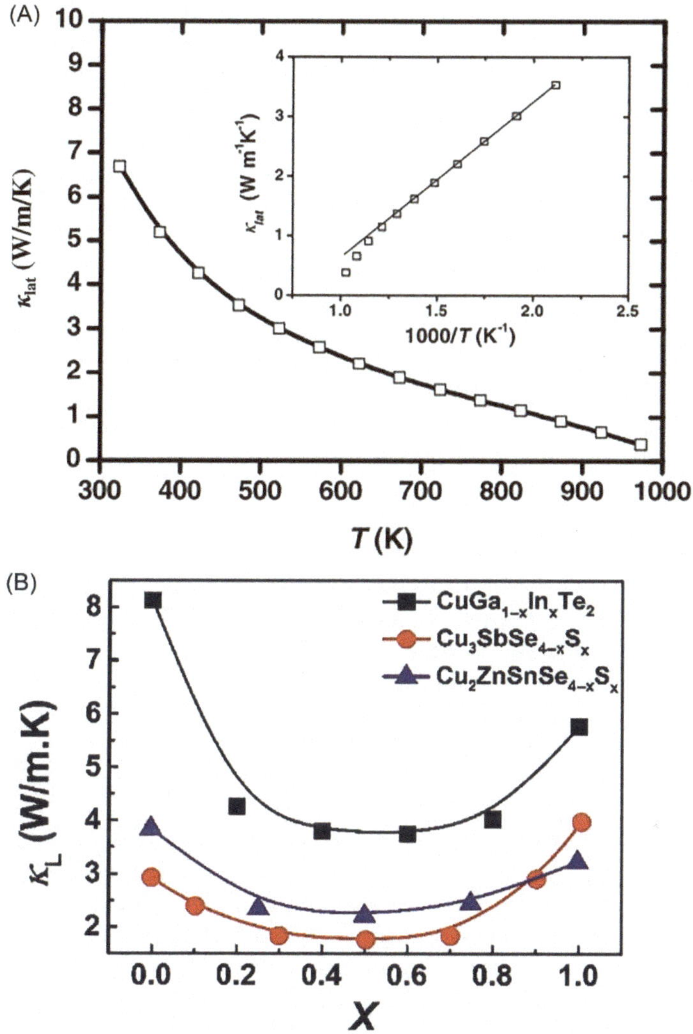


FIGURE 4.38 (A) Temperature dependence of the lattice thermal conductivity for $CuGaTe_2$. The insert shows the lattice thermal conductivity as a function of 1000/T. (B) the lattice thermal conductivity as a function of solid solubility $x$ for $CuGa_{1-x}In_xTe_2$, $Cu_3SbSe_{4-x}S_x$, and $Cu_2ZnSnSe_{1-x}S_x$.

introducing point defects via element doping such as $CuIn_{1-x}Cd_xTe_2$ [144], $Cu_2ZnSn_{0.9}In_{0.1}Se_4$ [137], and $Cu_2Sn_{0.9}In_{0.1}Se_3$ [138].

The characteristics of electronic structure of derived diamond-like structure compounds are basically determined by the distortion of crystalline structure. Generally, cubic diamond structure has the triply degenerate valence band $\Gamma_{15v}$ at valence band maximum. However, in the derived diamond-like structures, the crystal field effect and the tetragonal distortions shall lead to the valence band splitting. For example, in the tetragonal compounds, the triply degenerate valence band splits into a nondegenerate band and a doubly degenerate band. Zhang et al. [142] established a successful approach to realize cubic-like or pseudocubic band structure in noncubic diamond-like compounds through rationally tuning crystal structures. Taking the ternary tetragonal diamond-like structure compounds as an example, the triply degenerate valence band $\Gamma_{15v}$ splits into a nondegenerate band $\Gamma_{4v}$ and a doubly degenerate band $\Gamma_{5v}$, and the energy difference between $\Gamma_{5v}$ and $\Gamma_{4v}$ has a strong linear dependence on the distortion parameter. As mentioned above, when substitution is implemented, distortion is inevitably induced and the distortion parameter $\eta(=c/2a)$ changes and deviates from 1.0. Fortunately, it is possible to move the distortion parameter $\eta$ back close to 1.0, when alloying two kinds of substitutional diamond-like compounds with the distortion parameters $\eta$ deviating from 1.0 in different directions, that is, one is over 1.0 and the other is less 1.0. When $\eta$ approaches to 1.0, the splitting bands are overlapped again, leading to the zero splitting energy $\Delta_{CF}$. This is called the pseudocubic structure, which can significantly enhance the power factor and *ZT* values in these diamond-like compounds [142–146] (Fig. 4.39).

### 4.8.3 SnSe

SnSe is a well-known semiconductor with potential applications in photovoltaics, infrared optoelectronic technology, and memory-switching devices [147–150]. At low temperature, SnSe crystallizes in layered primitive orthorhombic crystal structure with a space group of Pnma, which can be derived from a three-dimensional distortion of the NaCl structure [151]. There are double SnSe layers along the b–c plane with a zig-zag folded accordion-like projection (along the b-axis) as shown in Fig. 4.40A [152,153]. Within these two layers, Sn atoms and Se atoms are held by strong covalent bonds, while the double layers are held together primarily by van der Waals forces leading to the easy cleavage along the b–c plane in this material. The structure contains highly distorted $SnSe_7$ coordination polyhedral with three short and four very long Sn–Se bonds and a lone pair ($5s^2$) from the $Sn^{2+}$ atoms sterically accommodated in between the four long Sn–Se bonds (see Fig. 4.40B) [152]. As temperature increasing to ~800K, SnSe undergoes a displacive (shear) phase transition from the low-symmetry orthorhombic phase to the high-symmetry orthorhombic phase with the space group of Cmcm [154,155]. Compared with the low-temperature orthorhombic structure, the bilayer stacking

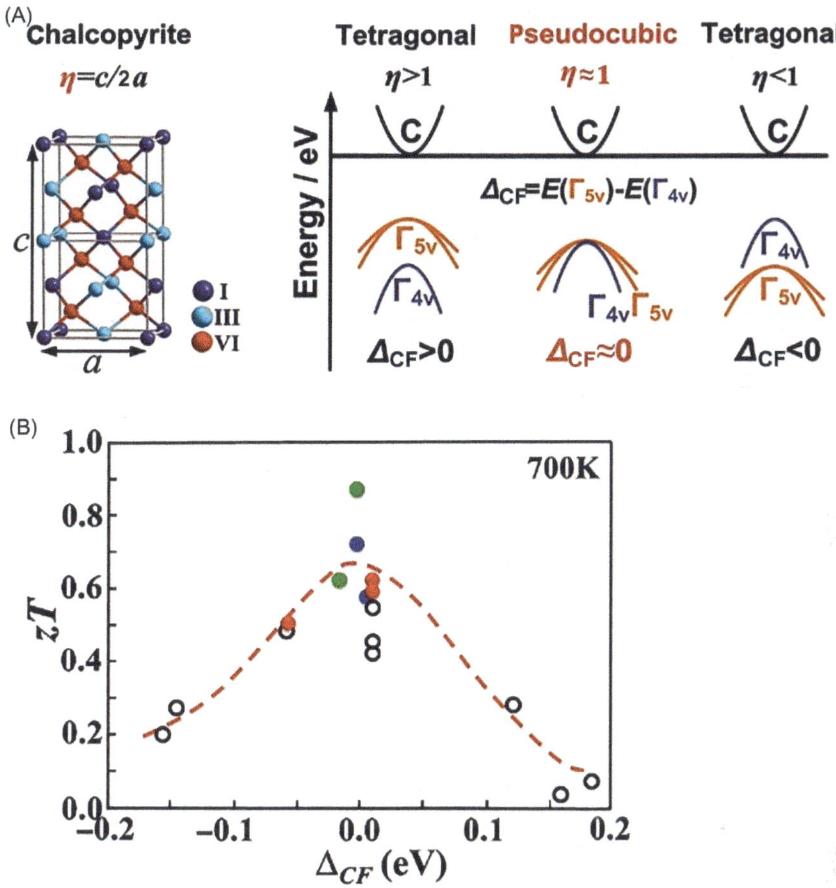


**FIGURE 4.39** (A) Crystal structure and electronic bands of ternary chalcopyrites. $\Gamma_{4v}$ is a nondegenerate band and $\Gamma_{5v}$ is a doubly degenerate band. $\Delta_{CF}$ is the energy split between the top of $\Gamma_{4v}$ and $\Gamma_{5v}$ bands. (B) *ZT* values at 700K in noncubic tetragonal chalcopyrites with calculated $\Delta_{CF}$ values.

still exists and each Sn (or Se) atom is coordinated to four neighboring Se (or Sn) atoms at equal distances in the b−c plane (see Fig. 4.40C and D) in the high-temperature phase [152].

SnSe has a complex electronic structure as indicated by first principles calculations. The band structures in both the low-temperature orthorhombic structure and the high-temperature orthorhombic structure for SnSe are given in Fig. 4.41 [151]. As shown in Fig. 4.41A, the valence band maximum (VBM) and conduction band minimum (CBM) in Pnma structure occur at different points of the Brillouin zone, which

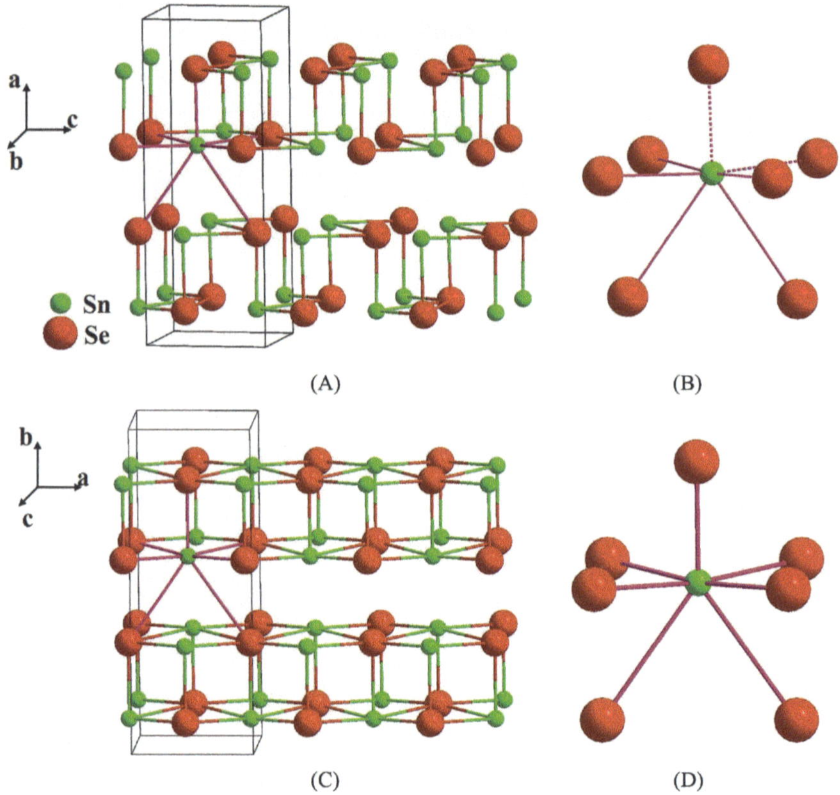


**FIGURE 4.40** (A) Pnma crystal structure in SnSe at room temperature, (B) highly distorted $SnSe_7$ coordination polyhedron with three short and four long Sn–Se bonds, (C) Cmcm crystal structure in SnSe at high temperature, and (D) Sn coordination polyhedron in the orthorhombic structure.

makes SnSe an indirect band gap semiconductor with a band gap $E_g$ of 0.86 eV [151]. There are multiple valence bands along the different direction in the Brillouin zone. The calculation shows a very small energy difference of ~0.06 eV between the first two valence band edges along the Γ–Z direction. Such a small energy gap is easily crossed by the Fermi level as the hole doping approaches $\sim 4 \times 10^{19}/\text{cm}^3$. In addition, the energy difference between the first and the third band edges (i.e., the maximum of U–X to the maximum Γ–Z) is only 0.13 eV. Furthermore, the Fermi level of SnSe even approaches the fourth, fifth, and sixth valence bands when the doping levels in the material are as high as $5 \times 10^{20}/\text{cm}^3$. In the high-temperature Cmcm-SnSe, the valance bands are different from those of the low-temperature structure. Both

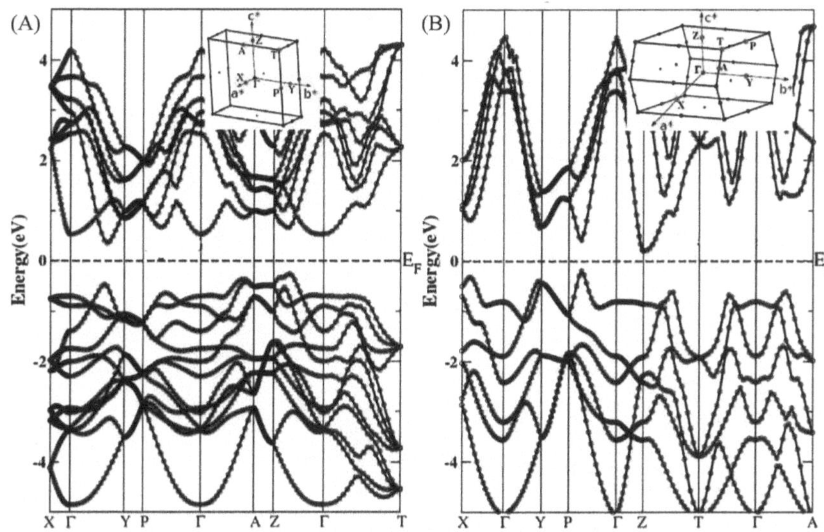


**FIGURE 4.41** Electronic band structure of (A) the low-temperature Pnma-SnSe and (B) the high-temperature Cmcm-SnSe.

the VBM and CBM lie in (0.34, 0.50, 0.00) along the X−A direction, showing a direct band gap with a lower $E_g$ of about 0.46 eV [150,155].

Since the initial report of single crystal SnSe as a high-performance thermoelectric material in 2014, there have been many reports on this compound [151,156−159]. Both p-type Na-doped SnSe and n-type Bi-doped SnSe single crystals have been reported exhibiting extremely high *ZT* value [156,158]. However, the reported high *ZT* values in SnSe single crystals are difficult to be realized in polycrystalline samples, as shown in Fig. 4.42 [151,152]. Relatively high *ZT* values ranging from 0.6 to 1.7 at 750K−873K have been obtained in polycrystalline SnSe samples by alloying [160−162], doping [163−165], and microstructure modulation [166−168], but most of the maximum *ZT* values are under 1.0. Firstly, it is a challenge to obtain comparable electrical transport properties in polycrystalline materials. It is said that the sintering or densification process may lead to inhomogeneous distribution of Sn and Se, and the segregation of Sn is frequently observed. Such inhomogeneity has a strong impact on the carrier concentrations and thermoelectric properties [155]. Secondly, there seem diverse reports on the thermal conductivity of SnSe. Generally, most of the reported thermal conductivity of single crystals is smaller than those of polycrystals (Table 4.7). Furthermore, the easy cleavage of SnSe single crystal does raise the difficulty of measurement of thermoelectric properties.

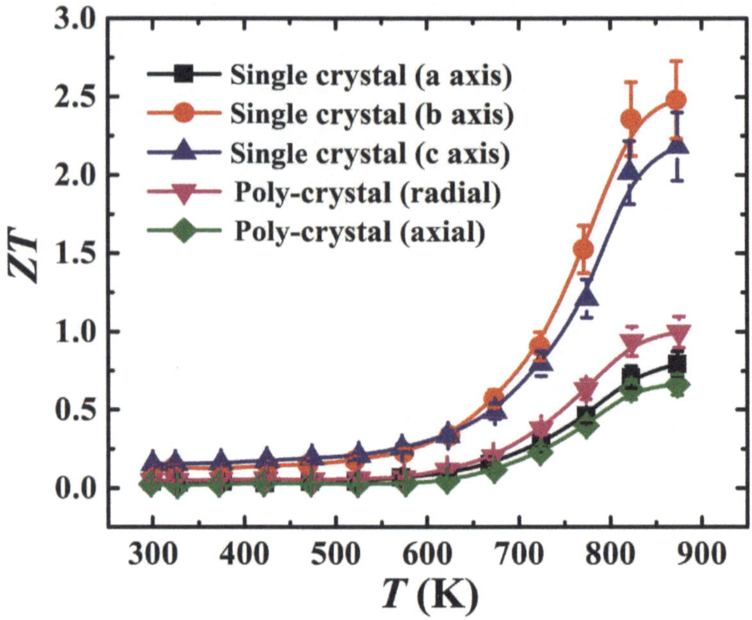


FIGURE 4.42 Temperature dependence of $ZT$ for reported single- and poly-crystal SnSe.

## 4.8.4 Zintl phases

Among thermoelectric materials, Zintl phases have attracted a great attention because of their excellent TE performance and fascinating physical properties [183–186]. The term "Zintl Phase" was proposed by Laves in 1941 and it was named after the famous German chemist Edward Zintl who was the first one to investigate the structure and properties of these compounds in 1930s [187]. Zintl studied the classical Zintl phase NaTl and illuminated the difference of electron transport behavior between NaTl and genuine ionic salts NaCl. This gave the initial description for Zintl phases, that is, they are valence compounds with typical chemical bonds between metallic and alloy-type intermetallic atoms. Based on crystallographic characters of these intermetallic compounds, Zintl described the compounds according to the coordination numbers, coordination polyhedral, atomic radii ratios, and their electronic situation [188]. Basically, Zintl phase compounds shall obey the octet rule and valence balance rule [189]. Most of Zintl compounds are semiconductors with some of them presenting excellent thermoelectric performance [187].

**TABLE 4.7** Comparisons of the total thermal conductivities for SnSe samples reported in the literature.

| References | Single/ingot/ polycrystalline | $\kappa$ (W/mK) at ~300K | $\kappa$ (W/mK) at ~800K |
|---|---|---|---|
| Zhao et al. [151] | Single | 0.46–0.70 | 0.23–0.34 |
| Zhao et al. [152] | Polycrystalline | 0.52–0.62 | 0.29–0.40 |
| Chen et al. [169] | Polycrystalline | 0.6–0.7 | 0.2–0.3 |
| Serrano-Sanchez et al. [170] | Ingot | 0.1 | 0.1–0.2 (at ~400K) |
| Bera et al. [171] | Polycrystalline | 0.2 | 0.24 (at ~400K) |
| Ju et al. [172] | Polycrystalline | 0.65 | - |
| Kim et al. [173] | Polycrystalline | 0.61–0.76 | 0.3–0.4 |
| Gharsallah et al. [174] | Ingot Ge–SnSe | 0.2–0.5 | 0.2–0.5 (at ~400K) |
| Zhang et al. [163] | Polycrystalline | 0.49–1.1 | 0.25–0.51 |
| Wei et al. [161] | Polycrystalline | 0.7–1.3 | 0.4–0.6 |
| Leng et al. [175] | Polycrystalline | 0.7–1.1 | 0.4–0.5 |
| Sassi et al. [176] | Polycrystalline | 0.75–1.25 | 0.4–0.6 |
| Han et al. [160] | Polycrystalline | 0.5–1.26 | 0.3–0.65 |
| Li et al. [177] | Polycrystalline | 0.9–1.2 | 0.3–0.4 |
| Chen et al. [165] | Polycrystalline | 0.9–1.35 | 0.5–0.8 |
| Leng et al. [178] | Polycrystalline | 0.7–1.0 | 0.3–0.4 |
| Chere et al. [179] | Polycrystalline | 0.85–1.0 | 0.4–0.5 |
| Peng et al. [159] | Single Na–SnSe | 1.2–2.0 | 0.4–0.45 |
| Zhao et al. [156] | Single Na–SnSe | 0.45–0.7 | 0.2–0.3 |
| Li et al. [180] | Polycrystalline | 1.0 | 0.35 |
| Popuri et al. [167] | Polycrystalline | 0.6–1.5 | 0.38–0.71 |
| Wasscher et al. [181] | Single (c axis) | 1.9 | - |
| Wei et al. [164] | Polycrystalline | 1.2–1.4 | 0.4–0.5 |
| Gao et al. [182] | Polycrystalline | 1.2–1.4 | 0.3–0.4 |

#### 4.8.4.1 *$AB_2C_2$-type Zintl phases*

$AB_2C_2$-type compounds refer to a family of Zintl phases with various elements, where A can be alkaline-earth or rare-earth elements (e.g.,

Mg, Ca, Sr, Ba, Eu, Yb), B can be transition metals or main-group elements (e.g., Zn, Cd, Mg, Mn, Al), and C can be group IV/V elements (e.g., Si, Sb, Bi). All these compounds crystallize in the $CaAl_2Si_2$ crystal structure (space group $P\bar{3}m1$) [190]. The complex crystal structure and composition result in a very low lattice thermal conductivity and usually a small bandgap and large density of states at the Fermi level. Changing chemical compositions or forming solid solutions is often used to adjust the carrier concentrations, mobility, effective mass, and lattice thermal conductivity in a wide range. Typically, the thermoelectric performance of $AB_2C_2$-type Zintl phases has been significantly improved by doping or substituting, such as $YbZn_{0.4}Cd_{1.6}Sb_2$ ($ZT = 1.26$ at 700K) [191], $Eu_{0.2}Yb_{0.2}Ca_{0.6}Mg_2Bi_2$ ($ZT = 1.3$ at 873K) [192], and $YbCd_{1.85}Mn_{0.15}Sb_2$ ($ZT = 1.14$ at 650K) [191]. Another typical thermoelectric material with $AB_2C_2$-type structure is the binary compound $Mg_3Sb_2$ (A = B = Mg, C = Sb), in which the covalently bonded Mg-Sb polyanion framework $[(Mg_2Sb_2)^{2-}]$ is considered to enable a high hole mobility due to its weak polarity. By introducing point defect, its $\kappa_L$ is reduced to 0.4 W/m/K resulting in a high $ZT$ of 1.7 for n-type $Mg_3Sb_2$-$Mg_3Bi_2$ solid solution [193,194] (Figs. 4.43 and 4.44).

#### 4.8.4.2 $A_{14}MPn_{11}$-type Zintl phases

$A_{14}MPn_{11}$-type compounds is another family of Zintl phases where A can be alkaline-earth or rare-earth elements (Ca, Sr, Ba, Eu, Yb), M can be transition metals or main-group elements (Mn, Ga, Nb, Cd, Zn, Mg, Al), and Pn is group V elements (P, As, Sb, Bi) [202]. $Yb_{14}MnSb_{11}$ is the most well-known thermoelectric material among $A_{14}MPn_{11}$-type Zintl phases. This compound crystallizes in tetragonal symmetry with a space group of $I4_1/acd$. Each unit cell has eight $Yb_{14}MnSb_{11}$ groups and 208 atoms in total. Each formula unit of $Yb_{14}MnSb_{11}$ is charge balanced by

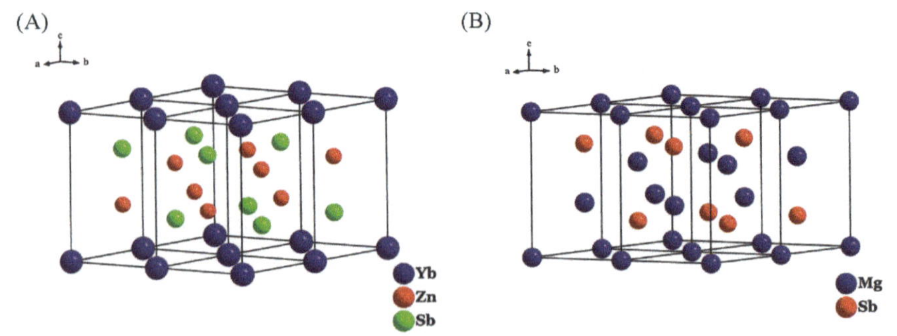


**FIGURE 4.43** Crystal structure of $AB_2C_2$-type Zintl compounds: (A) $YbZn_2Sb_2$ and (B) $Mg_3Sb_2$.

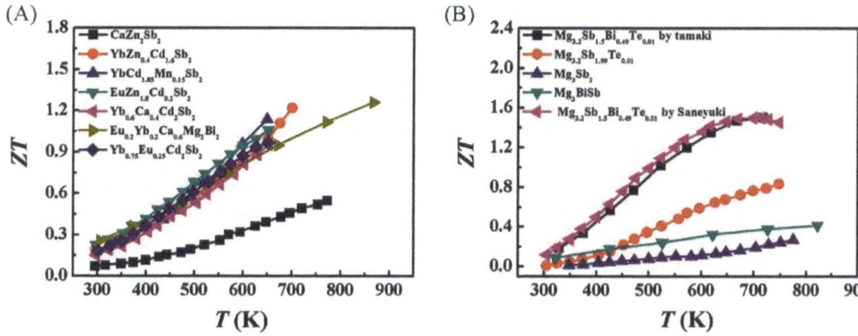


**FIGURE 4.44** Temperature dependence of *ZT* of $AB_2C_2$-based Zintl thermoelectric materials [195-201].

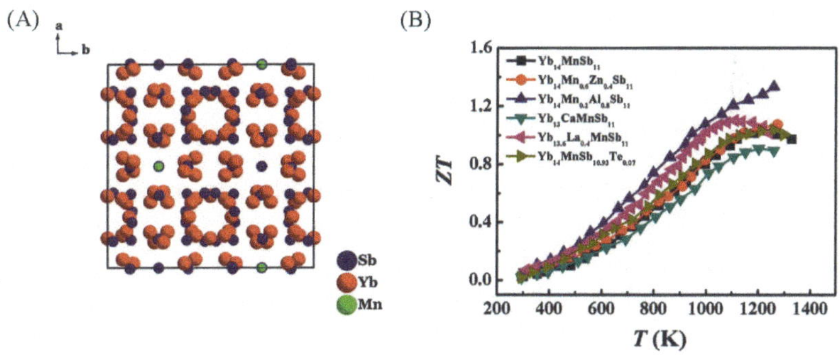


**FIGURE 4.45** (A) Crystal structure and (B) temperature dependence of *ZT* of $A_{14}MPn_{11}$ type Zintl thermoelectric materials.

14 $Yb^{2+}$ cations, one $[MnSb_4]^{9-}$ tetrahedron, one linear $Sb_3{}^{7-}$ anion, and four isolated $Sb^{3-}$ anions. The divalent Mn in the $[MnSb_4]^{9-}$ tetrahedron produces one hole and thus $Yb_{14}MnSb_{11}$ exhibits p-type conduction. $Yb_{14}MnSb_{11}$ does not strictly follow the Zintl-Klemm rule because of the complexity of both the crystal structure and the charge balance. $Yb_{14}MnSb_{11}$ is recognized as a typical PGEC material. Because of the large complex unit cells and larger atomic mass, a very low lattice thermal conductivity $\kappa_L$ (0.4 W/m/K at 300K) is observed in $Yb_{14}MnSb_{11}$. The electrical transport properties can be optimized by element doping or introducing nonstoichiometric atomic ratios on Mn and/or Yb sites. High *ZT* values of 1.1–1.3 have been obtained in various compositions at high temperatures as shown in Fig. 4.45B [203-208].

#### *4.8.4.3 $Zn_4Sb_3$-based materials*

$Zn_4Sb_3$-based materials are also one of the typical Zintl phases and have attracted great interest as promising thermoelectric material

(A) (B)

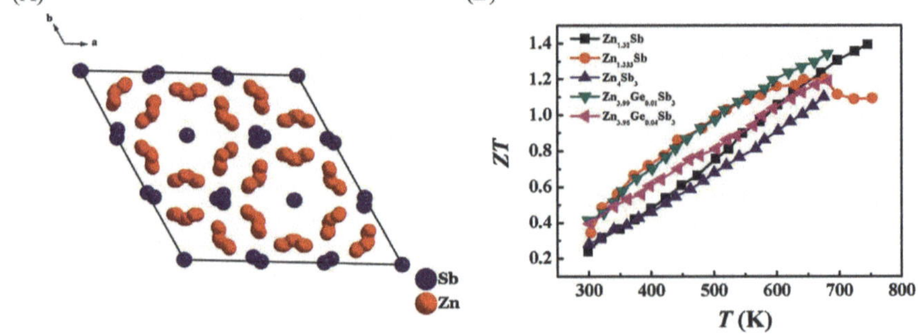


**FIGURE 4.46** (A) Crystal structure and (B) temperature dependence of *ZT* of $Zn_4Sb_3$-based thermoelectric materials.

because of its excellent TE performance as well as the abundance of component elements [209]. At room temperature, β-$Zn_4Sb_3$ belongs to rhombohedral symmetry with the space group of $R\bar{3}c$. The unit cell consists of 10 [$Zn_4Sb_3$] groups, where the 30 Sb atoms build up the anionic framework composed of 6 $Sb_2^{4-}$ dimers and 18 isolated $Sb^{3-}$, while the Zn atoms locate at multiple positions. The large unit cell and the extraordinarily disordered distribution of Zn atoms lead to a very low lattice thermal conductivity with the values about 1 W/m/K at room temperature and 0.7 W/m/K at 650K. Many synthetic routes have been employed to fabricate β-$Zn_4Sb_3$, such as melting/quenching, mechanical alloying, flux-growth, zone-melting, reactive hot-pressing, sinter-forging, and gradient freeze methods. $Zn_4Sb_3$-based thermoelectric materials exhibit high *ZT*s at medium temperatures. The typical experimental results are shown in Fig. 4.46B [210,211].

## References

[1] A.F. Ioffe, Semiconductor Thermoelements and Thermoelectric Cooling, Infosearch Limited, London, 1956.

[2] G.A. Slack, V.G. Tsoukala, Some properties of semiconducting $IrSb_3$, Journal of Applied Physics 76 (1994) 1665.

[3] B.L. Huang, M. Kaviany, Ab initio and molecular dynamics predictions for electron and phonon transport in bismuth telluride, Physical Review B 77 (2008) 125209.

[4] J. Horák, K. Čermák, L. Koudelka, Energy formation of antisite defects in doped $Sb_2Te_3$ and $Bi_2Te_3$ crystals, Journal of Physics and Chemistry of Solids 47 (1986) 805–809.

[5] L.P. Hu, T.J. Zhu, X.H. Liu, et al., Point defect engineering of high-performance bismuth-telluride-based thermoelectric materials, Advanced Functional Materials 24 (2014) 5211–5218.

[6] J.R. Sootsman, D.Y. Chung, M.G. Kanatzidis, New and old concepts in thermoelectric materials, Angewandte Chemie International Edition 48 (2009) 8616–8639.

[7] J.F. Dumas, G. Brun, B. Liautard, et al., New contribution in the study of the $Bi_2Te_3$-$Bi_2Se_3$ system, Thermochimica Acta 122 (1987) 135–141.
[8] R.T. Delves, A.E. Bowley, D.W. Hazelden, et al., Anisotropy of the electrical conductivity in bismuth telluride, Proceedings of the Physical Society 78 (1961) 838.
[9] H.J. Goldsmid, The thermal conductivity of bismuth telluride, Proceedings of the Physical Society(Section B) 69 (1956) 203.
[10] Z. Tang, L. Hu, T. Zhu, et al., High performance n-type bismuth telluride based alloys for mid-temperature power generation, Journal of Materials Chemistry C 3 (2015) 10597–10603.
[11] L.P. Hu, T.J. Zhu, Y.G. Wang, et al., Shifting up the optimum figure of merit of p-type bismuth telluride-based thermoelectric materials for power generation by suppressing intrinsic conduction, NPG Asia Mater 6 (2014) 88.
[12] X. Tang, W. Xie, H. Li, et al., Preparation and thermoelectric transport properties of high-performance p-type $Bi_2Te_3$ with layered nanostructure, Applied Physics Letters 90 (2007) 012102.
[13] W. Xie, J. He, H.J. Kang, et al., Identifying the specific nanostructures responsible for the high thermoelectric performance of $(Bi, Sb)_2Te_3$ nanocomposites, Nano Letters 10 (2010) 3283–3289.
[14] B. Poudel, Q. Hao, Y. Ma, et al., High-thermoelectric performance of nanostructured bismuth antimony telluride bulk alloys, Science 320 (2008) 634–638.
[15] Y. Ma, Q. Hao, B. Poudel, et al., Enhanced thermoelectric figure-of-merit in p-type nanostructured bismuth antimony tellurium alloys made from elemental chunks, Nano Letters 8 (2008) 2580–2584.
[16] J. Li, Q. Tan, J.F. Li, et al., BiSbTe-based nanocomposites with high ZT: The effect of SiC nanodispersion on thermoelectric properties, Advanced Functional Materials 23 (2013) 4317–4323.
[17] Y.I. Ravich, B.A. Efimova, I.A. Smirnov, Semiconducting Lead Chalcogenides, Plenum, New York, 1970, p. 91.
[18] H. Wang, E. Schechtel, Y.Z. Pei, et al., High thermoelectric efficiency of n-type PbS. Advanced Energy, Materials 3 (2013) 488.
[19] J.Q. He, L.D. Zhao, J.C. Zheng, et al., Role of sodium doping in lead chalcogenide thermoelectrics, Journal of the American Chemical Society 135 (2013) 4624.
[20] Q.Y. Zhang, H. Wang, W.S. Liu, et al., Enhancement of thermoelectric figure-of-merit by resonant states of aluminium doping in lead selenide, Energy Environmental Science 5 (2012) 5246.
[21] A.D. LaLonde, Y.Z. Pei, G.J. Snyder, Reevaluation of $PbTe_{1-x}I_x$ as high performance n-type thermoelectric material, Energy Environmental Science 4 (2011) 2090.
[22] J. Androulakis, I. Todorov, J.Q. He, et al., Thermoelectrics from abundant chemical elements: High-performance nanostructured PbSe-PbS, Journal of the American Chemical Society 133 (2011) 10920.
[23] S.N. Girard, J.Q. He, X.Y. Zhou, et al., High performance Na-doped PbTe-PbS thermoelectric materials: Electronic density of states modification and shape-controlled nanostructures, Journal of the American Chemical Society 133 (2011) 16588.
[24] Y.Z. Pei, X.Y. Shi, A. LaLonde, et al., Convergence of electronic bands for high performance bulk thermoelectrics, Nature 473 (2011) 66.
[25] R.J. Korkosz, T.C. Chasapis, S.H. Lo, et al., High ZT in p-type (PbTe)(1-2x)(PbSe)(x)(PbS)(x)thermoelectric materials, Journal of the American Chemical Society 136 (2014) 3225.
[26] C. Han, Q. Sun, Z. Li, et al., Thermoelectric enhancement of different kinds of metal chalcogenides, Advanced Energy Mateirals (2016) 6051600498.
[27] D. Wu, L.D. Zhao, X. Tong, et al., Superior thermoelectric performance in PbTe-PbS pseudo-binary: Extremely low thermal conductivity and modulated carrier concentration, Energy Environmental Science 8 (2015) 2056.

[28] G.A. Slack, M.A. Hussain, The maximum possible conversion efficiency of silicon-germanium thermoelectric generators, Journal of Applied Physics 70 (5) (1991) 2694–2718.
[29] D.M. Rowe, V.S. Shukla, The effect of phonon-grain boundary scattering on the lattice thermal-conductivity and thermoelectric conversion efficiency of heavily doped fine-grained, hot-pressed silicon germanium alloy, Journal of Applied Physics 52 (12) (1981) 7421–7426.
[30] B. Abeles, Lattice thermal conductivity of disordered semiconductor alloys at high temperatures, Physical Review 131 (5) (1963) 1906.
[31] J.P. Dismukes, E. Ekstrom, D.S. Beers, et al., Thermal + electrical properties of heavily doped Ge-Si alloys up to 1300 degrees K, Journal of Applied Physics 35 (10) (1964) 2899.
[32] M.S. Dresselhaus, G. Chen, M.Y. Tang, et al., New directions for low-dimensional thermoelectric materials, Advanced Materials 19 (8) (2007) 1043–1053.
[33] G. Joshi, H. Lee, Y. Lan, et al., Enhanced thermoelectric figure-of-merit in nanostructured p-type silicon germanium bulk alloys, Nano Letters 8 (12) (2008) 4670–4674.
[34] S. Bathula, M. Jayasimhadri, N. Singh, et al., Enhanced thermoelectric figure-of-merit in spark plasma sintered nanostructured n-type SiGe alloys, Applied Physics Letters 101 (21) (2012) 213902.
[35] S. Bathula, M. Jayasimhadri, B. Gahtori, et al., The role of nanoscale defect features in enhancing the thermoelectric performance of p-type nanostructured SiGe alloys, Nanoscale 7 (29) (2015) 12474–12483.
[36] N. Mingo, D. Hauser, N.P. Kobayashi, et al., "Nanoparticle-in-alloy" approach to efficient thermoelectrics: Silicides in SiGe, Nano Letters 9 (2) (2009) 711–715.
[37] K. Favier, G. Bernard-Granger, C. Navone, et al., Influence of in situ formed $MoSi_2$ inclusions on the thermoelectrical properties of an n-type silicon-germanium alloy, Acta Materialia 64 (2014) 429–442.
[38] B. Yu, M. Zebarjadi, H. Wang, et al., Enhancement of thermoelectric properties by modulation-doping in silicon germanium alloy nanocomposites, Nano Letters 12 (4) (2012) 2077–2082.
[39] S. Onari, M. Cardona, Resonant raman-scattering in II-IV semiconductors $Mg_2Si$, $Mg_2Ge$, and $Mg_2Sn$, Physical Review B condensed Matter 14 (8) (1976) 3520–3531.
[40] W. Liu, Q. Zhang, X.F. Tang, et al., Thermoelectric properties of Sb-doped $Mg_2Si_{0.3}Sn_{0.7}$, Journal of Electronic Materials 40 (5) (2011) 1062–1066.
[41] W.S. Liu, H.S. Kim, S. Chen, et al., n-type thermoelectric material $Mg_2Sn_{0.75}Ge_{0.25}$ for high power generation, Proceedings of the National Academy of Sciences of the USA 112 (11) (2015) 3269–3274.
[42] J.M. Higgins, A.L. Schmitt, I.A. Guzei, et al., Higher manganese silicide nanowires of nowotny chimney ladder phase, Journal of the American Chemical Society 130 (2008) 16086–16094.
[43] D.Y.N. Truong, D. Berthebaud, F. Gascoin, et al., Molybdenum, tungsten, and aluminium substitution for enhancement of the thermoelectric performance of higher manganese silicides, Journal of Eelctronic Materials 44 (2015) 3603–3611.
[44] G. Flieher, H. Völlenkle, H. Nowotny, Die kristallstruktur von $Mn_{15}Si_{26}$, Monatsheftefur Chemie 98 (1967) 2173–2179.
[45] R. Zhao, F. Guo, Y. Shu, et al., Improvement of thermoelectric properties via combination of nanostructurization and elemental doping, Jom 66 (2014) 2298–2308.
[46] X. Shi, et al., Enhanced power factor of higher manganese silicide via melt spin synthesis method, Journal of Applied Physics 116 (2014) 245104.
[47] X. She, X. Su, H. Du, et al., High thermoelectric performance of higher manganese silicides prepared by ultra-fast thermal explosion, Journal of Materials Chemistry C 3 (2015) 12116.

[48] X. Chen, A. Weathers, D. Salta, et al., Effects of (Al, Ge) double doping on the thermoelectric properties of higher manganese silicides, Journal of Applied Physics 114 (2013) 173705.
[49] R.M. Ware, D.J. McNeill, Iron disilicide as a thermoelectric generator material, Proceedings of the Institution of Electrical Engineers 111 (1) (1964) 178–182.
[50] T. Higashi, T. Nagase, I. Yamauchi, Long period structure of β -FeSi2, Journal of Alloys and Compounds 339 (1-2) (2002) 96–99.
[51] Y. Makita, Y. Nakayama, Y. Fukuzawa, et al., Important research targets to be explored for β -FeSi2 device making, Thin Solid Films 461 (1) (2004) 202–208.
[52] I. Yamauchi, A. Suganuma, T. Okamoto, I. Ohnaka, Effect of copper addition on the $\beta$-phase formation rate in $FeSi_2$ thermoelectric materials, Journal of Materials Sciences 32 (1997) 4603–4611.
[53] I. Yamauchi, T. Okamoto, A. Suganuma, I. Ohnaka, Effect of Cu addition on the $\beta$-phase formation rate in $Fe_2Si_5$ thermoelectric materials, Journal of Materials Sciences 33 (1998) 385–394.
[54] H.Y. Chen, X.B. Zhao, C. Stiewe, D. Platzek, E. Mueller, Microstructures and thermoelectric properties of Co-doped iron disilicides prepared by rapid solidification and hot pressing, Journal of Alloys and Compounds 433 (1-2) (2007) 338–344.
[55] J. Tani, H. Kido, Mechanism of electrical conduction of Mn-doped $\beta$-$FeSi_2$, Journal of Applied Physics 86 (1) (1999) 464–467.
[56] H.Y. Chen, X.B. Zhao, T.J. Zhu, Y.F. Lu, H.L. Ni, E. Müller, A. Mrotzek, Influence of nitrogenizing and Al-doping on microstructures and thermoelectric properties of iron disilicide materials, Intermetallics 13 (7) (2005) 704–709.
[57] S.W. Kim, M.K. Cho, Y. Mishima, D.C. Choi, High temperature thermoelectric properties of p- and n-type $\beta$-$FeSi_2$ with some dopants, Intermetallics 11 (5) (2003) 399–405.
[58] M. Komabayashi, K. Hijikata, S. Ido, Effect of some additives on thermoelectric properties of $FeSi_2$ thin films, Japanese Journal of Applied Physics 30 (2) (1991) 331–334.
[59] J. Tani, H. Kido, Thermoelectric properties of $\beta$-$Fe_{1-x}Co_xSi_2$ semiconductors, Japanese Journal of Applied Physics 40 (2001) 3236–3239.
[60] J. Tani, H. Kido, Thermoelectric properties of Pt-doped $\beta$-$FeSi_2$, Journal of Applied Physics 88 (10) (2000) 5810–5813.
[61] X. Du, P. Hu, T. Mao, Q. Song, P. Qiu, X. Shi, L. Chen, Ru alloying induced enhanced thermoelectric performance in $FeSi_2$-based compounds, ACS Applied Materials & Interfaces 11 (35) (2019) 32151–32158.
[62] M. Ito, T. Tada, S. Katsuyama, Thermoelectric properties of $Fe_{0.98}Co_{0.02}Si_2$ with $ZrO_2$ and rare-earth oxide dispersion by mechanical alloying, Journal of Alloys and Compounds 350 (2003) 296–302.
[63] K. Morikawa, H. Chikauchi, H. Mizoguchi, S. Sugihara, Improvement of thermoelectric properties of $\beta$-$FeSi_2$ by the Addition of $Ta_2O_5$, Materials Transactions 48 (8) (2007) 2100–2103.
[64] X. Qu, W. Wang, W. Liu, Z. Yang, X. Duan, D. Jia, Antioxidation and thermoelectric properties of ZnO nanoparticles-coated $\beta$-$FeSi_2$, Materials Chemistry and Physics 129 (1-2) (2011) 331–336.
[65] U. Stöhrer, R. Voggesberger, G. Wagner, et al., Sintered $FeSi_2$ for thermoelectric power generation, Energy Conversion and Management 30 (2) (1990) 143–147.
[66] H. Nagai, Effects of mechanical alloying and grinding on the preparation and thermoelectric properties of $\beta$-$FeSi_2$, Materials Transactions 36 (2) (1995) 365–372.
[67] M. Ito, H. Nagai, D. Harimoto, S. Katsuyama, K. Majima, Effects of Cu addition on the thermoelectric properties of hot-pressed $\beta$-$FeSi_2$ with SiC dispersion, Journal of Alloys and Compounds 322 (2001) 226–232.

[68] M. Ito, H. Nagai, E. Oda, S. Katsuyama, K. Majima, Thermoelectric properties of $\beta$-$FeSi_2$ with $B_4C$ and BN dispersion by mechanical alloying, Journal of Materials Science 37 (2002) 2609–2614.
[69] M. Ito, H. Nagai, E. Oda, S. Katsuyama, K. Majima, Effects of P doping on the thermoelectric properties of $\beta$-$FeSi_2$, Journal of Applied Physics 91 (4) (2002) 2138–2142.
[70] X.B. Zhao, H.Y. Chen, E. Müller, C. Drasar, Thermoelectric properties of Mn doped $FeSi_x$ alloys hot-pressed from nitrided rapidly solidified powders, Applied Physics A 80 (5) (2005) 1123–1127.
[71] O. Yamashita, S. Tomiyoshi, Thermoelectric properties of p- and n-type $FeSi_2$ prepared by spray drying, compaction and sintering technique, Journal of Materials Science 38 (2003) 1623–1629.
[72] M. Ito, Y. Takiguchi, Thermoelectric properties of p-type $Fe_{0.9}Mn_{0.1}Si_2$ with rare-earth oxide addition, Materials Transactions 46 (7) (2005) 1497–1501.
[73] G.S. Nolas, D.T. Morelli, T.M. Tritt, Skutterudites: A phonon-glass-electron crystal approach to advanced thermoelectric energy conversion applications, Annual Review of Materials Science 29 (1999) 89–116.
[74] C. Uher, Skutterudites: Prospective novel thermoelectrics, Semiconductors and Semimetals 69 (2001) 139–253.
[75] C. Uher, Recent trends in thermoelectric materials research II, in, Semiconductors and Semimetals, San Diego, 69, Academic Press, 2000, pp. 139–253.
[76] B.C. Sales, D. Mandrus, R.K. Williams, Filled skutterudite antimonides: A new class of thermoelectric materials, Science 272 (1996) 1325.
[77] G.S. Nolas, M. Kaeser, R.T. Littleton, et al., High figure of merit in partially filled ytterbium skutterudite materials.Applied Physics Letters 77 (2000) 1855.
[78] L.D. Chen, T. Kawahara, X.F. Tang, et al., Anomalous barium filling fraction and n-type thermoelectric performanceof $Ba_yCo_4Sb_{12}$, Journal of Applied Physics 90 (4) (2001) 1864–1868.
[79] X. Shi, W. Zhang, L.D. Chen, et al., Filling fraction limit for intrinsic voids in crystals: Doping in skutterudites, Physical Review Letters 95 (2005) 185503.
[80] J.H. Yang, W.Q. Zhang, S.Q. Bai, et al., Dual-frequency resonant phonon scattering in $Ba_xR_yCo_4Sb_{12}$ (R = La, Ce, Sr), Applied Physics Letters 90 (2007) 192111.
[81] X. Shi, J. Yang, J.R. Salvador, et al., Multiple-filled skutterudites: High thermoelectric figure of merit throughseparately optimizing electrical and thermal transports, Journal of the American Chemical Society 134 (5) (2012) 2842.
[82] X. Shi, H. Kong, C.P. Li, et al., Low thermal conductivity and high thermoelectric figure of merit in n-type $Ba_xYb_yCo_4Sb_{12}$ double-filled skutterudites, Applied Physics Letters 92 (2008) 182101.
[83] R.H. Liu, J. Yang, X.H. Chen, et al., p-Type skutterudites $R_xM_yFe_3CoSb_{12}$ (R, M = Ba, Ce, Nd, Yb): Effectiveness of double-filling for the lattice thermal conductivity reduction, Intermetallics 19 (2011) 1747–1751.
[84] G.S. Nolas, The Physics and Chemistry of Inorganic Clathrates, Springer, New York, 2014, pp. 1–4.
[85] Wang Li, Synthesis and thermoelectric performance of $Sr_8Ga_{16}Ge_{30}$-based type-I clathrate compounds. Shanghai, Doctor's Thesis of Shanghai Institute of Ceramics, Chinese Academy of Sciences, 2009.
[86] A. Saramat, G. Svensson, A.E.C. Palmqvist, et al., Large thermoelectric figure of merit at high temperature in Czochralski-grown clathrate $Ba_8Ga_{16}Ge_{30}$, Journal of Applied Physics 99 (2006) 023708.
[87] A. Prokofiev, A. Sidorenko, K. Hradil, et al., Thermopower enhancement by encapsulating cerium in clathrate cages, Nature materials 12 (2013) 1096.
[88] X. Shi, J. Yang, S.Q. Bai, et al., On the design of high-efficiency thermoelectric clathrates through a systematic cross-substitution of framework elements, Advanced Functional Materials 20 (2010) 755–763.

[89] H.L. Liu, X. Shi, F.F. Xu, et al., Copper ion liquid-like thermoelectrics, Nature Materials 11 (5) (2012) 422–425.
[90] P. Lu, H.L. Liu, X. Yuan, et al., Multiformity and fluctuation of Cu ordering in $Cu_2Se$ thermoelectric materials, Journal of Materials Chemistry A 3 (13) (2015) 6901–6908.
[91] K. Trachenko, Heat capacity of liquids: An approach from the solid phase, Physical Review B 78 (10) (2008) 104021.
[92] X.L. Su, F. Fu, Y.G. Yan, et al., Self-propagating high-temperature synthesis for compound thermoelectrics and new criterion for combustion processing, Nature Communications 5 (2014) 4908.
[93] L.L. Zhao, X.L. Wang, J.Y. Wang, et al., Superior intrinsic thermoelectric performance with ZT of 1.8 in single-crystal and melt-quenched highly dense $Cu_{2-x}Se$ bulks, Scientific Reports 5 (2015) 7671.
[94] B. Gahtori, S. Bathul, K. Tyagi, et al., Giant enhancement in thermoelectric performance of copper selenide by incorporation of different nanoscale dimensional defect features, Nano Energy 13 (2015) 36–46.
[95] Y. He, T. Day, T.S. Zhang, et al., High thermoelectric performance in non-toxic earth-abundant copper sulfide, Advanced Materials 26 (23) (2014) 3974–3978.
[96] Y. He, T.S. Zhang, X. Shi, et al., High thermoelectric performance in copper telluride, NPG Asia Mater 7 (8) (2015) e210.
[97] H.L. Liu, X. Yuan, P. Lu, et al., Ultrahigh thermoelectric performance by electron and phonon critical scattering in Cu2Se1-xIx, Advanced Materials 25 (45) (2013) 6607–6612.
[98] W.L. Mi, P.F. Qiu, T.S. Zhang, et al., Thermoelectric transports of Se-rich $Ag_2Se$ in normal phases and phase transitions, Applied Physics Letters 104 (13) (2014) 133903.
[99] Y. He, P. Lu, X. Shi, et al., Ultrahigh thermoelectric performance in mosaic crystals, Advanced Materials 27 (24) (2015) 3639–3644.
[100] S. Bhattacharya, R. Basu, R. Bhatt, et al., $CuCrSe_2$: A high performance phonon glass and electroncrystal thermoelectric material, Journal of Materials Chemistry A 1 (37) (2013) 11289–11294.
[101] S. Ishiwata, Y. Shiomi, J.S. Lee, et al., Extremely high electron mobility in a phonon-glass semimetal, Nature Materials 12 (6) (2013) 512–517.
[102] A.J. Hong, L. Li, H.X. Zhu, et al., Anomalous transport and thermoelectric performances of CuAgSe compounds, Solid State Ionics 261 (2014) 21–25.
[103] K.S. Weldert, W.G. Zeier, T.W. Day, et al., Thermoelectric transport in $Cu_7PSe_6$ with high copper ionic mobility, Journal of American Chemical Society 136 (34) (2014) 12035–12040.
[104] P.F. Qiu, T.S. Zhang, Y.T. Qiu, et al., Sulfide bornite thermoelectric material: A natural mineral with ultralow thermal conductivity, Energy and Environ Science 7 (12) (2014) 4000–4006.
[105] T.J. Zhu, S.N. Zhang, S.H. Yang, et al., Improved thermoelectric figure of merit of self-doped $Ag_{8-x}GeTe_6$ compounds with glass-like thermal conductivity, Physica Status Solidi-Rapid Research Letters 4 (11) (2010) 317–319.
[106] W. Li, S.Q. Lin, B.H. Ge, et al., Low sound velocity contributing to the high thermoelectric performance of $Ag_8SnSe_6$. Advanced, Science 3 (11) (2016) 1600196.
[107] H. Lin, H. Chen, J.N. Shen, et al., Chemical modification and energetically favorable atomic disorder of a layered thermoelectric material $TmCuTe_2$ leading to high performance, Chemistry—A European Journal 20 (47) (2014) 15401–15408.
[108] K. Fujita, T. Mochida, K. Nakamura, High-temperature thermoelectric properties of $Na_xCoO_{2-\delta}$ single crystals, Japanese Journal of Applied Physics 40 (7) (2001) 4644–4647.

[109] A.C. Masset, C. Michel, A. Maignan, et al., Misfit-layered cobaltite with an anisotropic giant magnetoresistance:$Ca_3Co_4O_9$, Physical Review B 62 (1) (2000) 166–175.
[110] Y.Y. Wang, N.S. Rogado, R.J. Cava, et al., Spin entropy as the likely source of enhanced thermopower in $Na_xCo_2O_4$, Nature 423 (6938) (2003) 425–428.
[111] M. Lee, L. Viciu, L. Li, et al., Large enhancement of the thermopower in $Na_xCoO_2$ at high Na doping, Nature Materials 5 (7) (2006) 537–540.
[112] Y. Miyazaki, K. Kudo, M. Akoshima, et al., Low-temperature thermoelectric properties of the composite crystal $[Ca_2CoO_{3.34}]_{0.614}$ $[CoO_2]$, Japanese Journal of Applied Physics, Part 2—Letters 39 (6A) (2000) L531–L533.
[113] J. Liu, X. Huang, Z. Sun, et al., Topotactic synthesis of alternately stacked $Ca_3Co_4O_9/\gamma$-$Na_{0.66}CoO_2$ composite with nanoscale layer structure, CrystEngComm 12 (2010) 4080–4083.
[114] S. Ohta, T. Nomura, H. Ohta, et al., High-temperature carrier transport and thermoelectric properties of heavily La- or Nb-doped $SrTiO_3$ single crystals, Journal of Applied Physics 97 (2005) 3.
[115] H.C. Wang, C.L. Wang, W.B. Su, et al., Enhancement of thermoelectric figure of merit by doping Dy in $La_{0.1}Sr_{0.9}TiO^3$ ceramic, Materials Research Bulletin 45 (7) (2010) 809–812.
[116] D. Flahaut, T. Mihara, R. Funahashi, et al., Thermoelectrical properties of A-site substituted $Ca_{1-x}Re_xMnO_3$ system, Journal of Applied Physics 100 (2006) 084911.
[117] L. Bocher, M.H. Aguirre, D. Logvinovich, et al., CaMn1-xNbxO3 ($x \leq 0.08$) perovskite-type phases as promising new high-temperature n-type thermoelectric materials, Inorganic Chemistry 47 (18) (2008) 8077–8085.
[118] L.D. Zhao, D. Berardan, Y.L. Pei, et al., $Bi_{1-x}Sr_xCuSeO$ oxyselenides as promising thermoelectric materials, Applied Physics Letters 97 (2010) 9.
[119] J. Li, J. Sui, C. Barreteau, et al., Thermoelectric properties of Mg doped p-type BiCuSeO oxyselenides, Journal of Alloys and Compounds 551 (2013) 649–653.
[120] J. Li, J. Sui, Y. Pei, et al., A high thermoelectric figure of merit ZT $>1$ in Ba heavily doped BiCuSeO oxyselenides, Energy and Environmental Science 5 (9) (2012) 8543–8547.
[121] F.G. Aliev, V. Kozyrkov, V. Mashchalkov, et al., Narrow-band in the intermetallic compounds tinisn, zrnisn, hfnisn, Zeitschrift Fur Physik B-Condensed Matter 80 (3) (1990) 353–357.
[122] J. Yang, H. Li, T. Wu, et al., Evaluation of half-heusler compounds as thermoelectric materials based on the calculated electrical transport properties, Advanced Functional Materials 18 (19) (2008) 2880–2888.
[123] Q. Shen, L. Chen, T. Goto, et al., Effects of partial substitution of Ni by Pd on the thermoelectric properties of ZrNiSn-based half-Heusler compounds, Applied Physics Letters 79 (25) (2001) 4165–4167.
[124] C. Yu, T.J. Zhu, R.Z. Shi, et al., High-performance half-Heusler thermoelectric materials $Hf_{1-x}$ $Zr_xNiSn_{1-y}Sb_y$ prepared by levitation melting and spark plasma sintering, Acta Materialia 57 (9) (2009) 2757–2764.
[125] G. Joshi, X. Yan, H. Wang, et al., Enhancement in thermoelectric figure-of-merit of an n-type half-heusler compound by the nanocomposite approach, Advanced Energy Materials 1 (4) (2011) 643–647.
[126] S. Chen, K. Lukas, W. Liu, et al., Effect of Hf concentration on thermoelectric properties of nanostructured n-type half-heusler materials $Hf_xZr_{1-x}NiSn_{0.99}Sb_{0.01}$, Advanced Energy Materials 3 (9) (2013) 1210–1214.
[127] Takeyuki Sekimoto, Ken Kurosaki, Hiroaki Muta, Shinsuke Yamanaka, Japanese Journal of Applied Physics 27 (46) (2007) L673–L675.
[128] S.R. Culp, J.W. Simonson, S.J. Poon, et al., (Zr, Hf)Co(Sb,Sn) half-Heusler phases as high-temperature ($>$ 700 degrees C) p-type thermoelectric materials, Applied Physics Letters 93 (2) (2008) 2436.

[129] X.A. Yan, G. Joshi, W. Liu, et al., Enhanced thermoelectric figure of merit of p-type half-heuslers, Nano Letters 11 (2) (2011) 556–560.

[130] X. Yan, W. Liu, S. Chen, et al., Thermoelectric property study of nanostructured p-type half-heuslers (Hf, Zr, Ti)$CoSb_{0.8}Sn_{0.2}$, Advanced Energy Materials 3 (9) (2013) 1195–1200.

[131] M.M. Zou, J.F. Li, P.J. Guo, et al., Synthesis and thermoelectric properties of fine-grained FeNbSb system half-heusler compound polycrystals with high phase purity, Journal of Physics D—Applied Physics 43 (41) (2010) 415403.

[132] C.G. Fu, T.J. Zhu, Y.T. Liu, et al., Band engineering of high performance p-type FeNbSb based half-heusler thermoelectric materials for figure of merit $ZT > 1$, Energy Environmental Science 8 (1) (2015) 216–220.

[133] C.G. Fu, T.J. Zhu, Y.Z. Pei, et al., High band degeneracy contributes to high thermoelectric performance in p-type half-heusler compounds. Advanced Energy, Materials 4 (18) (2014) 1400600.

[134] G. Joshi, R. He, M. Engber, et al., NbFeSb-based p-type half-heuslers for power generation applications, Energy and Environmental Science 7 (12) (2014) 4070–4076.

[135] C.G. Fu, S.Q. Bai, Y.T. Liu, et al., Realizing high figure of merit in heavy-band p-type half-heusler thermoelectric materials, Nature Communications 6 (2015) 8144.

[136] X.Y. Shi, F.Q. Huang, L.D. Chen, et al., A wide-band-gap p-type thermoelectric material based on quaternary chalcogenides of Cu(2)ZnSnQ(4) (Q = S, Se), Applied Physics Letters 94 (20) (2009) 202103.

[137] M.L. Liu, I.W. Chen, F.Q. Huang, et al., Improved thermoelectric properties of Cu-doped quaternary chalcogenides of $Cu_2CdSnSe_4$, Advanced Materials 21 (37) (2009) 3808–3812.

[138] X.Y. Shi, L.L. Xi, J. Fan, et al., Cu-Se bond network and thermoelectric compounds with complex diamondlike structure, Chemistry of Materials 22 (22) (2010) 6029–6031.

[139] A. Yusufu, K. Kurosaki, A. Kosuga, et al., Thermoelectric properties of $Ag_{1-x}GaTe_2$ with chalcopyrite structure, Applied Physics Letters 99 (2011) 061902.

[140] R.H. Liu, L.L. Xi, H.L. Liu, et al., Ternary compound $CuInTe_2$: A promising thermoelectric material with diamond-like structure, Chemical Commmications 48 (32) (2012) 3818–3820.

[141] T. Plirdpring, K. Kurosaki, A. Kosuga, et al., Chalcopyrite $CuGaTe_2$: A high-efficiency bulk thermoelectric material, Advanced Materials 24 (27) (2012) 3622–3626.

[142] Y. Qin, P. Qiu, R. Liu, et al., Optimized thermoelectric properties in pseudocubic diamond-like $CuGaTe_2$ compounds, Journal of Materials Chemisty A 4 (2016) 1277–1289.

[143] E.J. Skoug, J.D. Cain, D.T. Morelli, High thermoelectric figure of merit in the Cu(3) SbSe(4)-Cu(3)SbS(4) solid solution, Applied Physics Letters 98 (26) (2011) 261911.

[144] N. Cheng, R.H. Liu, S.Q. Bai, et al., Enhanced thermoelectric performance in Cd doped $CuInTe_2$ compounds, Journal of Applied Physics 115 (2014) 163705.

[145] J. Zhang, R. Liu, N. Cheng, et al., High-performance pseudocubic thermoelectric materials from non-cubic chalcopyrite compounds, Advanced Materials 26 (2014) 3848.

[146] W.G. Zeier, H. Zhu, Z.M. Gibbs, et al., Band convergence in the non-cubic chalcopyrite compounds $Cu_2MGeSe_4$, Journal Materials Chemistry C 2 (2014) 10189–10194.

[147] K.-M. Chung, D. Wamwangi, M. Woda, M. Wuttig, W. Bensch, Investigation of SnSe, $SnSe_2$, and $Sn_2Se_3$ alloys for phase change memory applications, Journal of Applied Physics 103 (8) (2008) 083523.

[148] A. Agarwal, M.N. Vashi, D. Lakshminarayana, N.M. Batra, Electrical resistivity anisotropy in layered p-SnSe single crystals, Journal of Materials Science: Materials in Electronics 11 (1) (2000) 67–71.

[149] C. Guillén, J. Montero, J. Herrero, Characteristics of SnSe and SnSe2 thin films grown onto polycrystalline $SnO_2$-coated glass substrates, physica status solidi (a) 208 (3) (2011) 679–683.
[150] S. Chen, K. Cai, W. Zhao, The effect of Te doping on the electronic structure and thermoelectric properties of SnSe, Physica B: Condensed Matter 407 (21) (2012) 4154–4159.
[151] L.-D. Zhao, S.-H. Lo, Y. Zhang, H. Sun, G. Tan, C. Uher, C. Wolverton, V.P. Dravid, M.G. Kanatzidis, Ultralow thermal conductivity and high thermoelectric figure of merit in SnSe crystals, Nature 508 (7496) (2014) 373–377.
[152] L.-D. Zhao, C. Chang, G. Tan, M.G. Kanatzidis, SnSe: a remarkable new thermoelectric material, Energ. Environ. Sci. 9 (10) (2016) 3044–3060.
[153] M.J. Peters, L.E. McNeil, High-pressure Mössbauer study of SnSe, Physical Review B 41 (9) (1990) 5893–5897.
[154] C.W. Li, J. Hong, A.F. May, D. Bansal, S. Chi, T. Hong, G. Ehlers, O. Delaire, Orbitally driven giant phonon anharmonicity in SnSe, Nat. Phys. 11 (12) (2015) 1063–1069.
[155] S. Li, X. Li, Z. Ren, Q. Zhang, Recent progress towards high performance of tin chalcogenide thermoelectric materials, Journal of Materials Chemistry A 6 (6) (2018) 2432–2448.
[156] L.-D. Zhao, G. Tan, S. Hao, J. He, Y. Pei, H. Chi, H. Wang, S. Gong, H. Xu, V.P. Dravid, C. Uher, G.J. Snyder, C. Wolverton, M.G. Kanatzidis, Ultrahigh power factor and thermoelectric performance in hole-doped single-crystal SnSe, Science 351 (6269) (2016) 141–144.
[157] M. Jin, H. Shao, H. Hu, D. Li, J. Xu, G. Liu, H. Shen, J. Xu, H. Jiang, J. Jiang, Single crystal growth of $Sn_{0.97}Ag_{0.03}Se$ by a novel horizontal Bridgman method and its thermoelectric properties, Journal of Crystal Growth 460 (2017) 112–116.
[158] A.T. Duong, V.Q. Nguyen, G. Duvjir, V.T. Duong, S. Kwon, J.Y. Song, J.K. Lee, J.E. Lee, S. Park, T. Min, J. Lee, J. Kim, S. Cho, Achieving $ZT = 2.2$ with Bi-doped n-type SnSe single crystals, Nat. Commun. 7 (1) (2016) 13713.
[159] K. Peng, X. Lu, H. Zhan, S. Hui, X. Tang, G. Wang, J. Dai, C. Uher, G. Wang, X. Zhou, Broad temperature plateau for high ZTs in heavily doped p-type SnSe single crystals. Energ, Environ. Sci. 9 (2016) 2.
[160] Y.-M. Han, J. Zhao, M. Zhou, X.-X. Jiang, H.-Q. Leng, L.-F. Li, Thermoelectric performance of SnS and SnS–SnSe solid solution, Journal of Materials Chemistry A 3 (8) (2015) 4555–4559.
[161] T.-R. Wei, C.-F. Wu, X. Zhang, Q. Tan, L. Sun, Y. Pan, J.-F. Li, Thermoelectric transport properties of pristine and Na-doped $SnSe_{1-x}Te_x$ polycrystals, Phys Chem Chem Phys 17 (44) (2015) 30102–30109.
[162] T.-R. Wei, G. Tan, C.-F. Wu, C. Chang, L.-D. Zhao, J.-F. Li, G.J. Snyder, M.G. Kanatzidis, Thermoelectric transport properties of polycrystalline SnSe alloyed with PbSe, Applied Physics Letters 110 (5) (2017) 053901.
[163] Q. Zhang, E.K. Chere, J. Sun, F. Cao, K. Dahal, S. Chen, G. Chen, Z. Ren, Studies on Thermoelectric Properties of n-type Polycrystalline $SnSe_{1-x}S_x$ by Iodine Doping, Adv Energy Mater 5 (12) (2015) 1500360.
[164] T.-R. Wei, G. Tan, X. Zhang, C.-F. Wu, J.-F. Li, V.P. Dravid, G.J. Snyder, M.G. Kanatzidis, Distinct Impact of Alkali-Ion Doping on Electrical Transport Properties of Thermoelectric p-Type Polycrystalline SnSe, Journal of the American Chemical Society 138 (28) (2016) 8875–8882.
[165] C.-L. Chen, H. Wang, Y.-Y. Chen, T. Day, G.J. Snyder, Thermoelectric properties of p-type polycrystalline SnSe doped with Ag, Journal of Materials Chemistry A 2 (29) (2014) 11171–11176.

[166] Y. Fu, J. Xu, G.-Q. Liu, J. Yang, X. Tan, Z. Liu, H. Qin, H. Shao, H. Jiang, B. Liang, J. Jiang, Enhanced thermoelectric performance in p-type polycrystalline SnSe benefiting from texture modulation, J. Mater. Chem. C 4 (6) (2016) 1201–1207.
[167] S.R. Popuri, M. Pollet, R. Decourt, F.D. Morrison, N.S. Bennett, J.W.G. Bos, Large thermoelectric power factors and impact of texturing on the thermal conductivity in polycrystalline SnSe, J. Mater. Chem. C 4 (8) (2016) 1685–1691.
[168] D. Feng, Z.-H. Ge, D. Wu, Y.-X. Chen, T. Wu, J. Li, J. He, Enhanced thermoelectric properties of SnSe polycrystals via texture control, Phys. Chem. Chem. Phys. 18 (46) (2016) 31821–31827.
[169] Y.-X. Chen, Z.-H. Ge, M. Yin, D. Feng, X.-Q. Huang, W. Zhao, J. He, Understanding of the Extremely Low Thermal Conductivity in High-Performance Polycrystalline SnSe through Potassium Doping, Adv. Funct. Mater. 26 (37) (2016) 6836–6845.
[170] F. Serrano-Sánchez, M. Gharsallah, N.M. Nemes, F.J. Mompean, J.L. Martínez, J.A. Alonso, Record Seebeck coefficient and extremely low thermal conductivity in nanostructured SnSe, Applied Physics Letters 106 (8) (2015) 083902.
[171] Tutul Bera, A. V. S., C. V. Tomy, Ajay D. Thakur. n-type $SnSe_{1-x}$ for Thermoelectric Application. arXiv:1601.00753 [cond-mat.mtrl-sci] 2016.
[172] H. Ju, J. Kim, Effect of SiC ceramics on thermoelectric properties of SiC/SnSe composites for solid-state thermoelectric applications, Ceram. Int. 42 (8) (2016) 9550–9556.
[173] J.H. Kim, S. Oh, Y.M. Kim, H.S. So, H. Lee, J.-S. Rhyee, S.-D. Park, S.-J. Kim, Indium substitution effect on thermoelectric and optical properties of $Sn_{1-x}In_xSe$ compounds, Journal of Alloys and Compounds 682 (2016) 785–790.
[174] M. Gharsallah, F. Serrano-Sánchez, N.M. Nemes, F.J. Mompeán, J.L. Martínez, M.T. Fernández-Díaz, F. Elhalouani, J.A. Alonso, Giant Seebeck effect in Ge-doped SnSe, Scientific Reports 6 (1) (2016) 26774.
[175] H. Leng, M. Zhou, J. Zhao, Y. Han, L. Li, Optimization of Thermoelectric Performance of Anisotropic $Ag_xSn_{1-x}Se$ Compounds, J. Electron. Mater 45 (1) (2016) 527–534.
[176] S. Sassi, C. Candolfi, J.-B. Vaney, V. Ohorodniichuk, P. Masschelein, A. Dauscher, B. Lenoir, Assessment of the thermoelectric performance of polycrystalline p-type SnSe, Applied Physics Letters 104 (21) (2014) 212105.
[177] J. Ding, B. Xu, Y. Lin, C. Nan, W. Liu, Lattice vibration modes of the layered material BiCuSeO and first principles study of its thermoelectric properties, New J. Phys. 17 (8) (2015) 083012.
[178] H.-Q. Leng, M. Zhou, J. Zhao, Y.-M. Han, L.-F. Li, The thermoelectric performance of anisotropic SnSe doped with Na, RSC Adv. 6 (11) (2016) 9112–9116.
[179] E.K. Chere, Q. Zhang, K. Dahal, F. Cao, J. Mao, Z. Ren, Studies on thermoelectric figure of merit of Na-doped p-type polycrystalline SnSe, Journal of Materials Chemistry A 4 (5) (2016) 1848–1854.
[180] Y. Li, F. Li, J. Dong, Z. Ge, F. Kang, J. He, H. Du, B. Li, J.-F. Li, Enhanced mid-temperature thermoelectric performance of textured SnSe polycrystals made of solvothermally synthesized powders, J. Mater. Chem. C 4 (10) (2016) 2047–2055.
[181] J.D. Wasscher, W. Albers, C. Haas, Simple evaluation of the maximum thermoelectric figure of merit, with application to mixed crystals $SnS_{1-x}Se_x$, Solid-State Electronics 6 (3) (1963) 261–264.
[182] H. Guo, H. Xin, X. Qin, J. Zhang, D. Li, Y. Li, C. Song, C. Li, Enhanced thermoelectric performance of highly oriented polycrystalline SnSe based composites incorporated with SnTe nanoinclusions, Journal of Alloys and Compounds 689 (2016) 87–93.
[183] S.M. Kauzlarich, S.R. Brown, G.J. Snyder, Zintl phases for thermoelectric devices, Dalton Transactions 21 (2007) 2099–2107.

[184] E.S. Toberer, C.A. Cox, S.R. Brown, T. Ikeda, A.F. May, S.M. Kauzlarich, G.J. Snyder, Traversing the Metal-Insulator Transition in a Zintl Phase: Rational Enhancement of Thermoelectric Efficiency in $Yb_{14}Mn_{1-x}Al_xSb_{11}$, Advanced Functional Materials 18 (18) (2008) 2795–2800.

[185] E.S. Toberer, A.F. May, G.J. Snyder, Zintl Chemistry for Designing High Efficiency Thermoelectric Materials, Chemistry of Materials 22 (3) (2010) 624–634.

[186] W.G. Zeier, A. Zevalkink, Z.M. Gibbs, G. Hautier, M.G. Kanatzidis, G.J. Snyder, Thinking Like a Chemist: Intuition in Thermoelectric Materials, Angewandte Chemie-International Edition 55 (24) (2016) 6826–6841.

[187] Thomas F. Zintl Phases: Principles and Recent Developments, (2011) Spriner.

[188] Slavi. Intermetallic Compounds - Principles and Practice: Progress, Vol (3), Chapter 6: Zintl Phases. Edited by J. H. Westbrook, R. L. Fleischer, 2002, John Wiley & Sons Ltd.

[189] Y. Hu, G. Cerretti, E.L.K. Wille, S.K. Bux, S.M. Kauzlarich, The remarkable crystal chemistry of the $Ca_{14}AlSb_{11}$ structure type, magnetic and thermoelectric properties, Journal of Solid State Chemistry 271 (2019) 88–102.

[190] X. Shi, X. Wang, W. Li, Y. Pei, Advances in Thermoelectric $Mg_3Sb_2$ and Its Derivatives, Small Methods 800022 (2018) 1–10.

[191] K. Guo, Q. Cao, J. Zhao, Zintl phase compounds $AM_2Sb_2$ (A = Ca, Sr, Ba, Eu, Yb; M = Zn, Cd) and their substitution variants: a class of potential thermoelectric materials, Journal of Rare Earths 31 (11) (2013) 1029–1038.

[192] J. Shuai, H. Geng, Y. Lan, Z. Zhu, C. Wang, Z. Liu, J. Bao, C.W. Chu, J. Sui, Z. Ren, Higher thermoelectric performance of Zintl phases $(Eu_{0.5}Yb_{0.5})_{1-x}CaxMg_2Bi_2$ by band engineering and strain fluctuation, Proceedings of the National Academy of Sciences of the United States of America 113 (29) (2016) E4125–E4132.

[193] X. Shi, L. Chen, C. Uher, Recent advances in high-performance bulk thermoelectric materials, International Materials Reviews 61 (6) (2016) 379–415.

[194] J. Zhang, L. Song, A. Mamakhel, M.R.V. Jorgensen, B.B. Iversen, High-Performance Low-Cost n-Type Se-Doped $Mg_3Sb_2$-Based Zintl Compounds for Thermoelectric Application, Chemistry of Materials 29 (12) (2017) 5371–5383.

[195] F. Gascoin, S. Ottensmann, D. Stark, S.M. Haile, G.J. Snyder, Zintl phases as thermoelectric materials: Tuned transport properties of the compounds $Ca_xYb_{1-x}Zn_2Sb_2$, Advanced Functional Materials 15 (11) (2005) 1860–1864.

[196] H. Zhang, M. Baitinger, M.B. Tang, Z.Y. Man, H.H. Chen, X.X. Yang, Y. Liu, L. Chen, Y. Grin, J.T. Zhao, Thermoelectric properties of $Eu(Zn_{1-x}Cd_x)_2Sb_2$, Dalton Transactions 39 (4) (2010) 1101–1104.

[197] H. Zhang, L. Fang, M.B. Tang, Z.Y. Man, H.H. Chen, X.X. Yang, M. Baitinger, Y. Grin, J.T. Zhao, Thermoelectric properties of $Yb_xEu_{1-x}Cd_2Sb_2$. The, Journal of Chemical Physics 133 (19) (2010) 194701.

[198] K. Guo, Q.G. Cao, X.J. Feng, M.B. Tang, H.H. Chen, X.X. Guo, L. Chen, Y. Grin, J.T. Zhao, Enhanced Thermoelectric Figure of Merit of Zintl Phase $YbCd_{2-x}Mn_xSb_2$ by Chemical Substitution, European Journal of Inorganic Chemistry 26 (2011) 4043–4048.

[199] H. Tamaki, H.K. Sato, T. Kanno, Isotropic Conduction Network and Defect Chemistry in $Mg_{3+\delta}Sb_2$-Based Layered Zintl Compounds with High Thermoelectric Performance, Advanced Materials 28 (2016) 10182–10187.

[200] Ohno, S., Imasato, K., Anand, S., Tamaki, H., Kang, S.D., Gorai, P., Sato, H.K., Toberer, E.S., Kanno, T., and Snyder, G.J., Phase Boundary Mapping to Obtain n-type $Mg_3Sb_2$-Based Thermoelectrics. Joule 2, 141-154.

[201] V. Ponnambalam, D.T. Morelli, On the thermoelectric properties of Zintl compounds $Mg_3Bi_{2-x}Pn_x$ (Pn = P and Sb), Journal of electronic materials 42 (2013) 1307–1312.

[202] Y. Hu, G. Cerretti, E.L.K. Wille, S.K. Bux, S.M. Kauzlarich, The remarkable crystal chemistry of the $Ca_{14}AlSb_{11}$ structure type, magnetic and thermoelectric properties, Journal of Solid State Chemistry 271 (2019) 88–102.

[203] C.A. Uvarov, F.O. Alvarez, S.M. Kauzlarich, Enhanced High-Temperature Thermoelectric Performance of $Yb_{14-x}Ca_xMnSb_{11}$, Inorganic Chemistry 51 (14) (2012) 7617–7624.

[204] E.S. Toberer, C.A. Cox, S.R. Brown, T. Ikeda, A.F. May, S.M. Kauzlarich, G.J. Snyder, Traversing the Metal-Insulator Transition in a Zintl Phase: Rational Enhancement of Thermoelectric Efficiency in $Yb_{14}Mn_{1-x}Al_xSb_{11}$, Advanced Functional Materials 18 (18) (2008) 2795–2800.

[205] B.C. Sales, P. Khalifah, T.P. Enck, E.J. Nagler, R.E. Sykora, R. Jin, D. Mandrus, Kondo lattice behavior in the ordered dilute magnetic semiconductor $Yb_{14-x}La_xMnSb_{11}$, Physical Review B 72 (2005) 20.

[206] S.R. Brown, S. Kauzlarich, M.F. Gascoin, G.J. Snyder, $Yb_{14}MnSb_{11}$: New high efficiency thermoelectric material for power generation, Chemistry of Materials 18 (7) (2006) 1873–1877.

[207] S.R. Brown, E.S. Toberer, T. Ikeda, C.A. Cox, F. Gascoin, S.M. Kauzlarich, G.J. Snyder, Improved thermoelectric performance in $Yb_{14}Mn_{1-x}Zn_xSb_{11}$ by the reduction of spin-disorder scattering, Chemistry of Materials 20 (10) (2008) 3412–3419.

[208] Y. Hu, S.K. Bux, J.H. Grebenkemper, S.M. Kauzlarich, The effect of light rare earth element substitution in $Yb_{14}MnSb_{11}$ on thermoelectric properties, Journal of Materials Chemistry C (2015).

[209] E.S. Toberer, P. Rauwel, S. Gariel, J. Tafto, G.J. Snyder, Composition and the thermoelectric performance of beta-$Zn_4Sb_3$, Journal of Materials Chemistry 20 (44) (2010) 9877–9885.

[210] J. Lin, X. Li, G. Qiao, Z. Wang, J. Carrete, Y. Ren, L. Ma, Y. Fei, B. Yang, L. Lei, J. Li, Unexpected High-Temperature Stability of beta-Zn4Sb3 Opens the Door to Enhanced Thermoelectric Performance, Journal of the American Chemical Society 136 (4) (2014) 1497–1504.

[211] S. Wang, X. Tan, G. Tan, X. She, W. Liu, H. Li, H. Liu, X. Tang, The realization of a high thermoelectric figure of merit in Ge-substituted beta-$Zn_4Sb_3$ through band structure modification, Journal of Materials Chemistry 22 (28) (2012) 13977–13985.

CHAPTER

# 5

# Low-dimensional and nanocomposite thermoelectric materials

## 5.1 Introduction

In early 1990s Hick and Dresselhaus et al. proposed the research direction of low-dimensional thermoelectrics [1–3]. They theoretically demonstrated that when the dimension of material is lowered to nanometer scale, the electronic density of state (DOS) around the Fermi level shall increase due to the quantum confinement effect, and meanwhile the phonon transport shall be greatly affected by interfacial scattering. It provides an additional freedom to tune material's thermal/electronic transport properties. The calculations predicted that the *ZT* value could be greatly enhanced when one of the dimensions is reduced to the level of the mean free path of phonon and/or electron. About 10 years later, the effectiveness of low-dimensionality for optimizing thermoelectric transport properties was preliminarily confirmed by the successful synthesis of superlattice films [4–7]. Thereafter significant improvements in thermoelectric performance were also accomplished in subsequent reports on quantum dot superlattice materials and one-dimensional nanowires. In the follow-up efforts, the idea of reducing material's dimension has been extendedly applied to the optimization of bulk materials through forming nano-grained structure or nanocomposite. This chapter focuses on the synthesis and structure control of both low-dimensional materials and bulk thermoelectric materials with nanostructures. The influences of nanostructure on the transport properties are also discussed.

*Thermoelectric Materials and Devices*
DOI: https://doi.org/10.1016/B978-0-12-818413-4.00005-3

# 5.2 Superlattice thermoelectric films

## 5.2.1 Synthesis of superlattice thermoelectric films

Superlattice thin films are the films with two or more lattice-matched materials being deposited alternatively in a certain period along a specific growth direction. The thickness of each layer can be several nanometers or more. As a typical example, Fig. 5.1 shows the schematic map of A/B superlattice structure, where A and B are two semiconductors with different conduction type or different bandgaps. In such a superlattice structure, periodic potential barriers and wells are formed along the growth direction. For example, when Si and $Si_{1-x}Ge_x$ are deposited to form a two-consecutive heterojunctions $Si/Si_{1-x}Ge_x/Si$, a potential well shall form. If the width of the potential well is less than the mean free path of electron, a single quantum well is formed as shown in Fig. 5.2A. Periodically multiple $Si/Si_{1-x}Ge_x$ layers construct multiple quantum well structures, as shown in Fig. 5.2A. If the thickness of potential barrier layer in the multiquantum wells (i.e., Si layer in $Si/Si_{1-x}Ge_x/Si$) is thin enough, the wave functions of electron/hole are no longer limited within the quantum layers, and the coupling between the electronic states of the adjacent quantum wells shall not be avoided. As a result, subenergy bands with certain widths ($\Delta E_1$, $\Delta E_2$) in the potential well will be produced, as shown in Fig. 5.2B. Similarly, when the barrier energy is low enough, the wave function of electron/hole shall spread into the potential barrier layer, resulting in effective overlapping of wave function of electron/hole in the potential well. The abovementioned thin barrier structure and low barrier height are the two of the typical superlattice structures.

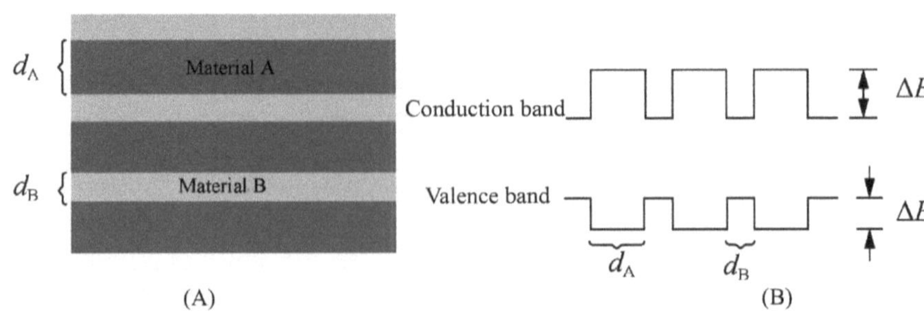


**FIGURE 5.1** (A) Schematic map of A/B superlattice structure with a period of ($d_A + d_B$), where $d_A$ and $d_B$ are the thickness of A and B, respectively. (B) Schematic map of the band structure of A/B superlattice, supposing A is an acceptor while B is a donor. $\Delta E_C$ and $\Delta E_V$ are the conduction and valence bands corresponding to A and B, respectively.

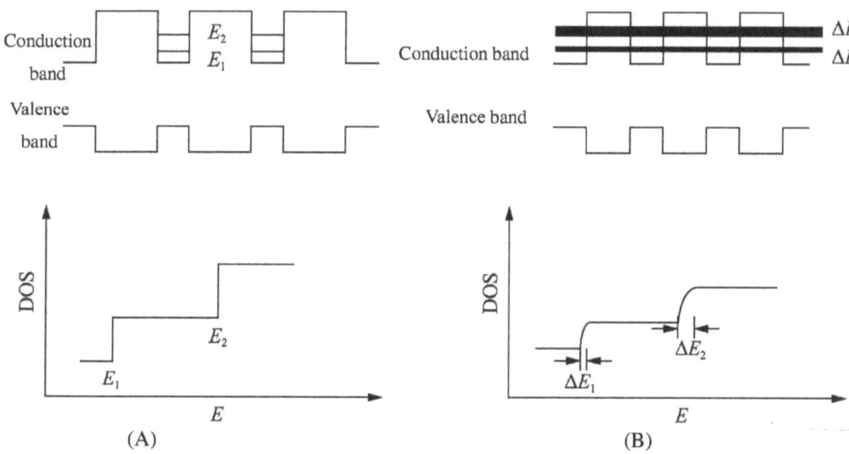


**FIGURE 5.2** (A) Band structure and DOS of multiple quantum wells. (B) The band structure and DOS of superlattice.

Two critical issues should be taken into account when designing and fabricating superlattice films [4]. The first one is the crystallographic compatibility between the two component materials, that is, they should have similar lattice structure with matched lattice constants to prevent the formation of dislocation around the interface. The second one is the chemical compatibility to prevent significant interfacial diffusion or reaction and keep the mutation of both structure and composition near the interface. However, it is not possible to realize complete matching of lattice structures in practical materials. In a lattice-mismatched system, two different manners of crystal growth are often observed: cogrowth (pseudocrystalline growth) and noncogrowth. The former occurs when the epitaxial layer exhibits same atomic arrangements with the substrate material (under-layer material), while the noncogrowth occurs when the epitaxial layer forms according to its own lattice constant and atomic arrangement. In the real material system, the difference of lattice constant between the neighbor layers shall lead to the formation of dislocations at the interface, and the interfacial dislocations as well as their extensions to the epitaxial layers will result in the deterioration of electric properties.

Molecular beam epitaxy (MBE), chemical vapor deposition (CVD), pulsed laser deposition, atomic layer deposition, etc. are usually applied for the fabrication of superlattice materials. MBE is featured with low growth temperature and good controllability, which enable the growth of ultrathin superlattice with steep interfaces. The real-time characterization is easily conducted by using integrated instruments into MBE process in one ultrahigh vacuum system. CVD is usually featured with relatively high growth temperature and excellent growth quality. The high growth

TABLE 5.1 Fabrication process of typical superlattice thermoelectric films.

| Superlattice | Substrate | Technique |
|---|---|---|
| $Bi_2Te_3/Sb_2Te_3$ [5] | GaAs (100) | MOCVD |
| $Bi_2Te_3/Bi_2Te_{2.83}Se_{0.17}$ [6] | GaAs (100) | MOCVD |
| $Bi_2Te_3/Bi_2(Te_{0.88}Se_{0.12})_3$ [7] | $BaF_2$ (111) | MBE |
| $PbTe/PbTe_{0.75}Se_{0.25}$ [8] | $BaF_2$ (111) | Thermal vapor deposition |
| $PbTe/Pb_{1-x}Eu_xTe$ [9] | $BaF_2$ (111) | MBE |
| Si/Ge [10] | SOI (001) | MBE |
| $Si/Si_{1-x}Ge_x$ [11] | SOI (001) | MBE |
| $Ge/Si_{1-x}Ge_x$ [12] | SOI (001) | MBE |

rate makes it capable for mass production. Table 5.1 summarizes the synthesis method of several typical superlattice thermoelectric films.

The film quality is determined by various thermodynamic factors such as chemical reaction, internal diffusion, growth rate, and growth direction as well. Keeping the substrate surface atomically clean is the first priority in order to prevent the formation of interfacial defect. Secondly, the lattice mismatch between the substrate material and superlattice material should not exceed 1% in order to suppress the dislocation. If the lattice mismatch between substrate and epitaxial layer is noticeable, epitaxial buffer layer is usually grown to relieve the lattice mismatch. In this case, the thickness of the epitaxial layer should not exceed the critical thickness of strain relaxation. However, in some exceptional cases, laminar growth of high quality films can be allowed on substrates with high lattice mismatch. For example, in the laminar structure compounds such as $Bi_2Te_3$, $Sb_2Te_3$, and their alloys, the weak bonding of Van der Waals' forces between layers plays roles in relaxing lattice strain during thin film growth, and therefore the growth method is called Van der Waals epitaxial growth [13]. A monocrystalline $Bi_2Te_3$ film is successfully deposited on a GaAs (100) monocrystalline substrate through metal-organic chemical vapor deposition (MOCVD) method despite a high lattice mismatch of 9.7% [14].

## 5.2.2 Phonon transport and thermal conductivity in superlattice films

The influence of superlattice structure on thermoelectric properties includes two possible positive effects: reducing lattice thermal conductivity by the high density of interfaces inside the superlattice and tuning

Seebeck coefficient and electrical conductivity through quantum confinement effect, energy filtering effect, or carrier pocket engineering. Meanwhile, the interfacial scattering to phonon and/or carriers along the vertical and horizontal directions is obviously different because the superlattice structure is highly anisotropic, which also makes the thermoelectric transport property of superlattice show significantly anisotropic.

The periodically arranged heterogeneous compositions inside superlattice not only enhance the interfacial scattering to phonons, but also induce the change in phonon spectrum, which also greatly affect the thermal conductivity. Based on Gang Chen's [15,16] theoretical analysis on interfacial scattering, the thermal conductivity in the in-plane direction is not only relevant to the thickness of superlattice layer, but also depends strongly on the scattering feature of interface. If the phonons reflect specularly at interface, the thermal conductivity of the superlattice shall be close to that of its bulk state. When there are slight diffuse reflections, the thermal conductivity shall be drastically reduced. Experimentally, Yang et al. [17] measured the in-plane thermal conductivity ($\kappa_{SL\text{-}in}$) of Sb-doped n-type Si/Ge superlattice with the thickness of 8 nm/2 nm and dopant concentration of $10^{16}\,cm^{-3}$ by $3\omega$ method, and found the weak temperature dependence of $\kappa_{SL\text{-}in}$. When increasing temperature, $\kappa_{SL\text{-}in}$ slightly increased and then turned to slowly decreasing afterward, while the thermal conductivity of elemental Si or Ge bulks obeys the temperature dependence of $\kappa \sim T^{-a}$ ($a_{Si} \sim 1.65$, $a_{Ge} \sim 1.25$) in the high-temperature range due to the predominant Umklapp scattering. This is explained by the contribution of extra interfacial scattering to phonons in addition to the Umklapp scattering. Beyer et al. [18] measured the in-plane thermal conductivity of n-PbTe/$PbSe_{0.20}Te_{0.80}$ superlattice at room temperature using the bridge method. The results show that the thermal conductivity is reduced by more than 25% (from 2.31 to 1.73 W/m K) when the PbTe layer thickness is changed from 13.4 to 2.3 nm while keeping the $PbSe_{0.20}Te_{0.80}$ thickness to 1.8 nm. Clearly, smaller superlattice thickness (period) means introducing more interfaces and therefore benefit to enhancing phonon scattering.

The cross-plane thermal conductivity ($\kappa_{SL\text{-}cr}$) of a superlattice is more sensitive to superlattice period. Generally, as decreasing superlattice period, $\kappa_{SL\text{-}cr}$ will decrease at first and then increase, which is observed in various superlattices, such as $Bi_2Te_3/Sb_2Te_3$ [19], PbTe/$PbTe_{0.75}Se_{0.75}$ [8], and Si/Ge [20]. Fig. 5.3 shows the change in lattice thermal conductivity along the cross-plane direction and the mean free path of phonons with varying periods of $Bi_2Te_3/Sb_2Te_3$ superlattices [19]. When the superlattice period is at about 5 nm, the $\kappa_{SL\text{-}cr}$ reaches the minimum value (0.22 W/m K), which is half of the bulk $Bi_2Te_3$-$Sb_2Te_3$ alloys. Further decreasing the superlattice period brings about the unexpected increase of $\kappa_{SL\text{-}cr}$ approaching to the lattice thermal conductivity of the

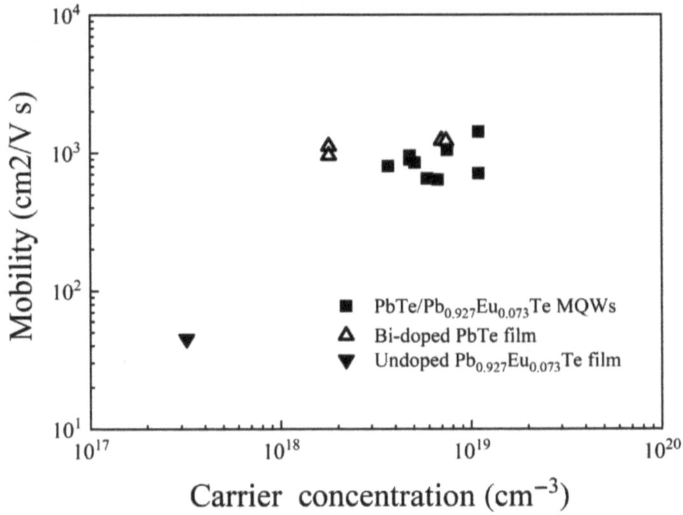


**FIGURE 5.3** The lattice thermal conductivities and mean free paths of $Bi_2Te_3/Sb_2Te_3$ superlattices with different periods [19].

bulk $Bi_2Te_3$-$Sb_2Te_3$ material. This is explained by the coexistence of phonon coherent conduction and interfacial diffuse reflection effects [21–23]. When the mean free path of phonon is less than the superlattice period, the phonon coherence effect is greatly reduced and the phonon transport obeys the Boltzmann transport equation. Then the thermal conductance is determined by the interfacial diffusion reflection of phonon as well as its relevant impedance of thermal boundary. Therefore, in a certain range, reducing superlattice period will increase the interface ratio and the phonon collision frequency, resulting in reduced thermal conductivity. However, when the superlattice period is further reduced to a critical value smaller than the mean free path of phonon, the tunneling of low-frequency phonon shall also be enhanced. The tunneled low-frequency phonons contribute the thermal conductance due to their high ability of heat-carrying. When the superlattice period is further reduced, the phonons with higher frequencies ($\lambda >$ superlattice period) shall also participate in the coherence heat conductance resulting in further increase of thermal conductivity.

## 5.2.3 Carrier transport in superlattice structure

The periodically arranged potential barriers and potential wells formed by the periodical heterogeneous structure in superlattice significantly affects the transport behavior of carriers. Apparently, the multipotential wells and the low potential barrier structures, which are the two typical types of superlattice, behave much differently on the carrier transport.

Harman et al. analyzed the electrical transport properties in a multipotential well structure of n-type PbTe/$Pb_{0.927}Eu_{0.073}Te$ [9]. In this multipotential well structure, the $Pb_{0.927}Eu_{0.073}Te$ serves as the potential barrier layer with a barrier height ($\triangle E_C$) of 0.17 eV and layer thicknesses of 36–54 nm, which is much thicker than that of the PbTe potential well layer (1.5–5.0 nm). Fig. 5.4 shows the carrier concentration dependences of mobility for n-type PbTe/$Pb_{0.927}Eu_{0.073}Te$ multipotential well, PbTe film, and $Pb_{0.927}Eu_{0.073}Te$ film. The PbTe/$Pb_{0.927}Eu_{0.073}Te$ superlattice presents a much higher mobility than that of $Pb_{0.927}Eu_{0.073}Te$ potential barrier layer, which indicates that the carrier is completely localized in PbTe quantum well layer at room temperature due to the relatively high barrier height in PbTe/$Pb_{0.927}Eu_{0.073}Te$ multipotential well structure and the PbTe layer becomes the only existing conducting tunnel.

In PbTe/$Pb_{1-x}Eu_xTe$ multipotential well, when the width of quantum well is about 4 nm, the Seebeck coefficient of the superlattice is similar to that of bulk PbTe material. When the quantum well width is reduced to about 2 nm, the Seebeck coefficient becomes twice that of bulk PbTe (Fig. 5.5). Fig. 5.6 shows the experimental and calculated results of equivalent electric power $S^2n$ of PbTe quantum well layer under various carrier concentrations and quantum well layer thicknesses [24]. The

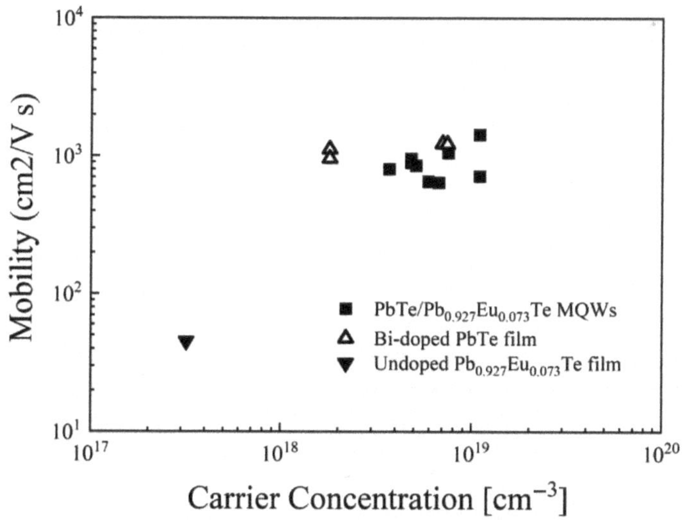


**FIGURE 5.4** The room-temperature carrier mobility of n-type PbTe/$Pb_{0.927}Eu_{0.073}Te$ multiple quantum well, PbTe film, and $Pb_{0.927}Eu_{0.073}Te$ films [9].

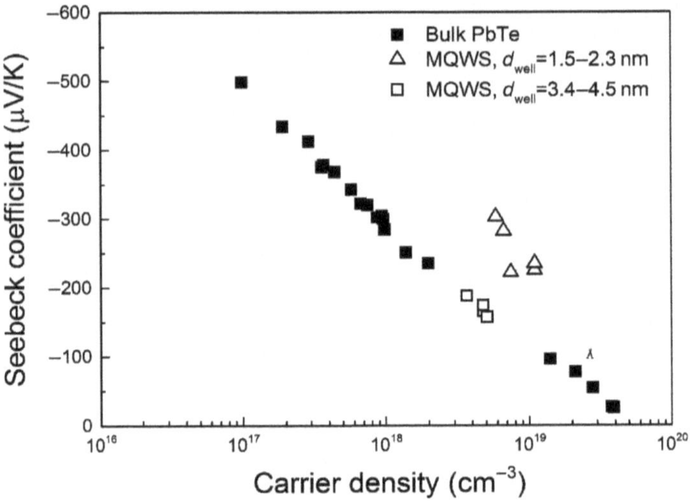


**FIGURE 5.5** Carrier concentration dependence of Seebeck coefficients for n-PbTe/$Pb_{0.927}Eu_{0.073}Te$ multiquantum well and bulk PbTe under 300K [9].

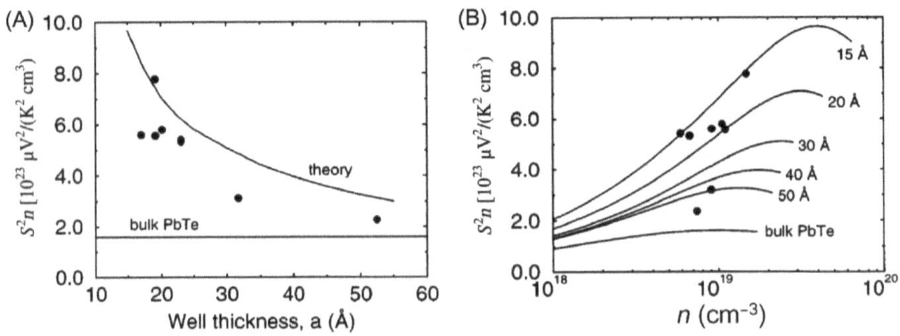


**FIGURE 5.6** Equivalent electric power ($S^2n$) of PbTe quantum well layer as a function of well thickness $a$ (A) and carrier density $n$ (B) different well depths under 300K [24].

Seebeck coefficient and equivalent electric power of superlattice material can be enhanced by increasing the DOS near the Fermi level through the quantum confinement effect. Typically, changing the quantum well width and/or carrier concentration and adjusting the Fermi level close to the subenergy levels can effectively increase the Seebeck coefficient and equivalent electric powers of superlattice films.

Differing from the charges in the quantum well structures, in the low potential barrier ($\triangle E \leq k_BT$) superlattice structure, charges are no longer localized in the quantum well layer. Some of the carriers are in the potential well layer and the others may locate in the potential barrier layer. Therefore both the potential well layers and potential barrier layers can be regarded as the transport tunnels for carriers. The electrical conductivity of superlattice shall be the sum of the electrical conductivities through both the potential barrier layers and potential well layers. Meanwhile, due to the low barrier height, the influence of quantum confinement effect on the Seebeck coefficient above room temperature is relatively weak. Therefore, in low potential barrier superlattice, the electrical performance parallel to the superlattice interface is usually worse as compared with those of component epitaxial films or bulk materials.

$PbTe/PbSe_xTe_{1-x}$ exhibits typical character of low-barrier superlattice structures ($\Delta E < 0.02$ eV). The electron potential well is located at the PbTe layer, while the hole potential well is located at $PbSe_xTe_{1-x}$ layer. The electrical conductivity and Hall measurement results of n-type $PbTe/PbSe_{0.20}Te_{0.80}$ (Fig. 5.7) reveal that when the thickness of $PbSe_{0.20}Te_{0.80}$ layer is fixed at 1.8 nm and PbTe layer thickness is reduced from 13.4 to 2.3 nm, the carrier mobility in the superlattice remains almost unchanged ($\sim 1100\ cm^2/V\ s$), which is lower than that of n-type PbTe ($\sim 1400\ cm^2/V\ s$) [18]. The electrical conductivity and Hall measurement results of n-type $PbTe/PbSe_{0.25}Te_{0.75}$ also show that the carrier mobility parallel to the superlattice interface ($\sim 750\ cm^2/V\ s$) is not sensitive to the change of superlattice period (2.5–12.5 nm) and is lower than that of n-type PbTe, which means that the interfacial scattering of carriers parallel to the superlattice interface is very weak [8]. Being made worse, due to the relatively low carrier mobility, the power factor of n-type $PbTe/PbSe_{0.20}Te_{0.80}$ superlattice is reduced by 20%–30% as compared with bulk n-type PbTe (Fig. 5.8). The reduced lattice thermal conductivity enables a trifle enhancement of $ZT$ as total. The maximum $ZT$ (300K) reaches 0.45 for n-type $PbTe/PbSe_{0.20}Te_{0.80}$ superlattice with the period of 4.1 nm, which is 20%–25% higher than that of bulk n-type PbTe material.

V and VI group elements are frequently used to fabricate low potential barrier superlattices. Peranio et al. [25] have grown epitaxial n-type $Bi_2Te_3/Bi_2(Te_{0.88}Se_{0.12})_3$ symmetrical superlattice ($d = 6$, 10, 12, and 20 nm) on $BaF_2$ substrate using MBE technique. The electrical transport properties of $Bi_2Te_3/Bi_2(Te_{0.88}Se_{0.12})_3$ superlattice at in-plane direction are similar to those of low potential barrier $PbTe/PbSe_xTe_{1-x}$ superlattice. Because both the $Bi_2Te_3$ and $Bi_2(Te_{0.88}Se_{0.12})_3$ layers contribute to the electrical conduction and the interfacial scattering to electron is not strong, the carrier mobility for different superlattice periods keeps almost unchanged ($100-110\ cm^2/V\ s$) as compared with the average value of $Bi_2Te_3$ ($120\ cm^2/V\ s$) and $Bi_2(Te_{0.88}Se_{0.12})_3$ ($\sim 80\ cm^2/V\ s$) films.

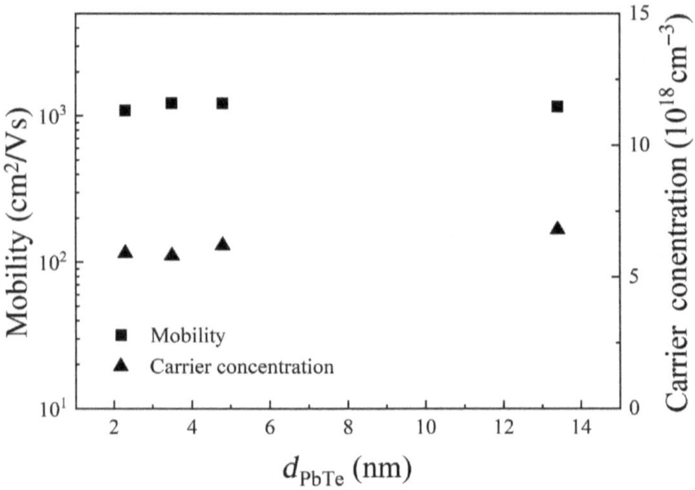


**FIGURE 5.7** Carrier mobility and carrier concentration of n-type $PbTe/PbSe_{0.20}Te_{0.80}$ under 300K ($d_{PbSe0.20Te0.80} = 1.8$ nm) [18].

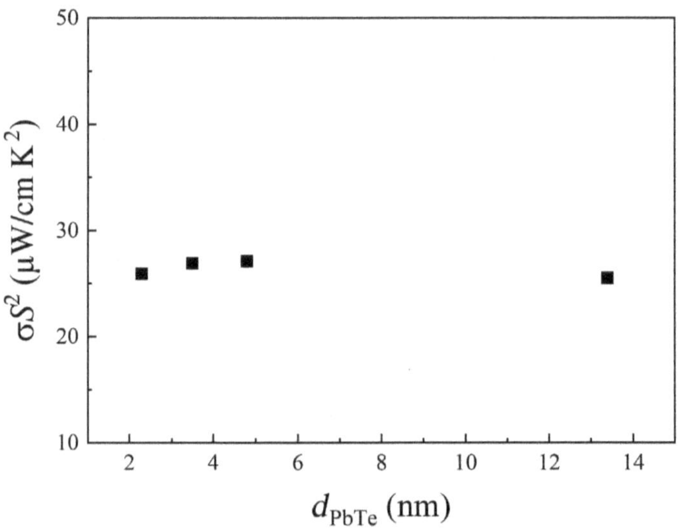


**FIGURE 5.8** The power factor ($S^2\sigma$) of n-type $PbTe/PbSe_{0.20}Te_{0.80}$ under 300K ($d_{PbSe0.20Te0.80} = 1.8$ nm) [18].

Therefore the Seebeck coefficient and power factor of the superlattice do not show obvious change as compared with $Bi_2Te_3$ films. These results further prove the weak quantum confinement and the weak influence of superlattice on the in-plane electrical transport properties in low potential barrier structures.

In addition, Koga et al. [26] proposed a concept—carrier pocket engineering—to improve the thermoelectric performance of superlattice. They constructed a dual quantum well structure by synergetically tuning the lattice period, thickness of potential barrier and/or potential well, and crystallographic orientation. This dual quantum well structure enables the two different types of carriers being selectively localized in the quantum well layer and quantum barrier layer, respectively. Therefore the two different types of carriers (hole and electron) participate together in the electric transportation resulting in enhancement of thermoelectric performance. This concept is successfully applied to fabricate GaAs/AlAs [27] and Si/Ge [28] superlattices.

## 5.3 Nanocrystalline thermoelectric films

Compared with superlattice films, nanocrystalline thermoelectric films are easily synthesized and more applicable for film devices. $Bi_2Te_3$-based V−VI family, such as $Bi_2Te_3$, $Sb_2Te_3$, $(Bi_{1-x}Sb_x)_2Te_3$, and $Bi_2(Te_{1-x}Se_x)_3$, has been widely investigated as thermoelectric thin films working under ambient temperatures. The essential strategy in nanocrystalline thermoelectric (TE) film is mainly reflected in the following aspects: (1) Selectively strengthen the scattering to phonons by modifying the size and distribution of nano-grains based on the different mean free paths between carrier and phonon. In fact, most of the reported experimental results show that the performance optimization of nanocrystalline film relies on the reduction of lattice thermal conductivity by constructing nanostructures. (2) Improve Seebeck coefficient by increasing the average energy of carriers through introducing energy filter and/or interfacial scattering to electrons. Although it is neither widely observed nor solidly proven, the positive contribution of energy filtering effect is continuously in high expectation.

Lots of thin-film fabrication techniques, for example, thermal evaporation, sputtering, flash evaporation, laser pulse deposition, CVD, MOCVD, electrochemical deposition, etc., are applied to synthesize $Bi_2Te_3$-based alloy films. Liao et al. [29] deposited p-type $(BiSb)_2Te_3$ polycrystalline films under either room temperature, 50°C, and 100°C via sputtering method. They found that the thermal conductivity increases from 0.45 to 0.8 W/m K when the average grain size changes from 26 to 85 nm. Takashiri et al. [30] deposited n-type $Bi_{2.0}Te_{2.7}Se_{0.3}$ nanocrystalline film on glass substrate via flash evaporation method and found that heat treatment is effective in improving the crystallinity and lowering the defect concentration in the film and therefore improving the carrier mobility and

**TABLE 5.2** Thermoelectric performance of n-type $Bi_{2.0}Te_{2.7}Se_{0.3}$ nanocrystalline films and the bulk material [30].

| Samples | Annealing temperature (°C) | Grain size | Conductivity ($10^4$ S/m) | Seebeck coefficient (μV/K) | Power factor [μW/(cm $K^2$)] | Thermal conductivity (W/m K) | $ZT$ (300K) |
|---|---|---|---|---|---|---|---|
| Nanocrystalline films | – | 10 nm | 5.5 | − 84.0 | 3.9 | 0.61 | 0.19 |
| | 150 | 27 nm | 5.4 | − 138.1 | 10.3 | 0.68 | 0.46 |
| | 250 | 60 nm | 5.4 | − 186.1 | 18.7 | 0.80 | 0.70 |
| Bulk sample | | 30 μm | 9.3 | − 177.5 | 29.3 | 1.6 | 0.55 |

**TABLE 5.3** The influence of annealing temperature on the chemical composition, grain size, and thickness of p-type $Bi_{0.45}Sb_{1.55}Te_3$ film [31].

| Annealing temperature (°C) | Bi (at.%) | Sb (at.%) | Te (at.%) | Grain size (nm) | Thickness (μm) |
|---|---|---|---|---|---|
| As-deposited | 9 | 31 | 60 | 30 | 1.2 |
| 150 | 9 | 31 | 60 | 36 | 1.19 |
| 200 | 9 | 31 | 60 | 41 | 1.17 |
| 250 | 9 | 32 | 59 | 48 | 1.31 |
| 300 | 10 | 32 | 58 | 60 | 1.42 |
| 350 | 10 | 33 | 57 | 64 | 1.46 |

Seebeck coefficient. The $Bi_{2.0}Te_{2.7}Se_{0.3}$ nanocrystalline film annealed under 250°C exhibits a *ZT* of 0.7 at room temperature (Table 5.2). Song et al. [31] synthesized p-type $Bi_{0.45}Sb_{1.55}Te_3$ polycrystalline film through magnetron sputtering at RT. Similarly, they found the effectiveness of heat treatment on improving crystallinity, lowing carrier concentration, and improving carrier mobility and Seebeck coefficient (Table 5.3 and Fig. 5.9).

## 5.4 Thermoelectric nanowires

Since the research direction of nano-thermoelectrics was proposed by Hick and Dresselhaus et al., the transport behavior of nanowires has extremely attracted wide attention, because theoretical analysis predicted that the band structure of nanowire thermoelectric materials could result in great enhancement of power factors as compared with bulk materials. Meanwhile, the strong phonon scattering caused by boundaries of nanowires can also effectively suppress the lattice thermal conductivity. The researches on thermoelectric nanowires have been mainly focused on the state-of-the-art TE materials, such as Bi-Te alloys, Pb-based chalcogenides, SiGe alloys, and III−V/II−IV compounds.

The bulk Si exhibits large phonon mean free path and high thermal conductivity of 150 W/m K, while the power factor of Si bulk is close to that of $Bi_2Te_3$ though the bandgap of Si is much larger than that of $Bi_2Te_3$. It is expected to achieve a high *ZT* in Si nanowires through reducing lattice thermal conductivity by phonon boundary scattering. Yang et al. [32] measured the thermal conductivity of single Si nanowires and found that the thermal conductivity of Si nanowires is much smaller than that of the bulk with the values decreased from 38 to 27 W/m K and to 18 W/m K when the nanowire diameter changes from 115 to 56 nm and to 37 nm, respectively. Especially, the thermal

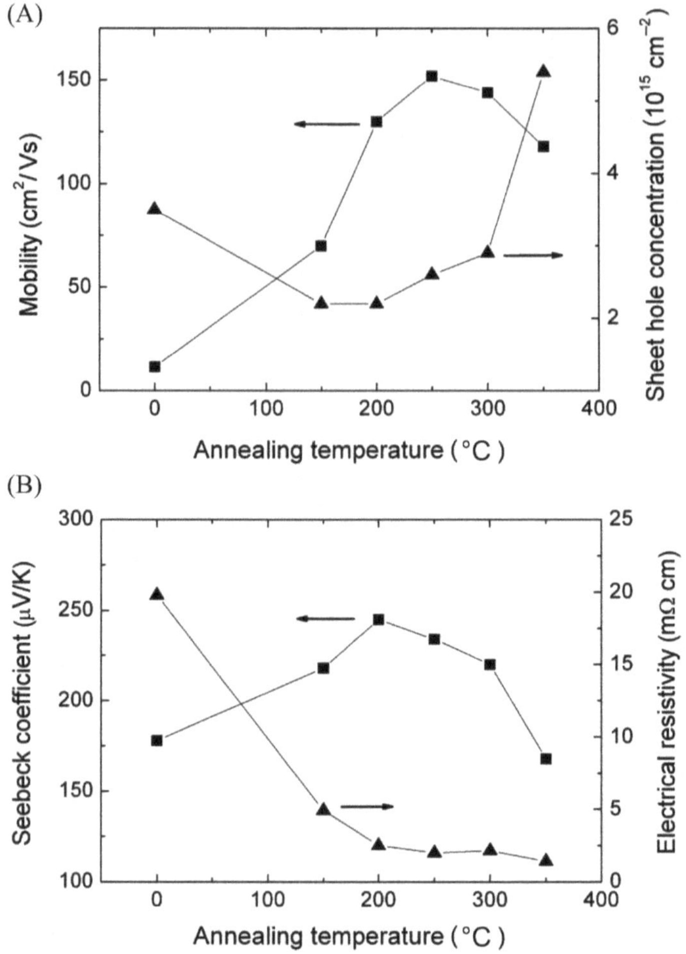


**FIGURE 5.9** The influence of annealing temperature on the room-temperature. (A) Mobility and sheet hole concentration. (B) Seebeck coefficient and electrical resistivity of $Bi_{0.45}Sb_{1.55}Te_3$ film.

conductivity of Si nanowire with rough surface is an order of magnitude lower than that of smooth Si nanowires synthesized by vapor-liquid-solid (VLS) method. Another typical example is the hexagonal hollow $Bi_2Te_3$-Te nanowire with a core–shell structure, which exhibits very low thermal conductivity of about 0.46 W/m K in the temperature range of 300K–400K [33].

Pb-based chalcogenides exhibit relatively low thermal conductivity less than 2.0 W/m K for the bulks. The thermal conductivity of PbTe and

PbSe nanowires fabricated by vapor phase methods can be further reduced due to the boundary scattering, reaching 1.29 and 0.8 W/m K, respectively. However, because of the existence of dense defects and the difficulties in adjusting doping concentration in the vapor phase deposited PbX nanowires, the electrical conductivities of PbTe and PbSe nanowires are very low, which results in lower *ZT* values as compared with the bulks. Similarly, although the nanowires of III–V and II–IV compounds are also theoretically predicted possessing excellent thermoelectric performance based on both the strengthened phonon scattering and quantum confinement effect, high *ZT* has not been experimentally realized in these nanowires, owing to the uncontrollable nonstoichiometry and carrier concentrations. The realization of optimized chemical compositions, doping concentrations, crystallinity, and suppressed defects is a big challenge for developing high-performance thermoelectric nanowires.

Vapor phase method, solution method, and templating method have been adopted to synthesize thermoelectric nanowires. There are two possible growth routines in vapor phase process: VLS and vapor-solid-solid. In vapor phase method, gaseous or vaporized precursors are used as raw materials, and the nucleation/growth process can be facilitated by raising temperature and/or using catalyst. In solution method, thermoelectric nanowires can be synthesized under mild conditions and low temperatures. Surfactants are usually required and the fresh surfaces of the grown nanowires are protected to enable the desired growth along a certain direction. By adopting metallic catalyst particles with different particle size and/or size distribution, the diameter and its distribution of the nanowires are able to be controlled.

## 5.5 Synthesis of nanopowders

The most commonly used approach to fabricate nano-grained and nanocomposite bulk thermoelectric materials is to densify nanopowders into bulk forms, and therefore the fabrication of nanopowders has attracted wide interests. Mechanical alloying (MA) is the most common method for smashing and alloying in the area of powder metallurgy. Through long-term ball milling, the mechanical energy is transferred to the powder under repeated rolling, shearing, rubbing, and impact, which results in alloyed particles with homogenous composition as schematically shown in Fig. 5.10. The commonly used MA devices include planetary, vibrating, stirring, high-energy (vibrating + stirring) and freezing ball mills. Quench-hardened steel, agate, corundum, and tungsten carbide are usually used as the grinding balls according to the milling condition and materials' chemical characters. The effectiveness of alloying is perplexingly dependent on all mill

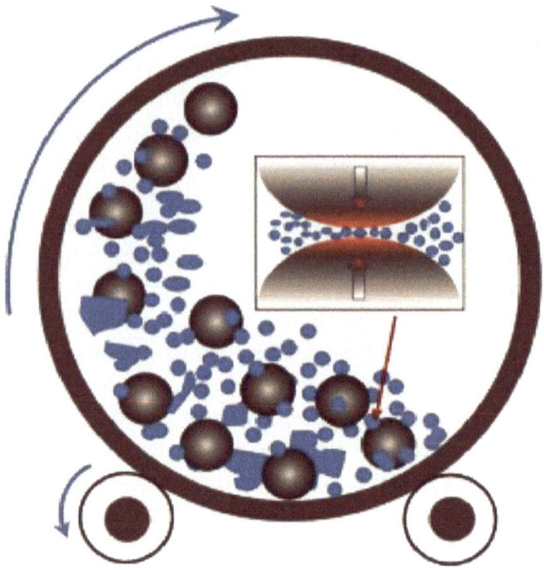

FIGURE 5.10 Synthesis of nanopowders through mechanical alloying.

hardware and milling conditions, such as rotation rate, ambient temperature, and atmosphere; the amount, size, density, and hardness of grinding balls; the shape and size of the raw particles to be alloyed; the grinding medium; as well as the volume and hardness of the milling tank. The milling energy is proportional to the ration of grinding balls and materials when the rotation rate is not exceeding the threshold value. Above the threshold value, the collision energy is reduced due to the limited accelerating space. In addition, the high-energy ball milling using quench-hardened steel grinding balls is the most effective one for the synthesis of nanopowders [34].

Ball milling is widely used for fabricating nanopowders of $Bi_2Te_3$-based compounds by either using their ingot or directly using elemental Bi, Sb, and Te as raw materials [35,36]. It is also adopted to synthesize SiGe nanopowder by directly milling elemental Si, Ge, and B (as dopants). By sintering these ball-milled nanopowders, a number of nano-grained thermoelectric materials with enhanced performance have been developed [37], in which the grain size usually distributes in wide range from tens to hundreds of nanometers with high crystallinity. For skutterudite compounds, single-phase compound is hard to be obtained by mechanical alloying owing to the peritectic reaction. Instead, it is convenient to synthesize the nanopowders composed of mixed phases by high-energy ball milling and then densify them into nano-grained bulk materials using rapid sintering techniques such as spark plasma sintering (SPS) [38–40].

Wet chemical approach is applied for the synthesis of $Bi_2Te_3$ nanopowders, which is firstly reported by Zhao and his colleagues [41] (Fig. 5.11A). They used water as solvent and ethylenediaminetetraacetic disodium salt (EDTA) (diethylenetriamine) as additive to prepare $Bi_2Te_3$ crystal with nano-capsule structure in an open system at 65°C and then densified them into nanocrystalline bulk material. Thereafter aqueous solution method becomes a typical routine for synthesizing thermoelectric nanopowders such as $Bi_2Te_3$ and PbTe. By controlling the reaction condition or using different reductants and/or additives, nanopowders with various morphologies such as lamellar nanocrystal, nanorod, and nanotube are obtained [42–47] (shown in Fig. 5.11). Their particle size and morphology are basically controllable. Furthermore, in order to facilitate mass production and structure control, a number of outfield-assisted solution methods have been developed, including microwave-assisted, ultrasound-assisted, or high-temperature/pressure wet chemical method. These newly developed wet chemical approaches are widely used for the synthesis of vast materials extending from $Bi_2Te_3$ to $Sb_2Te_3$, $Sb_2Se_3$, $Bi_2Se_3$, $AgSbTe_2$, $CoSb_3$, etc. [48–57] (Figs. 5.12 and 5.13). Table 5.4 lists the thermoelectric nanopowders synthesized via high-temperature/pressure wet chemical method.

Melt spinning method is a rapid melt/cooling technique, which had been widely used in the synthesis of amorphous metallic materials. In 2007, Tang et al. firstly reported the synthesis of $Bi_2Te_3$ nanopowders by melt spinning method [80] (Fig. 5.14). The main operation is to inject the melted raw material onto a rapidly rotating roller (usually Cu) which is cooled by liquid nitrogen, under protective atmosphere (Ar or $N_2$). The melts are thrown out along the tangent direct at a high speed and are rapidly solidified into flaky or banded particles [80–85]. The cooling rate can be as high as $10^4$–$10^7$K/s. Due to the difference in cooling rate between the contact and noncontact surfaces, the formed flaky nanoparticles exhibit different microstructures in their upper and lower surfaces. Amorphous structure is tended to form at the contact surface, while branched nanocrystalline structure is formed at the free surface. Such structural difference brings about different structure features into densified bulks, with apparently different grain sizes. Besides the abovementioned grain refinement function, this nonequivalent method can also benefit improving the homogeneity of powder composition as well as increasing the solid solubility limit. The melt spinning method has widely adopted in the synthesis of nanopowders for various kinds of thermoelectric materials including $Bi_2(Te_{1-x}Se_x)_3$ [86–88], β-$Zn_4Sb_3$ [89–91], $AgSbTe_2$ [92,93], Mn-Si compounds [94,95], skutterudites [96,97], SiGe alloys [98,99], half-Heusler alloys [100], clathrates [101–103], PbTe [104], etc.

**FIGURE 5.11** (A) Field emission scanning electron microscopy (FESEM) image of $Bi_2Te_3$ nanocapsules [41]. (B) FESEM image of $Bi_2Te_3$ nanosheets [42]. (C) TEM image of $Bi_2Te_3$ nanoparticles [42]. (D) TEM image of $Bi_2Te_3$ nanotubes [43]. (E) TEM image of $Bi_2Te_3$ nanoparticles [44]. (F) TEM image of $Bi_2Te_3$ nanoplates [45].

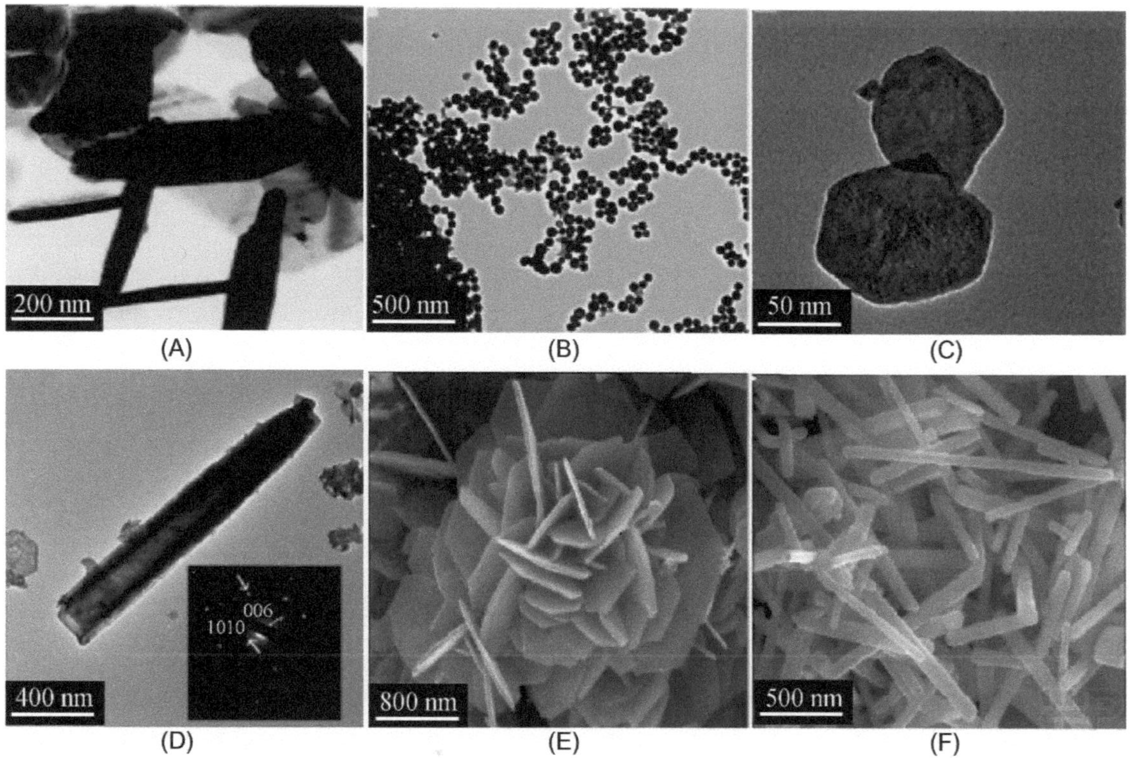


**FIGURE 5.12** (A) TEM image of $Bi_2Te_3$ nanopowder [48]. (B) TEM image of $Bi_2Te_3$ nanoball [49]. (C) TEM image of $Bi_2Te_3$ nanosheets [50]. (D) TEM image of $Bi_2Te_3$ nanotubes [50]. (E) Field emission scanning electron microscopy (FESEM) images of $Sb_2Te_3$ nanosheets [51]. (F) FESEM images of $Sb_2Te_3$ nanotubes [52].

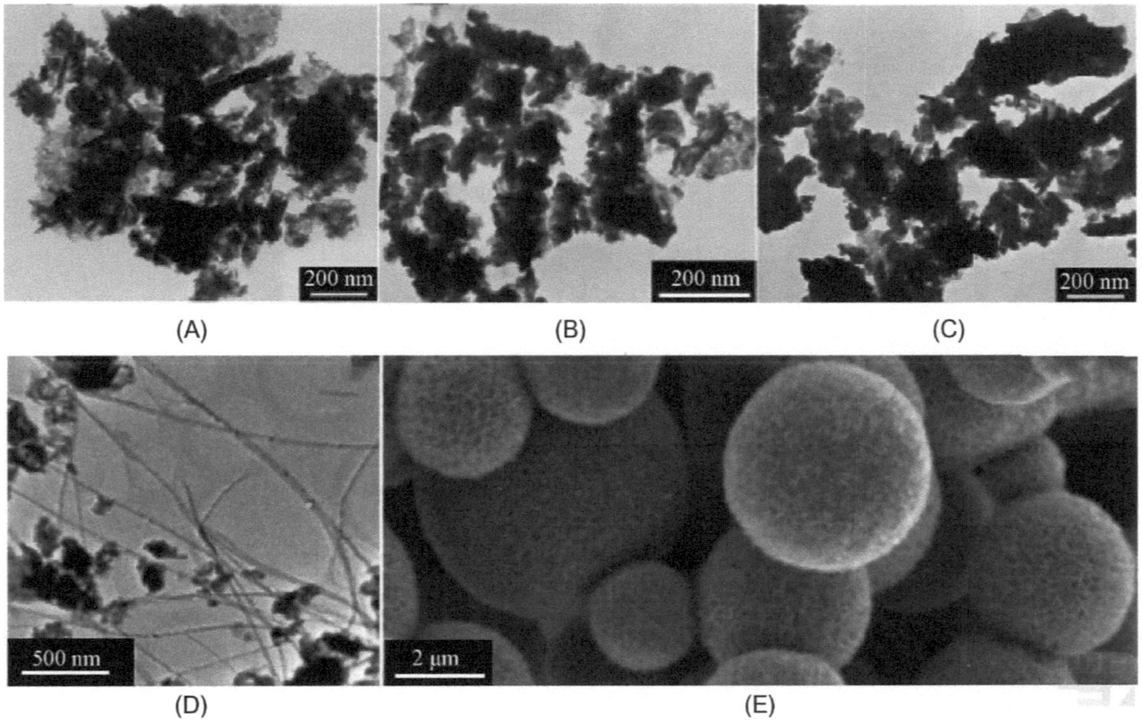


**FIGURE 5.13** TEM images of $Bi_2Te_3$ nanopowders obtained by ultrasonic well chemical method (A) without EDTA at70ºC for 16 h, (B) with EDTA at 70°C for 16 h, and (C) with EDTA at 40°C for 16 h [53]. (D) TEM image of $Bi_2Se_3$ nanobelts [54]. (E) FESEM image of spherical $Bi_2Te_3$ nanoplate aggregates [55]. *EDTA*, Ethylenediaminetetraacetic disodium salt.

**TABLE 5.4** The representative studies on the synthesis of thermoelectric nanopowders using high-temperature and high-pressure solution chemical methods.

| Compounds | Raw materials | Solvents | Temperature (°C) | Time (h) | Products morphology | References |
|---|---|---|---|---|---|---|
| $Bi_2Te_3$ | $BiCl_3$, Te, EDTA, KOH | DMF | 140 | 10–36 | Nanotube | [58] |
| $Bi_2Te_3$ | EDTA, $BiCl_3$, $K_2TeO_3$, Alginate, NaOH | Deionized water | 220 | 24 | Multiple nanostructure | [59] |
| $Bi_2Te_3$ | $Bi(NO_3)_3 \cdot 5H_2O$, $Na_2TeO_3$,<br>Glucose, NaOH, $N_2H_4 \cdot H_2O$ | Glycol | 180 | 8 | Nanoflower | [60] |
| $La_{0.2}Bi_{1.8}Te_3$ | $LaCl_3$, $BiCl_3$, Te $NaBH_4$,<br>NaOH, EDTA | Deionized water | 150 | 24 | Nanosheets | [61] |
| $CsBi_4Te_6$ | $CsCH_3COO$, $Bi(NO_3)_3 \cdot 5H_2O$, $Na_2TeO_3$, $NaBH_4$ | TetraEG | 200 | 16 | Nanosheets | [62] |
| $Sb_2Te_3$ | Te, $NaBH_4$, $SbCl_3$, dihydroxysuccinic acid, AOT | Deionized water | 200 | 24 | Nanobelts | [63] |
| $Sb_2Te_3$ | $SbCl_3$, dihydroxysuccinic acid, $NH_3H_2O$, $K_2TeO_3$, $N_2H_4 \cdot H_2O$ | Deionized water | 180 | 5 | Nanosheets | [64] |
| $Sb_2Te_3$ | $Sb_2O_3$, HCl, Te, polyethylene glycol, $HNO_3$ | Glycol | 150 | 36 | Nanofork | [65] |
| $Bi_{0.4}Sb_{1.6}Te_3$ | $BiCl_3$, $SbCl_3$, $TeO_2$, $NaBH_4$, NaOH | Deionized water | 200 | 20 | Nanopowder | [66] |
| $CoSb_3$ | $CoCl_2 \cdot 6H_2O$, $SbCl_3$, $NaBH_4$ | Ethyl alcohol | 250 | 72 | Nanopowder | [67] |
| $Co_{4-x}Fe_xSb_{12}$ | $SbCl_3$, $CoCl_2 \cdot 6H_2O$, $FeCl_3 \cdot 6H_2O$, $NaBH_4$ | Triethylene glycol | 290 | 12 | Nanopowder | [68] |
| $La_{0.3}Co_4Sb_{12}$ | La, Co, Sb, $NaBH_4$ | Ethyl alcohol | 240 | 48 | Nanopowder | [69] |

*(Continued)*

TABLE 5.4 (Continued)

| Compounds | Raw materials | Solvents | Temperature (°C) | Time (h) | Products morphology | References |
|---|---|---|---|---|---|---|
| $Yb_xCo_4Sb_{12}$ | $CoCl_2 \cdot 6H_2O$, $SbCl_3$, $YbCl_3$, $NaBH_4$ | Ethyl alcohol | 240 | 72 | Nanopowder | [70] |
| PbTe | $Pb(CH_3COO)_2 \cdot 3H_2O$, Te, NaOH, $NaBH_4$ | DMF/ethyl alcohol/ propyl alcohol/glycol | 150 | 12 | Nanocrystal | [71] |
| PbTe | $Pb(CH_3COO)_2$, $Na_2TeO_3$, β-cyclodextrin, DETA, $N_2H_4 \cdot H_2O$ | Deionized water | 180 | 6 | Nanoflower | [72] |
| PbTe/$Ag_2$Te | $SbCl_3$, $AgNO_3$, $Pb(CH_3COO)_2 \cdot 3H_2O$, $Na_2TeO_3$, NaOH, $NaBH_4$ | Glycol | 250 | 0.4 | Core–shell nanocube | [73] |
| $AgPb_{18}SbTe_{20}$ | $Pb(CH_2OCOO)_2 \cdot 3H_2O$, NaOH, Te, $AgNO_3$, $SbCl_3$, $NaBH_4$, Cetyltrimethylammonium bromide | Ethyl alcohol | 180 | 24 | Nanorod | [74] |
| PbSe | $Pb(CH_3COO)_2$, tetrahydrofuran, DMF, $Na_2SeO_3$, sodium tartrate | DMF | 180 | 24 | Nanoflower | [75] |
| PbS/PbSe | $Pb(CH_3COO)_2$, $Na_2SeO_3$, tetrahydrofuran, $N_2H_4 \cdot H_2O$, thioglycolic acid | Glycol | 120 | 12 | Hollow sphere | [76] |
| $Cu_2Se$ | Polyvinylpyrrolidone, CuO, $SeO_2$, NaOH | Glycol | 230 | 24 | Nanosheets | [77] |
| $FeSb_2$ | $Fe(CH_3COO)_2$, $Sb(CH_3COO)_3$, $NaBH_4$ | Ethyl alcohol | 220 | 16 | Nanopowder | [78] |
| $Cu_2NiSnS_4$ | $CuCl_2 \cdot 2H_2O$, $NiCl_2 \cdot 6H_2O$, $SnCl_4 \cdot 5H_2O$, $CH_4N_2S$ | Deionized water | 180 | 12 | Nanopowder | [79] |

AOT, (2-ethylhexyl) sodium sulfosuccinate; DETA, Diethylenetriamine; *DMF*, Dimethylformamide; *EDTA*, Ethylenediaminetetraacetic disodium salt.

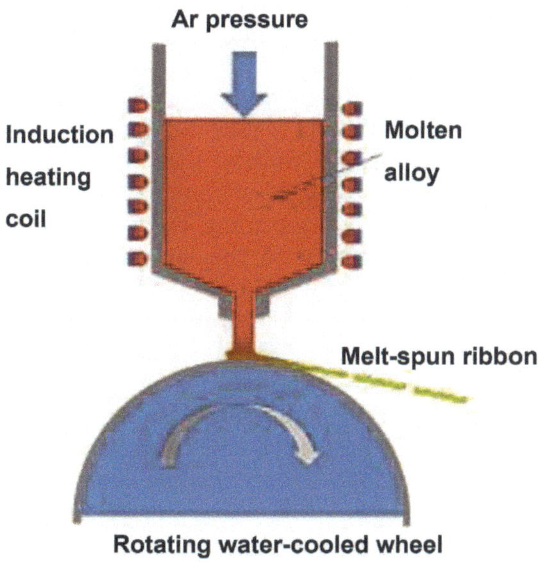


FIGURE 5.14 Schematic diagram of synthesis of thermoelectric nanoparticles through melt spinning.

## 5.6 Nano-grained and nanocomposite thermoelectric materials

### 5.6.1 Preparation techniques for nanostructured materials

Since the concept of nano-thermoelectrics was proposed, various bulk nanostructures, such as nano-grained and nanoparticle-dispersed bulk materials, have been prepared in an attempt to improve thermoelectric performance of bulk materials. As for thermoelectric nanocomposites, the particle size and its distribution should be well controlled for the tuning of electrical and thermal transports. This is a great challenge for the synthesis of nanocomposites. How to realize the uniform distribution of nanoparticles in nanocomposites is thus the first and key task for thermoelectric nanocomposites.

Solid-state mechanical mixing method is a conventional approach to prepare composite materials. Bulk composites can be conveniently obtained by mixing nanoparticles and matrix materials by using ball milling and then sintering the compact into bulk composites. This has been widely used in the fabrication of various nanocomposite thermoelectric materials such as skutterudites [105–109], PbTe [110–112], $Bi_2Te_3$ [113–119], and half-Heusler alloys [120] based composites. The temperature rising during ball milling and the high specific surface of the ball-milled particles often result in amorphization and/or oxidation on the particle surfaces, which may deteriorate the thermoelectric properties of the composite material. It is also difficult to achieve extremely homogenous dispersion of second phase particles using this method.

The solvent mixing method can realize better homogeneous dispersion of secondary phase in the matrix than the solid-state mixing method. The stirring operation (e.g., mechanical and ultrasonic stirring) and the in situ precipitation of dispersing components are commonly utilized to improve the dispersity of the components [121–131]. For example, Xiong et al. [132] suspended the as-synthesized $Ba_{0.22}Co_4Sb_{12}$ particles in a tetrabutyl titanate containing aqueous solution under mechanical stirring and then obtained $Ba_{0.22}Co_4Sb_{12}/TiO_2$ mixed powder by filtration. Here the $TiO_2$ nano-sized particles are formed by the in situ hydrolysis reaction. By using the obtained mixed powder, they densified $Ba_{0.22}Co_4Sb_{12}/TiO_2$ mixture powder into dense nanocomposite by SPS, in which $TiO_2$ particles with an average size of about 30 nm are evenly distributed in the grain boundary or inside the skutterudite grains. Shi et al. [121] developed a solid reaction/SPS process to prepare $Ba_{0.44-x}Co_4Sb_{12}/Ba_xC_{60}$ nanocomposite. In this process, uniformly mixed $BaSb_3/Co/Sb/C_{60}$ powder is acquired by suspending these constituent raw powders in ethanol under ultrasonication at first, and then the $Ba_{0.44-x}Co_4Sb_{12}/Ba_xC_{60}$ nanocomposite is synthesized through high-temperature solid-state reaction and SPS sintering. The nano-sized $Ba_xC_{60}$ particles are formed during the high-temperature solid-state reaction and are embedded into the $Ba_{0.44-x}Co_4Sb_{12}$ grain boundaries during SPS densification.

In situ growth method has been utilized as an effective approach to synthesize nanostructured and nanocomposite thermoelectric materials. It can be presented as in situ precipitation, in situ reaction, and in situ oxidation. It has significant advantages such as uniform dispersion and good size controllability of second phase. The spinodal decomposition is a comprehensive example to form uniformly distributed second phase by precipitating isomeric and heterogeneous nanoparticles or nanostructures. $CoSb_3/(Ir,Rh)Sb_3$ nanocomposites are synthesized through phase separation of $Co_{1-x}(Ir,Rh)_xSb_3$ solid solution by an appropriate heat treatment [125–127]. Similarly, long-term annealing at low temperature results in the phase separation of $Ba_{0.2}(Co_{0.9}Ir_{0.1})_4Sb_{12}$ and yielding Ir-rich and Co-rich skutterudite nanoparticles [128]. Eutectic reaction can be also used to in situ produce second phase having different chemical composition and/or different crystal structure with the matrix. Through the formation of solid solution at high temperature and phase separation at low temperature, a series of $PbTe/Sb_2Te_3$-based nanocomposites are successfully synthesized by selecting a designed composition and cooling process according to the pseudopotential binary phase diagram

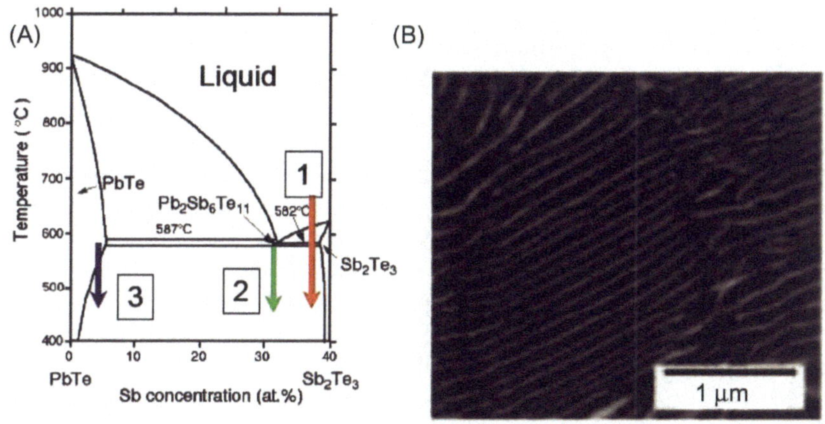


**FIGURE 5.15** (A) Phase diagram and designed route for precipitating second-phase nanoparticles. Route 1: lamellar structure composed of two phases, $Sb_2Te_3$ containing 5% Pb and $Pb_2Sb_6Te_{11}$, can be obtained; Route 2: Laminated structure of PbTe-$Sb_2Te_3$ with a period of tens of nanometers can be obtained near eutectic region through rapid cooling or decomposition of metastable $Pb_2Sb_6Te_{11}$ phase; Route 3: Nano-$Sb_2Te_3$-rich phase can be precipitated out of the oversaturated PbTe-$Sb_2Te_3$ matrix; (B) laminated nanostructure in PbTe-$Sb_2Te_3$ obtained by Route 2.

of PbTe-$Sb_2Te_3$ system as shown in Fig. 5.15 [122–124]. For example, near the eutectic composition (Route 2 in Fig. 5.15), laminated structure with a period of tens of nanometers can be obtained through rapid cooling. Li et al. [133] synthesized $Yb_{0.2}Co_4Sb_{12}$/Sb nanocomposite through melt spinning and rapid cooling technique. The excess Sb is in situ precipitated in nanometer scale and distributes in the grain boundaries. Li et al. in situ synthesized InSb nanoparticles in $In_xCe_xCo_4Sb_{12}$ system, in which a large number of InSb particles in 50–80 nm distribute on the boundaries of the matrix skutterudite grains. In addition, skutterudite-based nanocomposites dispersed with oxide nanoparticles are also prepared by combining the in situ precipitation and the subsequent oxidation of the precipitated rare earth elements [129]. The thermoelectric performance of presentative thermoelectric nanocomposites will be described in detail in Sections 5.6.2 and 5.6.3.

## 5.6.2 Skutterudite-based nanocomposites

Nanostructuring or nanocompositing has been applied to various state-of-the-art thermoelectric materials to improve their performance, such as filled skutterudites, PbTe-based chalcogenides, $Bi_2Te_3$-based compounds, half-Heusler alloys, and SiGe alloys. As well known, $CoSb_3$-based skutterudites exhibit large power factors, but high thermal conductivity. Besides the approach of filling guest atoms into the lattice intrinsic voids (forming filled skutterudite, see Chapter 4: Review of Inorganic Thermoelectric Materials), many other strategies have been used to reduce their thermal conductivity, among which dispersing nanoparticles into the skutterudite or filled skutterudite matrix has been widely tried. Shi et al [134] synthesized $CoSb_3/C_{60}$ nanocomposite containing evenly distributed fullerene. The thermal conductivity of $CoSb_3/C_{60}$ nanocomposite is greatly reduced owing to the phonon scattering by large $C_{60}$ defects distributed in the grain boundaries. Dispersing $Yb_2O_3$ nanoparticles into $Yb_xCo_4Sb_{12}$ (Fig. 5.16) by in situ oxidization of excess Yb is proven effective to suppress its lattice thermal conductivity while maintaining high power factors [129]. The lattice thermal conductivity is reduced to about 1.72 W/m K at room temperature and to 0.52 W/m K at 800K, which is almost half of the matrix.

The nanocompositing also give moderate influence on the electrical transport properties of the composites. Xiong et al. [135] synthesized $Yb_{0.26}Co_4Sb_{12}/GaSb$ nanocomposites through controlling excess Sb content and cooling process on the basis of understanding the thermodynamic characteristic of Ga in the voids of $CoSb_3$ at high temperature. They found that $Yb_{0.26}Co_4Sb_{12}/0.2GaSb$ composite has an enhanced Seebeck coefficient as compared with the matrix, which results in an enhanced *ZT* at high temperatures. The enhancement of thermal power is explained by the energy filtering effect caused by the semiconductor GaSb nanoparticles (Fig. 5.17B).

### 5.6.3 Multiscaling structures in PbTe-based materials

In PbTe-based IV–VI chalcogenide compounds, the $Pb^{2+}$ and $Te^{2-}$ form a NaCl type cubic lattice. The lattice thermal conductivity can be effectively reduced by introducing coherent or incoherent secondary phases into the PbTe matrix. There are obvious grain boundaries between incoherent nano-phases and matrix, and the inhibiting effect of incoherent grain boundary on the thermal conductivity is originated from the mismatch between the phonon modes. Although introducing nanoparticles of incoherent secondary phase can significantly reduce the lattice thermal conductivity, it has negative influence on the electron transport. Therefore the enhancement of *ZT* with the values of 1.4–1.5 in the PbTe-based nanocomposites with dispersing incoherent secondary phase (e.g., Sb [136], $Ag_2Te$ [137]) is not so significant. The coherent secondary phase has similar lattice parameters as matrix and exhibits good lattice matching. Usually, the internal stress of coherent grain boundaries is higher than that of incoherent grain boundaries. The phonon scattering is mainly originated from the concentrated stress caused

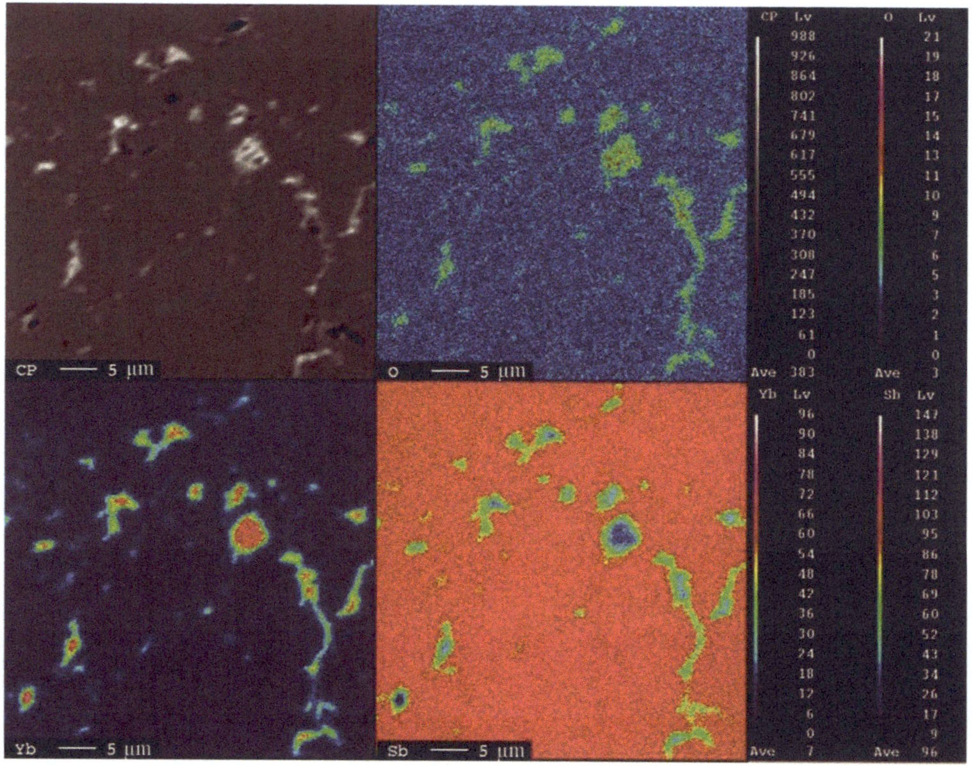


**FIGURE 5.16** Back-scattered image and the corresponding O, Yb, and Sb elemental mapping of $Yb_{0.25}Co_4Sb_{12}/Yb_2O_3$ nanocomposite.

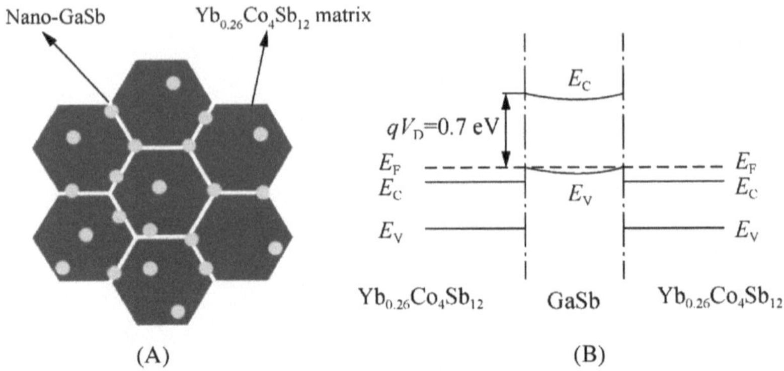


**FIGURE 5.17** (A) Schematic diagram of the microstructure of $Yb_{0.26}Co_4Sb_{12}$/GaSb nanocomposite. (B) The band diagram around the phase interface in $Yb_{0.26}Co_4Sb_{12}$/GaSb nanocomposite [135].

by the slight lattice mismatch, which is similar to the point effect. Significant enhancement of *ZT* (maximum values of 1.7–1.8 at high temperatures) is observed in the PbTe-based nanocomposites with dispersion of coherent nanoparticles (e.g., $AgSbTe_2$ [138], $NaSbTe_2$ [139], SrTe [140]). In these coherent nanocomposites, because of the very close lattice parameters between the matrix and dispersed phase, nanoscaled phase separation easily occurs, leading to the formation of composition fluctuation in nanoscale. For example, in $Ag_{1-x}Pb_{18}SbTe_{20}$, Ag-Sb-rich "domains" (or nano-inclusions) with the size of 2–3 nm are frequently observed homogeneously distributing inside the matrix [138,141,142]. Within the Ag-Sb-rich nano-inclusions are the $Ag^+$-$Sb^{3+}$ dipoles, which may also act as energy filter to enhance thermal power in addition to scattering phonons as nano-size defects [142]. Similar structure characteristics and extremely depressed lattice thermal conductivity are also unveiled in other PbTe-based nanocomposites such as Na-doped PbTe-SrTe. On the basis of understanding such complex defect systems, Biswas et al. [143] proposed an atomic-nanometer-mesoscopic-through full-scale phonon scattering model (Fig. 5.18B), in which the full-scale distributed structural defects scatter a wide wavelength range of phonons. Such multiscaling structure has brought extremely high *ZT* values in PbTe-based nanocomposites.

## 5.7 Summary

Nanostructuring and nanocompositing are effective for the optimization of thermoelectric performance for existing materials, especially

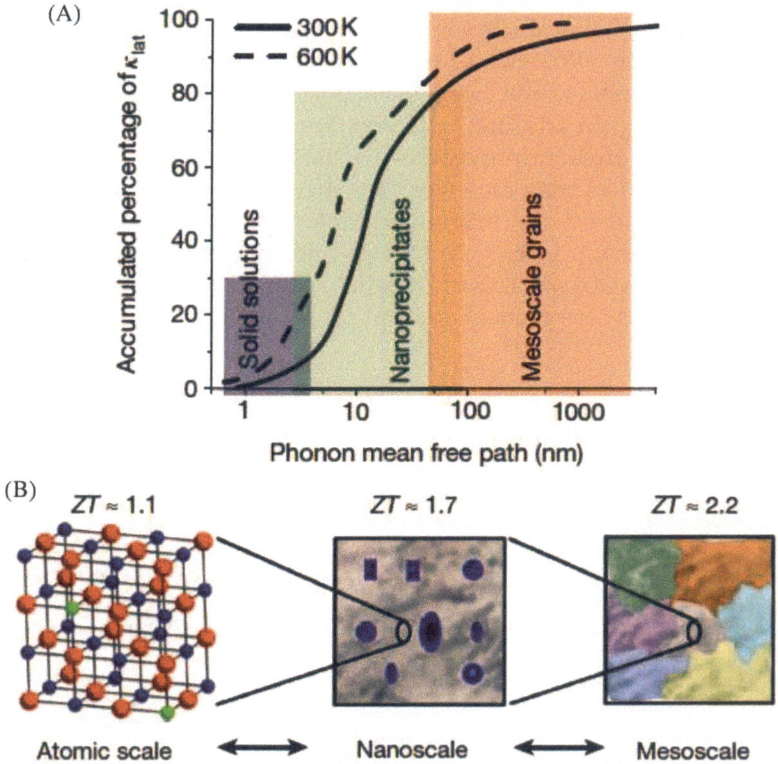


FIGURE 5.18 (A) Schematic diagram of short-range, medium-range, and long-range phonon scattering. (B) The atomic-nanometer-mesoscopic-through full-scale phonon scattering model.

through suppressing lattice thermal conductivity. Significant anisotropic feature and size confinement effect are undoubtedly observed in various thin films, which lead to distinctive electronic/phonon transportation characteristics. With the development of micro-device technologies, nanowires and thin films are expected with broad prospective for the applications in electronics and human health as well. As for the bulk materials, nanostructuring and nanocompositing exhibit great capability for enhancing *ZT* especially by strengthening interfacial scattering to phonons while maintaining good electrical transport properties. The nanostructure and nanoparticles dispersion may also increase the DOS near the Fermi level and/or filter low-energy electrons, which benefit to enhancing power factor ($S^2\sigma$). In order to effectively enhance thermoelectric performance by nanoparticles or nanostructures, lots of things should be severely considered and optimized, such as the size, content, and distribution of the dispersed particles. Furthermore, the following factors also need to be concerned: (1) band structure matching between nano-phase and matrix material; (2) the chemical inertness of nanophase under service environment; (3) a certain difference of the elastic modulus between nanophase and matrix material, which is believed one of the reasons to strengthen phonon scattering. In addition, it is well known that dispersing particles into matrix can often improve the mechanical properties in structural materials. This effect may also work for thermoelectric materials, that is, introducing nanoparticles into matrix compounds may improve the mechanical properties in a certain range and thus enhance its high temperature service stability. This is definitely helpful for the study on thermoelectric devices for practical applications.

## References

[1] L.D. Hicks, M.S. Dresselhaus, The effect of quantum-well structures on the thermoelectric figure of merit, Phys. Rev. B 47 (19) (1993) 12727–12731.
[2] L.D. Hicks, M.S. Dresselhaus, Thermoelectric figure of merit of a one-dimensional conductor, Phys. Rev. B 47 (24) (1993) 16631–16634.
[3] L.D. Hicks, T.C. Harman, M.S. Dresselhaus, Use of quantum-well superlattices to obtain a high figure of merit from nonconventional thermoelectric materials, Appl. Phys. Lett. 63 (23) (1993) 3230–3232.
[4] D. Chen, X. Zhang, X. Wang, Strain layer superlattice of $Ge_xSi_{1-x}$/Si and Ge/Si, Physics 18 (1) (1989) 21–26.
[5] R. Venkatasubramanian, T. Colpitts, B.O. Quinn, et al., Low-temperature organometallic epitaxy and its application to superlattice structures in thermoelectrics, Appl. Phys. Lett. 75 (8) (1999) 1104–1106.
[6] R. Venkatasubramanian, E. Siivola, T. Colpitts, et al., Thin-film thermoelectric devices with high room-temperature figures of merit, Nature 413 (6856) (2001) 597–602.
[7] J. Nurnus, H. Böttner, H. Beyer, et al., Proceedings of the 18th International Conference on Thermoelectrics (ICT, 1999), Baltimore, MD, IEEE, Piscataway, NJ, 1999, Catalog No. 99TH 8407, pp. 696–699.
[8] J.C. Caylor, K. Coonley, J. Stuart, et al., Enhanced thermoelectric performance in PbTe-based superlattice structures from reduction of lattice thermal conductivity, Appl. Phys. Lett. 87 (2005) 023105.
[9] T.C. Harman, D.L. Spears, M.P. Walsh, High thermoelectric figures of merit in PbTe quantum wells, J. Electron. Mater. 25 (7) (1996) 1121–1127.
[10] T. Koga, S.B. Cronin, M.S. Dresselhaus, et al., Experimental proof-of-principle investigation of enhanced $Z_{3D}T$ in (001) oriented Si/Ge superlattices, Appl. Phys. Lett. 77 (10) (2000) 1490–1492.
[11] S.T. Huxtable, A.R. Abramson, C. Tien, et al., Thermal conductivity of Si/SiGe and SiGe/SiGe superlattices, Appl. Phys. Lett. 80 (10) (2002) 1737–1739.
[12] A. Samarelli, L.F. Llin, S. Cecchi, et al., The thermoelectric properties of Ge/SiGe modulation doped superlattices, J. Appl. Phys. 113 (2013) 233704.
[13] A. Koma, K. Yoshimura, Ultrasharp interfaces grown with Van der Waals epitaxy, Surf. Sci. 174 (1986) 556–560.
[14] R. Venkatasubramanian, T. Colpttis, B. Watko, et al., MOCVD of $Bi_2Te_3$, $Sb_2Te_3$ and their superlattice structures for thin-film thermoelectric applications, J. Cryst. Growth 170 (1997) 817–821.
[15] G. Chen, Thermal conductivity and ballistic-phonon transport in the cross-plane direction of superlattices, Phys. Rev. B 57 (23) (1998) 14958–14973.

[16] G. Chen, M. Neagu, Thermal conductivity and heat transfer in superlattices, Appl. Phys. Lett. 71 (19) (1997) 2761–2763.
[17] B. Yang, W.L. Liu, J.L. Liu, et al., Measurements of anisotropic thermoelectric properties in superlattices, Appl. Phys. Lett. 81 (19) (2002) 3588–3590.
[18] H. Beyer, J. Nurnus, H. Böttner, et al., PbTe based superlattice structures with high thermoelectric efficiency, Appl. Phys. Lett. 80 (7) (2002) 1216–1218.
[19] R. Venkatasubramanian, Lattice thermal conductivity reduction and phonon localizationlike behavior in superlattice structures, Phys. Rev. B 61 (4) (2000) 3091–3097.
[20] S. Chakraborty, C.A. Kleint, A. Heinrich, et al., Thermal conductivity in strain symmetrized Si/Ge superlattices on Si(111), Appl. Phys. Lett. 83 (20) (2003) 4184–4186.
[21] M.V. Simkin, G.D. Mahan, Minimum thermal conductivity of superlattices, Phys. Rev. Lett. 84 (5) (2000) 927–930.
[22] M.N. Luckyanova, J. Garg, K. Esfarjani, et al., Coherent phonon heat conduction in superlattices, Science 338 (6109) (2012) 936–939.
[23] J. Garg, G. Chen, Minimum thermal conductivity in superlattices: a first-principles formalism, Phys. Rev. B 87 (2013). 140302(R).
[24] L.D. Hicks, T.C. Harman, X. Sun, et al., Experimental study of the effect of quantum-well structures on thethermoelectric figure of merit, Phys. Rev. B 53 (1996) R10493.
[25] N. Peranio, O. Eibl, J. Nurnus, Structural and thermoelectric properties of epitaxially grown $Bi_2Te_3$ thin films and superlattices, J. Appl. Phys. 100 (2006) 114306.
[26] T. Koga, X. Sun, S.B. Cronin, et al., Carrier pocket engineering to design superior thermoelectric materials using As/AlAs superlattices, Appl. Phys. Lett. 73 (20) (1998) 2950–2952.
[27] T. Koga, X. Sun, S.B. Cronin, et al., Carrier pocket engineering applied to "strained" Si/Ge superlattices to design useful thermoelectric materials, Appl. Phys. Lett. 75 (16) (1999) 2438–2440.
[28] D.J. Paul, Si/SiGe heterostructures: from material and physics to devices and circuits, Semicond. Sci. Technol. 19 (2004) R75.
[29] C.N. Liao, Y.C. Wang, H.S. Chu, Thermal transport properties of nanocrystalline Bi-Sb-Te thin films prepared by sputter deposition, J. Appl. Phys. 104 (2008) 104312.
[30] M. Takashiri, K. Miyazaki, S. Tanaka, et al., Effect of grain size on thermoelectric properties of n-type nanocrystalline bismuth-telluride based thin films, J. Appl. Phys. 104 (2008) 084302.
[31] J.Q. Song, X.H. Chen, Y.S. Tang, et al., Post-annealing effect on microstructures and thermoelectric properties of $Bi_{0.45}Sb_{1.55}Te_3$ thin films deposited by co-sputtering, J. Electron. Mater. 41 (11) (2012) 3068–3072.
[32] A.I. Hochbaum, R.K. Chen, R.D. Delgado, et al., Enhanced thermoelectric performance of rough silicon nanowires, Nature 451 (7175) (2008) 163–167.
[33] Z.L. Li, S.Q. Zheng, T. Huang, et al., Rational design, high-yield synthesis, and low thermal conductivity of Te/$Bi_2Te_3$ core/shell heterostructure nanotube composites, J. Alloy. Compd. 617 (2014) 247–252.
[34] X.S. Zhang, G. Ran, Review of the reactive mechanism in mechanical alloying, Heat. Treat. Met. 28 (2003) 28–31.
[35] B. Poudel, Q. Hao, Y. Ma, et al., High-thermoelectric performance of nanostructured bismuth antimony telluride bulk alloys, Science 320 (5876) (2008) 634–638.
[36] Y. Ma, Q. Hao, B. Poudel, et al., Enhanced thermoelectric figure-of-merit in p-type nanostructured bismuth antimonyntellurium alloys made from elemental chunks, Nano Lett. 8 (8) (2008) 2580–2584.
[37] X.W. Wang, H. Lee, Y.C. Lan, et al., Enhanced thermoelectric figure of merit in nanostructured n-type silicon germanium bulk alloy, Appl. Phys. Lett. 93 (2008) 193121.
[38] J. Yang, Q. Hao, H. Wang, et al., Solubility study of Yb in n-type skutterudites $Yb_xCo_4Sb_{12}$ and their enhanced thermoelectric properties, Phys. Rev. B 80 (11) (2009) 115329.

[39] B.-L. Yu, Q. Qi, X.-F. Tang, et al., Preparation and thermal transport properties of $CoSb_3$ nano-compounds, Acta Phys. Sin. 54 (12) (2005) 5763–5768.
[40] Y.C. Lan, A.J. Minnich, G. Chen, et al., Enhancement of thermoelectric figure-of-merit by a bulk nanostructuring approach, Adv. Funct. Mater. 20 (3) (2010) 357–376.
[41] X.B. Zhao, T. Sun, T.J. Zhu, et al., In-situ investigation and effect of additives on low temperature aqueous chemical synthesis of $Bi_2Te_3$ nanocapsules, J. Mater. Chem. 15 (16) (2005) 1621–1625.
[42] L.N. Zhou, X.B. Zhang, X.B. Zhao, et al., Influence of NaOH on the synthesis of $Bi_2Te_3$ via a low temperature aqueous chemical method, Mater. Sci. 44 (13) (2009) 3528–3532.
[43] Y.Q. Cao, T.J. Zhu, X.B. Zhao, Thermoelectric $Bi_2Te_3$ nanotubes synthesized by low-temperature aqueous chemicalnmethod, J. Alloy. Compd. 449 (1–2) (2008) 109–112.
[44] Y. Li, J. Jiang, G. Xu, et al., The reaction mechanism for the synthesis of $Bi_2Te_3$-based thermoelectric materials by aqueous chemical method, Rare Met. Mater. Eng. 36 (2) (2007) 233–235.
[45] Z.L. Sun, S.C. Liu, Q. Yao, et al., Low temperature synthesis of $Bi_2Te_3$ nanosheetsand thermal conductivity of nanosheet contained-composites, Mater. Chem. Phys. 121 (2010) 138–141.
[46] Y. Zheng, Wet-Chemical Method Preparation and Thermoelectric Properties of Nano $Bi_2Te_3$ Based Compound, Master's Thesis of Wuhan University of Technology, Wuhan, 2009.
[47] T.J. Zhu, Y.Q. Cao, Q. Zhang, et al., Bulk nanostructured thermoelectric materials: preparation, structure and properties, J. Electron. Mater. 39 (9) (2010) 1990–1995.
[48] B. Zhou, Y. Zhao, L. Pu, et al., Microwave-assisted synthesis of nanocrystalline $Bi_2Te_3$, Mater. Chem. Phys. 96 (2006) 192–196.
[49] Y. Jiang, Y.J. Zhu, L.D. Chen, Microwave-assisted preparation of $Bi_2Te_3$ hollow nanospheres, Chem. Lett. 36 (3) (2007) 382–383.
[50] Q. Yao, Y.J. Zhu, L.D. Chen, et al., Microwave-assisted synthesis and characterization of $Bi_2Te_3$ nanosheets and nanotubes, J. Alloy. Compd. 481 (2009) 91–95.
[51] X. Zhen, Synthesis of Nano $Bi_2Te_3$ Based Thermoelectric Materials by Wet Chemical Route, Master's Thesis of Huazhong University of Science and Technology, Wuhan, 2007.
[52] Kai Li, Study on Preparation of $Sb_2Te_3$ and $Sb_2Se_3$ Nano Materials by Microwave Heating and Wet-Chemical Method, Master's Thesis of Huazhong University of Science and Technology, Wuhan, 2008.
[53] Y. Zheng, T.J. Zhu, X.B. Zhao, et al., Sonochemical systhesis of nano-crystalline $Bi_2Te_3$ thermoelectric compound, Mater. Lett. 59 (2005) 286–288.
[54] H. Liu, H.M. Cui, F. Han, et al., Growth of $Bi_2Se_3$ nanobelts synthesized through a co-reduction method under ultrasonic irradiation at room temperature, Cryst. Growth Des. 5 (5) (2005) 1711–1714.
[55] T. Wang, R. Mehta, C. Karthik, et al., Microphere bouquets of bismuth telluride nanoplates: room temperature synthesis and thermoelectric properties, Phys. Chem. 114 (4) (2011) 1796–1799.
[56] J.J. Xu, H. Li, B.L. Du, et al., High thermoelectric figure of merit and nanostructureng in bulk $AgSbTe_2$, J. Mater. Chem. 20 (29) (2010) 6138–6143.
[57] J. Xie, X.B. Zhao, J.L. Mi, et al., Solvothermal synthesis of nanosized $CoSb_3$ skutterudite, J. Zhejiang Univ. Sci. 5 (12) (2004) 1504–1508.
[58] Y. Deng, C.W. Cui, N.L. Zhang, et al., Fabrication of bismuth telluride nanotubes via a simple solvothermal process, Solid State Commun. 138 (3) (2006) 111–113.
[59] J.L. Mi, N. Lock, T. Sun, et al., Biomolecule-assisted hydrothermal synthesis and self-assembly of $Bi_2Te_3$ nanostring-cluster hierarchical structure, ACS Nano 4 (5) (2010) 2523–2530.

[60] R.C. Jin, J.S. Liu, G.H. Li, Facile solvothermal synthesis, growth mechanism and thermoelectric property of flower-like $Bi_2Te_3$, Cryst. Res. Technol. 49 (7) (2014) 460–466.

[61] Y.H. Zhang, T.J. Zhu, J.P. Tu, et al., Flower-like nanostructure and thermoelectric properties of hydrothermally synthesized La-containing $Bi_2Te_3$ based alloys, Mater. Chem. Phys. 103 (2–3) (2007) 484–488.

[62] A. Datta, G.S. Nolas, Solution-based synthesis and low-temperature transport properties of CsBi4Te6, ACS Appl. Mater. Interfaces 4 (2) (2012) 772–776.

[63] W.D. Shi, J.B. Yu, H.S. Wang, et al., Hydrothermal synthesis of single-crystalline antimony telluride nanobelts, J. Am. Chem. Soc. 128 (51) (2006) 16490–16491.

[64] W.D. Shi, L. Zhou, S.Y. Song, et al., Hydrothermal synthesis and thermoelectric transport properties of impurity-free antimony telluride hexagonal nanoplates, Adv. Mater. 20 (10) (2008) 1892–1897.

[65] S.F. Shi, M.H. Cao, C.W. Hu, Controlled solvothermal synthesis and structural characterization of antimony telluride nanoforks, Cryst. Growth Des. 9 (5) (2009) 2057–2060.

[66] Z. Chen, M.Y. lin, G.D. Xu, et al., Hydrothermal synthesized nanostructure Bi-Sb-Te thermoelectric materials, J. Alloy. Compd. 588 (2014) 384–387.

[67] J.L. Mi, X.B. Zhao, T.J. Zhu, et al., Solvothermal synthesis and electrical transport properties of skutterudite $CoSb_3$, J. Alloy. Compd. 417 (36) (2006) 269–272.

[68] J.Q. Li, X.W. Feng, W.A. Sun, et al., Solvothermal synthesis of nano-sized skutterudite $Co_{4-x}Fe_xSb_{12}$ powders, Mater. Chem. Phys. 112 (1) (2008) 57–62.

[69] J.L. Mi, X.B. Zhao, T.J. Zhu, et al., Nanosized La filled $CoSb_3$ prepared by a solvothermal- annealing method, Mater. Lett. 62 (15) (2008) 2363–2365.

[70] K. Kadel, W.Z. Li, Solvothermal synthesis and structural characterization of unfilled and Yb-filled cobalt antimony skutterudite, Cryst. Res. Technol. 49 (2-3) (2014) 135–141.

[71] T.J. Zhu, Y.Q. Liu, X.B. Zhao, Synthesis of PbTe thermoelectric materials by alkaline reducing chemical routes, Mater. Res. Bull. 43 (11) (2008) 2850–2854.

[72] R.C. Jin, G. Chen, J. Pei, et al., Facile solvothermal synthesis and growth mechanism of flower-like PbTe dendrites assisted by cyclodextrin, Cryst. Eng. Commun. 14 (6) (2012) 2327–2332.

[73] G.H. Dong, Y.J. Zhu, One-step microwave-solvothermal rapid synthesis of Sb doped PbTe/$Ag_2Te$ core/shell composite nanocubes, Chem. Eng. J. 193-194 (2012) 227–233.

[74] J. Liu, X.G. Wang, L.M. Peng, Solvothermal synthesis and growth mechanism of Ag and Sb co-doped PbTe heterogeneous thermoelectric nanorods and nanocubes, Mater. Chem. Phys. 133 (1) (2012) 33–37.

[75] X.R. Shi, G. Chen, D.H. Chen, et al., PbSe hierarchical nanostructures: solvothermal synthesis, growth mechanism and their thermoelectric transportation properties, Cryst. Eng. Commun. 16 (41) (2014) 9704–9710.

[76] R.C. Jin, G. Chen, J. Pei, PbS/PbSe hollow spheres: solvothermal synthesis, growth mechanism, and thermoelectric transport property, J. Phys. Chem. C. 116 (2012) 16207–16216.

[77] L. Yang, Z.G. Chen, G. Han, et al., High-performance thermoelectric $Cu_2Se$ nanoplates through nanostructure engineering, Nano Energy 16 (2015) 367–374.

[78] A. Datta, G.S. Nolas, Synthesis and characterization of nanocrystalline $FeSb_2$ for thermoelectric applications, Eur. J. Inorg. Chem. 2012 (1) (2012) 55–58.

[79] S. Sarkar, B. Das, P.R. Midya, et al., Optical and thermoelectric properties of chalcogenide based $Cu_2NiSnS_4$ nanoparticles synthesized by a novel hydrothermal route, Mater. Lett. 152 (2015) 155–158.

[80] X. Tang, W. Xie, H. Li, et al., Preparation and thermoelectric transport properties of high-performance p-type $Bi_2Te_3$ with layered nanostructure, Appl. Phys. Lett. 90 (2007) 012102.

[81] W.J. Xie, X.F. Tang, Y.G. Yan, et al., Unique nanostructures and enhanced thermoelectric performance of melt-spun BiSbTe alloys, Appl. Phys. Lett. 94 (2009) 102111.

[82] W.J. Xie, X.F. Tang, Y.G. Yan, et al., High thermoelectric performance BiSbTe alloy with unique low-dimensional structure, J. Appl. Phys. 105 (2009) 113713.
[83] W. Xie, J. He, H.J. Kang, et al., Identifying the specific nanostructures responsible for the high thermoelectric performance of $(Bi, Sb)_2Te_3$ nanocomposites, Nano Lett. 10 (2010) 3283.
[84] H. Li, X. Tang, X. Su, et al., Preparation and thermoelectric properties of high-performance Sb additional $Yb_{0.2}Co_4Sb_{12+y}$ bulk materials with nanostructure, Appl. Phys. Lett. 92 (2008) 202114.
[85] H. Li, X. Tang, Q. Zhang, et al., Rapid preparation method of bulk nanostructured $Yb_{0.3}Co_4Sb_{12+y}$ compounds andtheir improved thermoelectric performance, Appl. Phys. Lett. 93 (2008) 252109.
[86] S.Y. Wang, W.J. Xie, H. Li, et al., Enhanced performances of melt spun $Bi_2(Te, Se)_3$ for n-type thermoelectric legs, Intermetallics 19 (7) (2011) 1024–1031.
[87] X.Z. Cai, X.A. Fan, Z.Z. Rong, et al., Improved thermoelectric properties of $Bi_2Te_{3-x}Se_x$ alloys by melt spinning and resistance pressing sintering, J. Phys. D Appl. Phys. 47 (2014) 115101.
[88] W.J. Xie, S.Y. Wang, S. Zhu, et al., High performance $Bi_2Te_3$ nanocomposites prepared by single-elementmelt-spinning spark-plasma sintering, J. Mater. Sci. 48 (2013) 2745–2760.
[89] S.Y. Wang, H. Li, D.K. Qi, et al., Enhancement of the thermoelectric performance of $\beta$-$Zn_4Sb_3$ by *in situ* nanostructures and minute Cd-doping, Acta Mater. 59 (12) (2011) 4805–4817.
[90] D. Qi, X.F. Tang, H. Li, et al., Improved thermoelectric performance and mechanical properties of nanostructured melt-spun $\beta$-$Zn_4Sb_3$, J. Electron. Mater. 39 (8) (2010) 1159–1165.
[91] M. Hasaka, T. Morimura, H. Nakashima, Thermoelectric properties of melt-spun $Zn_xSb_3$ ribbons, J. Electron. Mater. 41 (6) (2012) 1193–1198.
[92] Y. Chen, T.J. Zhu, S.H. Yang, et al., High-performance $(Ag_xSbTe_{x/2+1.5})_{15}$ $(GeTe)_{85}$ thermoelectric materials preparedby melt spinning, J. Electron. Mater. 39 (9) (2010) 1719–1723.
[93] B. Du, H. Li, J. Xu, et al., Enhanced thermoelectric performance and novel nanopores in $AgSbTe_2$ prepared by melt spinning, J. Solid State Chem. 184 (2011) 109–114.
[94] W. Luo, H. Li, Y. Yan, et al., Rapid synthesis of high thermoelectric performance higher manganese silicide with in-situ formed nano-phase of MnSi, Intermetallics 19 (3) (2011) 404–408.
[95] X.Y. Shi, X. Shi, Y.L. Li, et al., Enhanced power factor of higher manganese silicide via melt spin synthesis method, J. Appl. Phys. 116 (2014) 245104.
[96] J. Zhou, Q. Jie, L. Wu, et al., Nanostructures and defects in non-equilibrium synthesized filled skutterudite $CeFe_4Sb_{12}$, J. Mater. Res. 26 (2011) 1842.
[97] J.R. Salvador, R.A. Waldo, C.A. Wong, et al., Thermoelectric and mechanical properties of melt spun and spark plasma sintered n-type Yb- and Ba-filled skutterudites, Mater. Sci. Eng. B 178 (17) (2013) 1087–1096.
[98] G.H. Zhu, H. Lee, Y.C. Lan, et al., Increased phonon scattering by nanograins and point defects in nanostructured silicon with a low concentration of germanium, Phys. Rev. Lett. 102 (2009) 196803.
[99] P. Zhang, Z. Wang, H. Chen, et al., Effect of cooling rate on microstructural homogeneity and grain size of n-type Si-Ge thermoelectric alloy by melt spinning, J. Electron. Mater. 39 (10) (2010) 2251–2254.
[100] C. Yu, T.J. Zhu, K. Xiao, et al., Reduced grain size and improved thermoelectric properties of melt spun (Hf, Zr)NiSn half-Heusler alloys, J. Electron. Mater. 39 (9) (2010) 2008–2012.

[101] Y. Yan, X. Tang, P. Li, et al., Microstructure and thermoelectric transport properties of type I clathrates $Ba_8Sb_2Ga_{14}Ge_{30}$ prepared by ultrarapid solidification process, J. Electron. Mater. 38 (7) (2009) 1278–1281.
[102] A. Prokofiev, M. Ikeda, E. Makalkina, et al., Melt spinning of clathrates: electron microscopy study and effect of composition on grain size, J. Electron. Mater. 7 (42) (2013) 1628–1633.
[103] W.Q. Cao, S.K. Deng, X.F. Tang, et al., The effects of melt spinning process on microstructure and thermoelectric properties of Zn-doped type-I clathrates, Acta Phys. Sin. 1 (58) (2009) 612.
[104] G.C. Ding, J.X. Si, H.F. Wu, et al., Thermoelectric properties of melt spun PbTe with multi-scaled nanostructures, J. Alloy. Compd. 662 (2016) 368–373.
[105] S. Katsuyama, Y. Kanayama, M. Ito, et al., Thermoelectric properties of $CoSb_3$ with dispersed $FeSb_2$ particles, J. Appl. Phys. 88 (6) (2000) 3484.
[106] S. Katsuyama, M. Watanabe, M. Kuroki, et al., Effect of NiSb on the thermoelectric properties of skutterudite $CoSb_3$, J. Appl. Phys. 93 (5) (2003) 2758–2764.
[107] Z. He, C. Stiewe, D. Platzek, et al., Nano $ZrO_2/CoSb_3$ composites with improved thermoelectric figure of merit, Nanotechnology 18 (5) (2007) 235602.
[108] S. Katsuyama, H. Okada, K. Miyajima, Thermoelectric properties of $CeFe_3CoSb_{12}$-$MoO_2$ composite, Mater. Trans. 49 (8) (2008) 1731.
[109] P.N. Alboni, X. Ji, J. He, et al., Thermoelectric properties of $La_{0.9}CoFe_3Sb_{12}$-$CoSb_3$ skutterudite nanocomposites, J. Appl. Phys. 103 (2008) 113707.
[110] S.P. Li, J.Q. Li, Q.B. Wang, et al., Synthesis and thermoelectric properties of the $(GeTe)_{1-x}(PbTe)_x$ alloys, Solid State Sci. 2 (13) (2011) 399–403.
[111] J.Q. Li, X.X. Li, F.S. Liu, et al., Enhanced thermoelectric properties of $(PbTe)_{0.88}(PbS)_{0.12}$ composites by Sb doping, J. Electron. Mater. 3 (42) (2013) 366–371.
[112] O. Falkenbach, D. Hartung, P.J. Klar, et al., Thermoelectric properties of nanostructured bismuth-doped lead telluride $Bi_x(PbTe)_{1-x}$ prepared by Co-ball-milling, J. Electron. Mater. 43 (6) (2014) 1674–1680.
[113] S. Sumithra, N.J. Takas, W.M. Nolting, et al., Effect of NiTe nanoinclusions on thermoelectric properties of $Bi_2Te_3$, J. Electron. Mater. 41 (6) (2012) 1401–1407.
[114] S. Yoon, O. Kwon, S. Ahn, et al., The effect of grain size and density on the thermoelectric properties of $Bi_2Te_3$-PbTe compounds, J. Electron. Mater. 42 (12) (2013) 3390–3396.
[115] W.S. Liu, K.C. Lukas, K. McEnaney, et al., Studies on the $Bi_2Te_3$-$Bi_2Se_3$-$Bi_2S_3$ system for mid-temperature thermoelectric energy conversion, Energy Environ. Sci. 6 (2013) 552–560.
[116] S. Sumithra, N.J. Takas, D.K. Misra, et al., Enhancement in thermoelectric figure of merit in nanostructured $Bi_2Te_3$ with semimetal nanoinclusions, Adv. Energy Mater. 6 (1) (2011) 1141–1147.
[117] F.R. Sie, C.H. Kuo, C.S. Hwang, et al., Thermoelectric performance of n-type $Bi_2Te_3$/Cu composites fabricated by nanoparticle decoration and spark plasma sintering, J. Electron. Mater. 45 (3) (2016) 1927–1934.
[118] J. Hyun, K. Jooheon, Preparation and structure dependent thermoelectric properties of nanostructured bulk bismuth telluride with graphene, J. Alloy. Compd. 664 (2016) 639–647.
[119] D. Suh, S. Lee, H. Mun, et al., Enhanced thermoelectric performance of $Bi_{0.5}Sb_{1.5}Te_3$-expanded graphene composites by simultaneous modulation of electronic and thermal carrier transport, Nano Energy 13 (2015) 67–76.
[120] C.C. Hsu, Y.N. Liu, H.K. Ma, Effect of the $Zr_{0.5}Hf_{0.5}CoSb_{1-x}Sn_x/HfO_2$ half-heusler nanocomposites on the ZT value, J. Alloy. Compd. 597 (2014) 217–222.
[121] X. Shi, L.D. Chen, S.Q. Bai, et al., Influence of fullerene dispersion on high temperature thermoelectric properties of $Ba_yCo_4Sb_{12}$-based composites, J. Appl. Phys. 102 (2007) 103709.

[122] T. Ikeda, L.A. Collins, V.A. Ravi, et al., Self-assembled nanometer lamellae of thermoelectric PbTe and $Sb_2Te_3$ with epitaxy-like interfaces, Chem. Mater. 19 (4) (2007) 763–767.
[123] T. Ikeda, S.M. Haile, V.A. Ravi, et al., Solidification processing of alloys in the pseudo-binary PbTe-$Sb_2Te_3$ system, Acta Mater. 55 (4) (2007) 1227–1239.
[124] T. Ikeda, E.S. Toberer, V.A. Ravi, et al., In situ observation of eutectoid reaction forming a PbTe-$Sb_2Te_3$ thermoelectric nanocomposite by synchrotron X-ray diffraction, Scr. Mater. 60 (5) (2009) 321–324.
[125] X. Shi, Z. Zhou, W. Zhang, et al., Solid solubility of Ir and Rh at the Co sites of skutterudites, J. Appl. Phys. 101 (2007) 123525.
[126] X. Shi, W. Zhang, L.D. Chen, et al., Phase-diagram-related problems in thermoelectric materials: skutterudites as an example, Int. J. Mater. Res. 99 (6) (2008) 638.
[127] C. Uher, X. Shi, H. Kong, Filled $Ir_xCo_{1-x}Sb_3$-based skutterudite solid solutions, in: Proceedings of the 25th International Conference on Thermoelectrics, 2007, p. 189.
[128] Z. Xiong, X.Y. Huang, X.H. Chen, et al., Realizing phase segregation in the $Ba_{0.2}(Co_{1-x}Ir_x)_4Sb_{12}$ ($x = 0$, 0.1, 0.2) filled skutterudite system, Scr. Mater. 62 (2) (2010) 93–96.
[129] X.Y. Zhao, X. Shi, L.D. Chen, et al., Synthesis of $Yb_yCo_4Sb_{12}/Yb_2O_3$ composites and their thermoelectric properties, Appl. Phys. Lett. 89 (2006) 092121.
[130] T. Wu, W. Jiang, X. Li, et al., Thermoelectric properties of p-type Fe-doped TiCoSb half-heusler compounds, J. Appl. Phys. 102 (2007) 103705.
[131] J.P. Heremans, C.M. Thrush, D.T. Morelli, Thermopower enhancement in PbTe with Pb precipitates, J. Appl. Phys. 98 (2005) 063703.
[132] Z. Xiong, X.H. Chen, X.Y. Zhao, et al., Effects of nano-$TiO_2$ dispersion on the thermoelectric properties of filled-skutterudite $Ba_{0.22}Co_4Sb_{12}$, Solid State Sci. 11 (9) (2009) 1612–1616.
[133] H. Li, X. Tang, Q. Zhang, et al., High performance $In_xCe_yCo_4Sb_{12}$ thermoelectric materials with *in situ* forming nanostructured InSb phase, Appl. Phys. Lett. 94 (2009) 102114.
[134] X. Shi, L. Chen, J. Yang, et al., Enhanced thermoelectric figure of merit of $CoSb_3$ via large-defect scattering, Appl. Phys. Lett. 84 (13) (2004) 2301.
[135] Z. Xiong, X.H. Chen, X.Y. Huang, et al., High thermoelectric performance of $Yb_{0.26}Co_4Sb_{12}/y$GaSb nanocomposites originating from scattering electrons of low energy, Acta Mater. 58 (2010) 3995.
[136] J.R. Sootsman, H.J. Kong, C. Uher, et al., Large enhancements in the thermoelectric power factor of bulk PbTe at high temperature by synergistic nanostructuring, Angew. Chem. Int. Ed. 47 (45) (2008) 8618.
[137] Y.Z. Pei, J. Lensch-Falk, E.S. Toberer, et al., High thermoelectric performance in PbTe due to large nanoscale $Ag_2Te$ precipitates and La doping, Adv. Funct. Mater. 21 (2) (2011) 241–249.
[138] K.F. Hsu, S. Loo, F. Guo, et al., Cubic $AgPb_mSbTe_{2+m}$: bulk thermoelectric materials with high figure of merit, Science 303 (5659) (2004) 818–821.
[139] F.P. Poudeu, J.D. Angel, A.D. Downey, et al., High thermoelectric figure of merit and nanostructuring in bulk p-type $Na_{1-x}Pb_mSb_yTe_{m+2}$, Angew. Chem. Int. Ed. 45 (23) (2006) 3835–3839.
[140] K. Biswas, J.Q. He, Q.C. Zhang, et al., Strained endotaxial nanostructures with high thermoelectric figure of merit, Nat. Chem. 3 (2) (2011) 160–166.
[141] I.U. Arachchige, J. Wu, V.P. Dravid, et al., Nanocrystals of the quaternary thermoelectric materials: $AgPb_mSbTe_{m+2}$ ($m = 1 \sim$E;18): phase-segregated or solid solutions? Adv. Mater. 20 (19) (2008) 3638–3642.
[142] N. Chen, F. Gascoin, G.J. Snyder, et al., Macroscopic thermoelectric inhomogeneities in $(AgSbTe_2)_x(PbTe)_{1-x}$, Appl. Phys. Lett. 87 (2005) 171903.
[143] K. Biswas, J.Q. He, I.D. Blum, et al., High-performance bulk thermoelectrics with all-scale hierarchical architectures, Nature 489 (7416) (2012) 414–418.

CHAPTER

# 6

# Organic thermoelectric materials

## 6.1 Introduction

Since the thermoelectric effects were discovered, the research on thermoelectrics had been only concentrated on the exploration and optimization of inorganic thermoelectric materials (including inorganic semiconductors and semimetals) for a long time. At the end of 20th century, organic conductors began to attract attention as potential thermoelectric materials because of their good flexibility, processability, and tunable thermoelectric properties. Especially, the emerging trend for wearable and autonomous devices has accelerated the research of organic thermoelectric materials.

The discovery of conducting polyacetylene with high conductivity by Shirakawa, Heeger, and MacDiarmid in 1976 did trigger a research upsurge in the field of organic conducting polymers. In 1990s the thermoelectric transport behavior of conducting polymers began to enter the vision of researchers in the thermoelectric field, and a few conducting polymers (e.g., polyaniline and polyacetylene) were reported possessing tunable Seebeck coefficient by changing the doping concentrations. However, their $ZT$ values had lingered in the range of $10^{-6}$–$10^{-3}$ for a long time, which is far lower than those of state-of-the-art inorganic thermoelectric materials. Since 2000, with the exploration of thermoelectric properties being expanded to various organic materials, the recorded maximum $ZT$ values for organic materials have been successively renewed. Among them, the typical members are conducting polymers [such as polyaniline, poly(3,4-ethylenedioxythiophene), and poly(3-hexythiophene)], organic small molecules (such as tetracyano-*p*-phenylquinone-dimethylmethane), metal complexes (such as polyethylene tetramercaptan). Especially, remarkable progress has been made in conducting polymers, and their $ZT$ values have been improved


DOI: https://doi.org/10.1016/B978-0-12-818413-4.00006-5

from $10^{-3}$ to $10^{-1}$, almost by two orders of magnitude. Various novel approaches have been developed for optimizing the TE performance of polymers, such as doping regulation, molecular structure engineering, organic/inorganic nanocomposites or hybrids, and so on. Lots of elementary but guiding significant mechanisms for tuning thermoelectric transport of organic materials have been proposed, mainly including charge-transport bridging, interfacial scattering to electrons/phonons, interface engineering–induced ordering of polymer chain alignment and/or chain conformation, nanointercalated superlattice, and junctions-assisted abnormal charge transfer as well. In this chapter, the synthesis method and optimization of thermoelectric properties of conducting polymers and polymer-based nanocomposites will be particularly presented.

## 6.2 Doping and charge transport in organic semiconductors

Generally, doping in an inorganic material means "contamination" of host materials with dopants and the charge transport behavior is usually determined by the band structure. However, the doping mechanism in organic semiconductors remains somewhat ambiguous compared with the well-established band theory for inorganic counterparts, because of the complex relation between the charge transport and the complicated structure factors—chemical bonds, molecular structure, molecular chain conformation, etc. The organic semiconductors mostly consist of $sp^2$ and/or sp-hybridized carbon atoms and the remained p electrons form extended π-systems. Because of the localized charges and weak coupling effect of π-orbitals in molecules, a hopping mechanism is often used to explain the carrier transport in disordered or low-mobility organic semiconductors with low mobility. Chemical doping is an efficient way to increase the charge carrier concentrations. The chemical doping has two essential roles: (1) tune electrical conductivity by different doping levels; (2) change the position of Fermi levels to adjust the charge injection barriers. The former is widely used in organic thermoelectrics and bioelectronics [1,2], while the latter is popular in organic field-effect transistor (OFET), organic light-emitting diode, and solar cells [3–5]. In the chemical doping process, free charge carriers are generated by the electron transfer between host molecule and the dopant (Fig. 6.1). To facilitate the charge transfer, the lowest unoccupied molecular orbital (LUMO) of the p-dopant has to be aligned below or at least close to the highest occupied molecular orbital (HOMO) of the matrix molecule (for p-type doping, left in Fig. 6.1), or the HOMO of the n-dopant has to be aligned above or at least close to the LUMO of the matrix molecule (for n-type doping, right in Fig. 6.1).

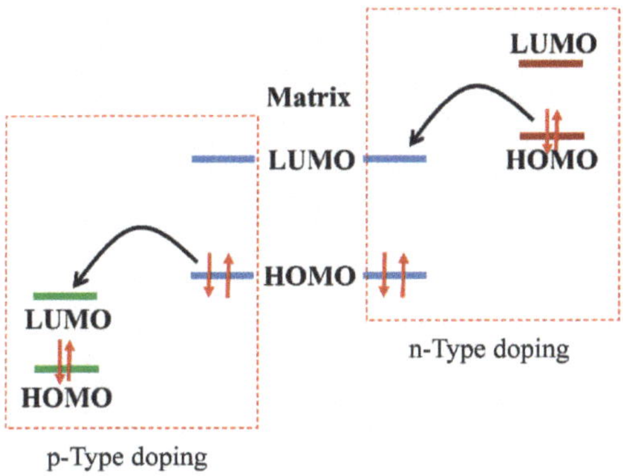


**FIGURE 6.1** The schematic model of chemical doping process. Free charge carriers are generated by the electron transfer between dopant and organic semiconductors (matrix).

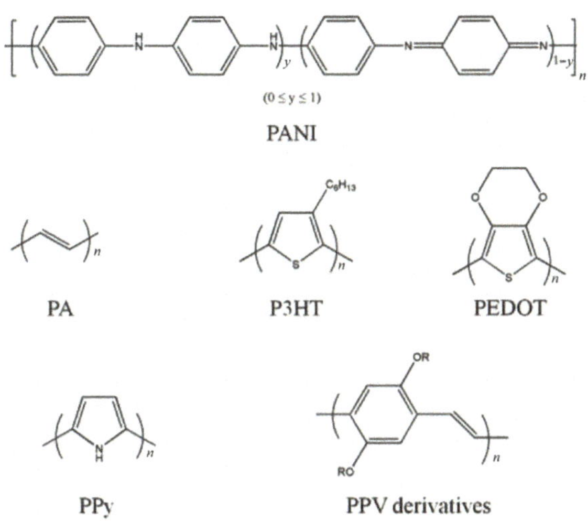


**FIGURE 6.2** Chemical structures of typical conducting polymers.

The typical semiconductor polymers generally have rigid and conjugated backbone to form weak van der Waals interaction in solid state. As shown in Fig. 6.2, most of semiconductors consist of alternative single bonds and double bonds or show aromatic character. The room-temperature electrical conductivities of these conducting polymers are usually very low, typically around $10^{-8}$ S/cm or lower. By doping, electrical conductivity can be improved by orders of magnitudes due to the significant increase of charge carrier concentrations in the form of polaron or bipolaron.

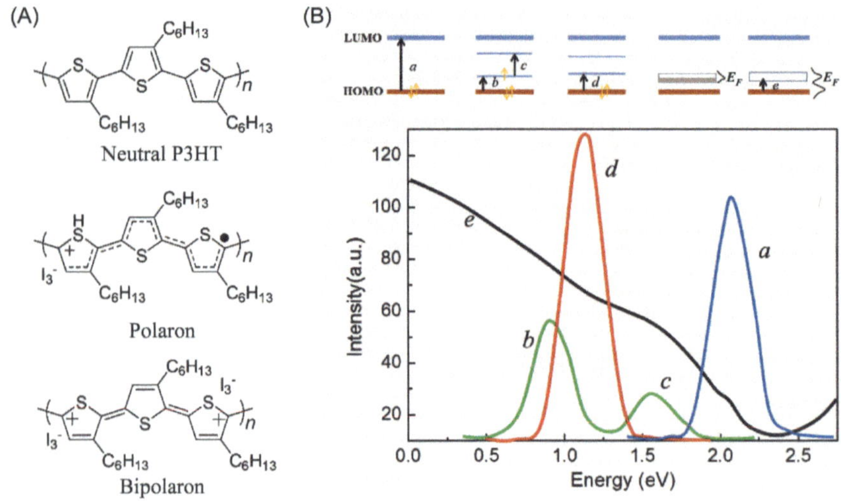


FIGURE 6.3 (A) Chemical structures of a neutral P3HT chain and P3HT chain carrying a polaron and a bipolaron. (B) Electronic structures of polymer chains at various oxidation levels and the corresponding optical spectra.

Taking the p-type doping as an example, a neutral polythiophene (P3HT) chain will carry a polaron or a bipolaron at a high doping level (Fig. 6.3A), while the reduced dopant transforms into a negative counter-ion upon doping. With the formation of (bi)polaron, the structure of the aromatic monomers tends to adopt quasiquinoid structures, which results in the change of bond length. The appearance of polarons or bipolarons can be detected by the absorption spectra when a conjugated polymer subjecting to dopants (Fig. 6.3B), which corresponds to the evolution of electronic structures of polymer chains at various oxidation levels. Transition *a* represents a first optical transition from the top of the HOMO level to the bottom of the LUMO level of a neutral polymer chain. Upon doping, new polaronic or bipolaronic states appear in the bandgap of the polymer (transitions *b* and *c* mean the polarons and a transition *d* means bipolaron on an isolated polymer chain). At high doping levels, (bi)polaron band shall form and shows a broad infrared absorption (transition *e*) [6].

It is generally agreed that the transport of charge carrier in organic solids follows thermally activated polaron hopping and percolation theories (Fig. 6.4) [7], owing to the weak intermolecular interaction and the feature of energetic disorder in organic polymers. So far, the concept of variable-range hopping is widely accepted to interpret the hopping mechanism, that is, a carrier could hop over a certain distance with proper activation energy (Fig. 6.4A) [8]. Therefore the conduction

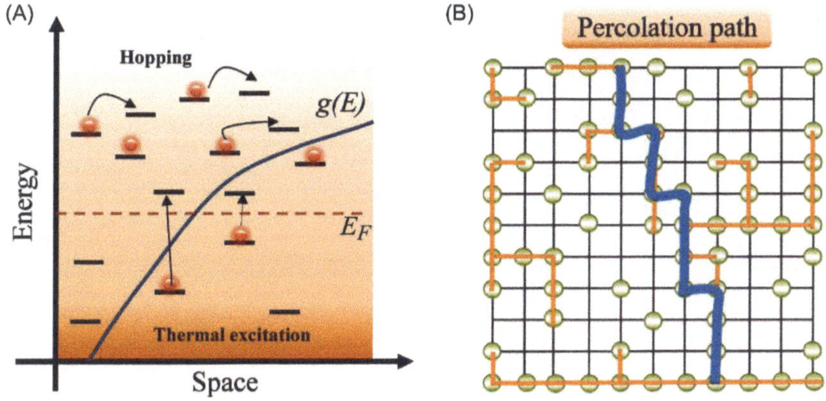


FIGURE 6.4 (A) The schematic diagram of the charge transport in a hopping system. (B) The percolation current path through the polymer matrix.

process in organic films can be described by percolation theory. It is generally believed that there are conductive networks in the polymer solids. The localized sites in the network that are interconnected to form transfer paths for the carrier to travel through shall contribute to the large electrical conductivity (Fig. 6.4B) [9]. The hopping and percolation theories are related to the electronic couplings between adjacent molecules and thus strongly dependent on the molecular packing and the morphology. The mode of the molecular packing can be affected by the degree of π-orbital overlaps, intermolecular interactions, the molecular rigidity, the conjugation length, etc. However, the introduced dopants will inevitably impact these factors and thus change the intrinsic packing and morphology of polymer molecules.

The complex microstructure of conjugated polymers in a solid is ascribed to the diverse intermediate status with order-disorder structures in multiple scales. Usually, the coexistence of polycrystalline and amorphous region can be observed in these materials on the macroscale (Fig. 6.5). These polycrystalline domains are largely responsible for charge transport to strongly affect the electrical conductivity. On smaller scale, the electronic coupling between polymeric chains results in highly anisotropic charge transport within individual crystalline domains. On the molecular scale, π-orbital overlaps and intermolecular interactions strongly affect the strength of coupling and the charge transport. Therefore the introduced of dopants not only induce charges into the assembly, but also interact with semiconductors in micro- and/or nano-size scale. These interactions give us opportunities to improve the doping efficiency and control the charge transport by the strategies of designing organic materials, developing processing methods, and

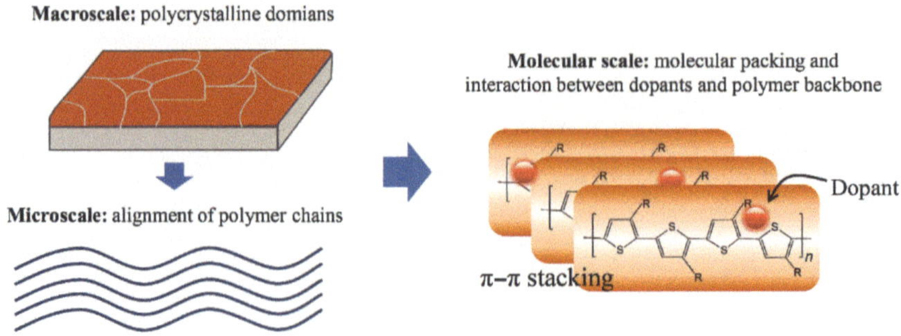


FIGURE 6.5 Size scales and relevant structure features in polymer semiconductor film.

forming organic-inorganic hybrids/composites. These strategies will be discussed in detail in the following sections.

## 6.3 Thermoelectric properties of typical conducting polymers

### 6.3.1 Polyaniline

Among numerous conducting polymers, polyaniline (PANI) is earlier investigated as thermoelectric materials owing to its unique properties such as straightforward protocol for preparation, easy control of electrical conductivity (e.g., reversible transformation between insulating and conducting), and low cost. According to the structural model proposed by MacDiarmid, the polyaniline molecular consists of two repeat units: reduced unit (leucoemeraldine base) and oxidized unit (permigraniline), with a head-tail link structure as shown in Fig. 6.6. With changing the ratio of reduced unit ($y$ value), polyanilines show three typical states. Polyaniline is in the fully reduced state when $y=1$, while it is in the fully oxidized state when $y=0$. When $y$ is about 0.5, polyaniline is in the half-oxidized state (emeraldine base, EB). The polyaniline presents electrically insulating in both the fully reduced and fully oxidized states. Polyanilines can be changed from insulating to conducting states by changing the oxidation degree and/or doping protonic acid.

Protonic acid is a common dopant for polyaniline. The doping process of polyaniline is quite different from that of other conducting polymers. Generally, the doping process of conducting polymer is always accompanied by the gain and loss of electrons on the backbone chain. However, when polyaniline is doped with protonic acid, the number of electrons keeps unchanged, while the protons enter the polyaniline chain to make the chain positively charged. The counter-ion enters the polyaniline molecular chain as a balance of positive or negative charges. The molecular size and electronegativity of the counter-ions will directly

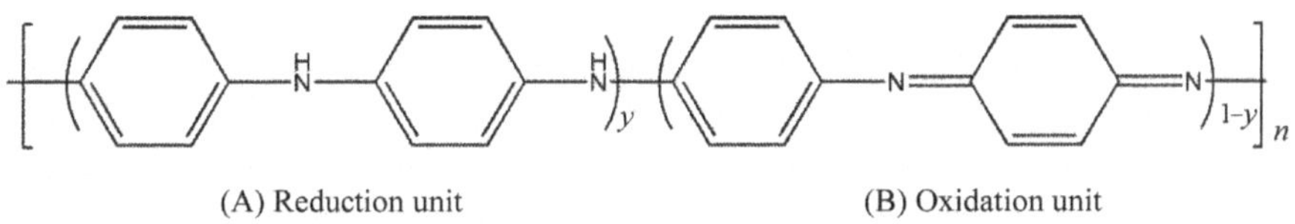


**FIGURE 6.6** The chemical structure of polyaniline. $y(0 \leq y \leq 1)$ is the degree of reduction of polyaniline, and $n$ is the degree of polymerization.

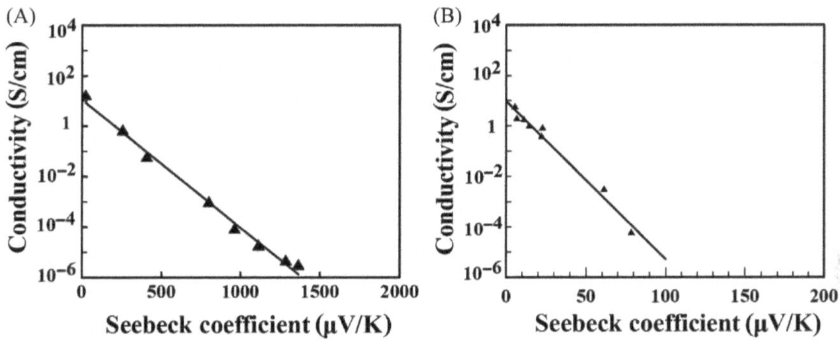


**FIGURE 6.7** Electrical conductivity versus Seebeck coefficient for (A) polyacetylene and (B) polyaniline with different doping degree.

affect the molecular conformation and arrangement of polyaniline chains and therefore affect the electrical transport properties.

Since the end of 1990s the thermoelectric properties of polyaniline have been widely studied. As a general concept, the major parameters (electrical conductivity, Seebeck coefficient, and thermal conductivity) are strongly correlated, making the enhancement of $ZT$ very difficult. In 1998 Mateeva et al. studied the effect of carrier concentrations on the thermoelectric properties of polyaniline by doping protonic acid at different doping levels. They found that increasing doping level resulted in the increase of carrier concentration and electrical conductivity, but decreasing Seebeck coefficients (Fig. 6.7) [10]. It is similar to the case of inorganic semiconductors. On the other hand, the thermal conductivity of polyaniline is found less dependent on the doping level and electrical conductivity [11].

Improving carrier mobility is the most effective way to simultaneously increase the electrical conductivity and Seebeck coefficient of conducting polymers. In general, an expanded chain conformation and an ordered chain arrangement would result in a reduced charge hopping barrier in both interchain and intra-chain paths and, therefore enhance the carrier mobility. Toshima et al. studied the effect of stretching on the thermoelectric properties of ( ± )-10-camphorsulfonic acid

(CSA)-doped polyaniline films [12]. They found that both the $\sigma$ and S of the stretched films in the direction parallel to the stretching direction are much higher than those of the unstretched counterpart and increase with increasing drawing ratio. X-ray diffraction (XRD) results revealed that the ordering degree of molecular chain packing in the PANI film is improved by the mechanical stretching as shown in Fig. 6.8, which is considered to contribute to the enhanced carrier mobility, electrical conductivity, and Seebeck coefficient.

The ordering degree of PANI molecular chain packing can be also modulated by polymerization condition or posttreatment. Yao et al. prepared PANI films with different ordering degrees of molecular chains packing by simply changing the m-cresol content in the solvent [13]. They found that the conformation of PANI molecules changes from a compacted coil to an expanded coil by the chemical interactions among PANI, dopant CSA and m-cresol (Fig. 6.9). The Raman spectrum and XRD analysis results confirm that the percentage of the ordered molecular packing region increases with increasing m-cresol content, which is benefit to decrease the hopping barrier and increase the carrier mobility.

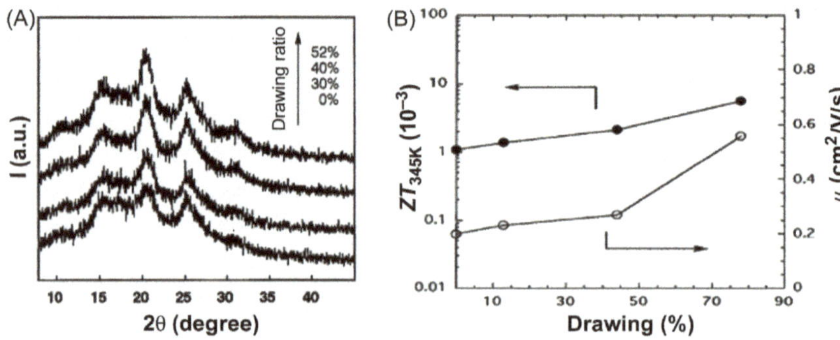


**FIGURE 6.8** (A) XRD of polyaniline films of various drawing ratios; (B) *ZT* value and carrier mobility of polyaniline films of various drawing ratios.

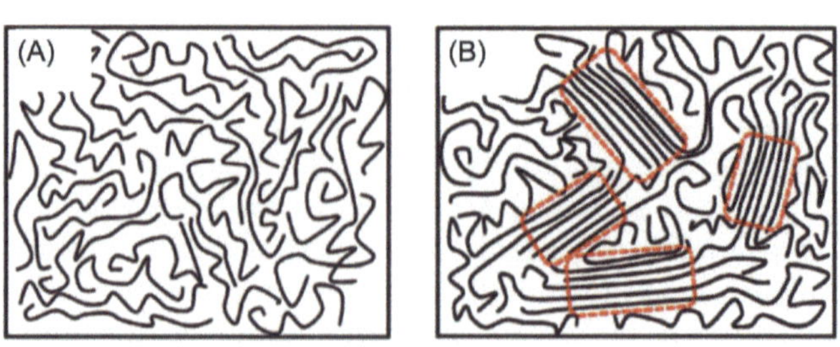


**FIGURE 6.9** Schematic illustration of the packing state of polyaniline molecules before (A) and after (B) m-cresol solution treatment.

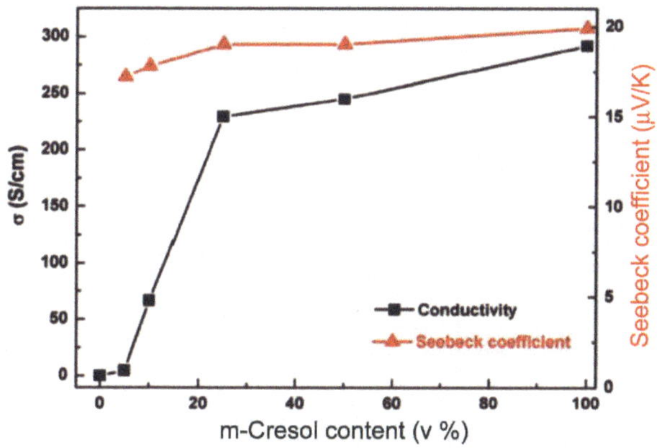


**FIGURE 6.10** The electrical conductivity and Seebeck coefficient of polyaniline films prepared from the mixed solvent with different m-cresol contents.

Therefore increasing m-cresol content in the posttreatment solvent results in remarkable improvement of electrical conductivity, while the Seebeck coefficient is maintained the same level or slightly increased, as shown in Fig. 6.10. The power factor of the polyaniline film treated by 100% m-cresol solvent is about 60 times higher than that of the as-prepared polyaniline films without m-cresol solvent treatment.

Template-induced ordering technique can be also used to construct highly ordered PANI molecular packing. Wang et al. achieved highly ordered chain structure of PANI by introducing self-assembled supramolecule (SAS) (3,6-dioctyldecyloxy-1,4-benzenedicarboxylicacid) as a template and found that the ordered packing regions in the PANI-SAS film increase with decreasing film thickness [14]. Consequently, both the electrical conductivity and Seebeck coefficient of PANI-SAS films are much higher than those of PANI films without SAS, as shown in Fig. 6.11. The maximum power factor of the PANI-SAS film reaches 31 $\mu W/m/K^2$, which is approximately six times higher than the power factor of a normal PANI film with similar thickness.

The multilayer structure is also an effective way to tune the thermoelectric properties of polyaniline films. Toshima et al. fabricated the multilayered PANI film, composed of electrically insulating emeraldine base layers and electrically conducting CSA-doped emeraldine salt layers [11]. This multilayer film exhibits six times higher *ZT* at 300K than that of a homogeneous film. Hye Jeong Lee et al. constructed organic/organic multilayer structures through a layer-by-layer deposition of PANI-CSA and PEDOT:PSS thin films (Fig. 6.12) [15]. They found that the electrical conductivity is increased by ~1.3 times and the power factor by ~2 times in comparison to those of single-layer PEDOT:PSS films. The enhancement of electrical conductivity is

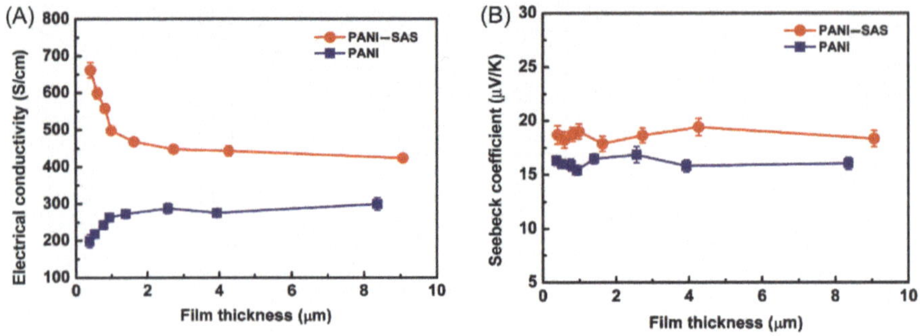


**FIGURE 6.11** The electrical conductivity and Seebeck coefficient of PANI-SAS and PANI films with different film thicknesses.

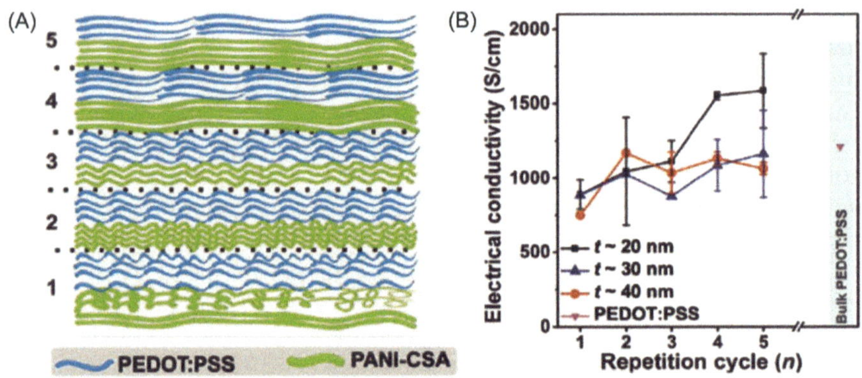


**FIGURE 6.12** (A) Schematic diagram illustrations explaining the stretching of the chain of both the PEDOT:PSS and PANI-CSA layer with increasing repetition cycles $n$; (B) electrical conductivity as a function of the repetition cycle ($n$) for multilayer films with various component layer thicknesses.

considered to be attributed to the stretching of PEDOT and PANI chains in multilayer films prepared via the alternative stacking of PEDOT:PSS and PANI:CSA.

## 6.3.2 P3HT

Poly (3-hexylthiophene) (P3HT) refers to one type of polythiophene derivatives. P3HT has high solubility in most organic solvents and high tunability of molecular structure and molecular weight (MW). Oxidation is generally used as doping process for P3HT, by being exposed to a gas or dipped in a solution containing an oxidizing agent (dopant). During the doping process, the reduced dopant transforms

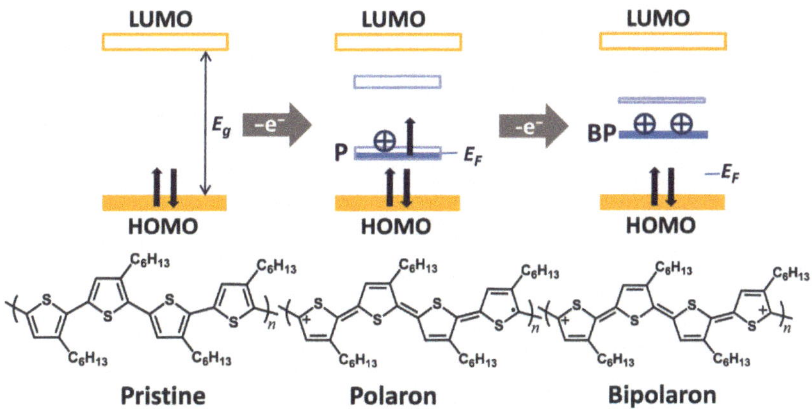


FIGURE 6.13 Electronic and chemical structure of pristine P3HT and positive charges of P3HT in polarons and bipolarons depending on the doping level.

into a negative counter-ion and neutralizes the positive charge introduced in the π-electron system of P3HT backbone. For example, when a solution of $NO^+PF_6^-$ salt is used as dopant, after dipping P3HT film into this solution, the NO gas shall be released, while the $PF_6^-$ anion becomes a counter-ion in the oxidized polymer chain. In P3HT, the charges can be found in polarons or bipolarons depending on the doping level. When an electron is removed from the top of the valence band (HOMO level), structure distortion is induced in the P3HT backbone and makes the P3HT backbone behave as quinoid character. This distortion shall extend over three to four monomer units depending on the counter-ion. The quasiparticles composed of a positive charge associated with the lattice distortion are called positive polarons. The unpaired electrons shall locate at the half-filled electronic level above the valence band edge. With further increasing doping level, the quasiparticles become carrying two positive charges associating with the same lattice defect, which is called positive bipolarons (as shown in Fig. 6.13). P3HT can be synthesized from 2-bromo-3-alkyl thiophene as raw material under existence of magnesium forming Grignard reagent and Ni (II) as catalyst (as shown in Fig. 6.14).

The focus research focuses at the early stage about the thermoelectric properties of P3HT had been on the exploration of effective dopants and optimization of doping level. In 2010 Xuan et al. [16] reported a heavily doped regioregular P3HT films by introducing negatively charged counter-ions, $PF_6^-$, in which the doping level reaches up to 34% (Fig. 6.15A). They found that, with increasing dopant content, the electrical conductivity of P3HT increases remarkably but the Seebeck coefficient exhibits slight decrease, which is similar to PANI system. As

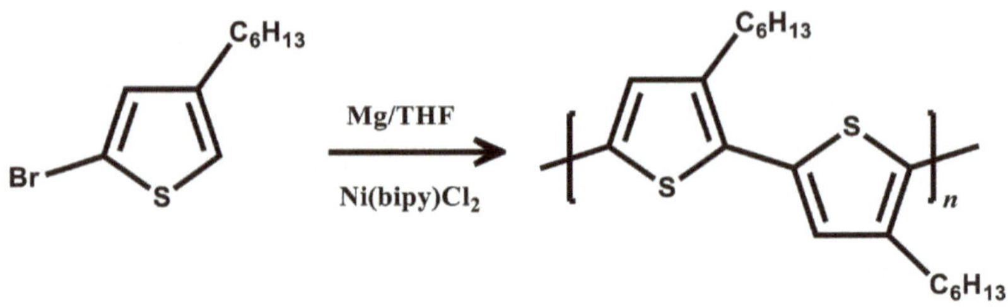


FIGURE 6.14 Synthesis of P3HT using 2-bromo-3-hexylthiophene as raw material.

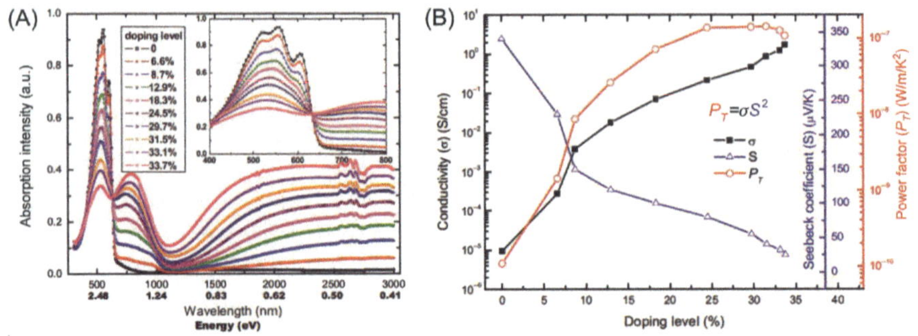


FIGURE 6.15 (A) Evolution of the optical-absorption spectra of P3HT as a function of doping level. (B) Doping dependence of the room-temperature Seebeck coefficient, electrical conductivity, and power factor.

a result, the power factor reaches a plateau-like maximum at the doping level range between 20% and 31% (Fig. 6.15). Similar behavior is also observed for other dopant systems such as F4TCNQ [17] and $FeCl_3$.

Similar to polyaniline, the molecular chain's conformation and arrangement of P3HT are also sensitive to its thermoelectric properties. Qu et al. [18] investigated the influence of molecular configuration on the ordering degree of the chain arrangement. They found that highly regular molecular configuration benefits to form highly ordered chain arrangement in P3HT film. Subsequently, they demonstrated that the ordered chain arrangement is not only helpful to improve the charge carrier mobility but also contribute to producing more charge carriers, thereby remarkably improve the electrical conductivity. Furthermore, Qu et al. [19] investigated the influence of molecular weights (MWS) of P3HT on the thermoelectric properties. They indicated that the MW of P3HT also has impact on the carrier-transport properties by affecting the molecular structure. Upon increasing the MW, the long molecular chain provides enough connectivity for the charge to move through the ordered regions, which is beneficial to decreasing the carrier barrier and

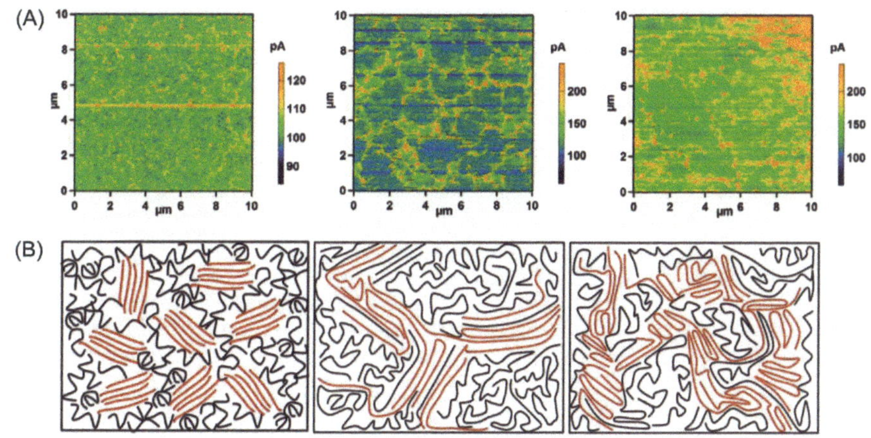


FIGURE 6.16 (A) Current images of the P3HT films mapped by c-AFM. The molecular weights are 10,000, 50,000, and 100,000 g/mol from left to right, respectively. (B) Schematic illustrations of the molecular chain conformation in the P3HT films with different molecular weights: low to high from left to right.

increasing carrier mobility. However, a too-high MW could cause more folding of the polymer chains, which would deteriorate the electrical properties (Fig. 6.16).

Template process is typically useful for constructing highly oriented molecular arrangement of P3HT. In 2016 Qu et al. developed a unique method to fabricate P3HT films with highly oriented molecular chains. In this process, P3HT films are epitaxially grown under a temperature gradient by using 1,3,5-trichlorobenzene (TCB) small-molecule as a template [20]. The lattice matching and $\pi-\pi$ conjugation between P3HT and TCB enable the backbone chains of P3HT polymer highly orderly arrange along the axial direction of columnar TCB crystals. The epitaxy crystallization results in markedly reduced defects along the polymer backbone, which is beneficial to effectively increasing the degree of electron delocalization. These combined advantages resulted in an efficient quasi-1D path for the carrier transport and therefore enhance the carrier mobility in the TCB-treated P3HT films. The highest electrical conductivity in parallel direction of the P3HT molecular chain reaches 320 S/cm. The power factor reaches 38 and 62.4 $\mu W/m/K^2$ at room temperature and at 365K, respectively, as shown in Fig. 6.17.

The carrier transport behavior in conducting polymers represents the average properties contributed from all carrier transport channels, including intra-chain, interchain, intergrain, and hopping between localized disordering sites. The fundamental understanding of the influence of molecular structure on the carrier transport properties is still a big challenge. In 2018 Qu et al. [21] investigated the carrier transport

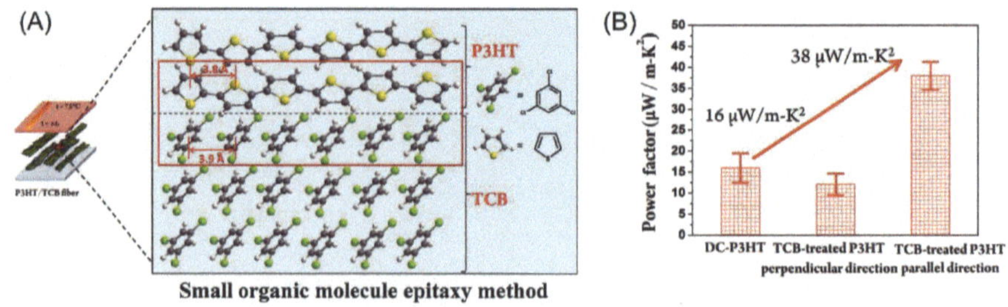


FIGURE 6.17 (A) Schematic diagram of ordered arrangement mechanism of P3HT molecule chain induced by TCB; (B) The power factor at room temperature of the DC-P3HT film and the TCB-treated P3HT film in the perpendicular direction and parallel direction.

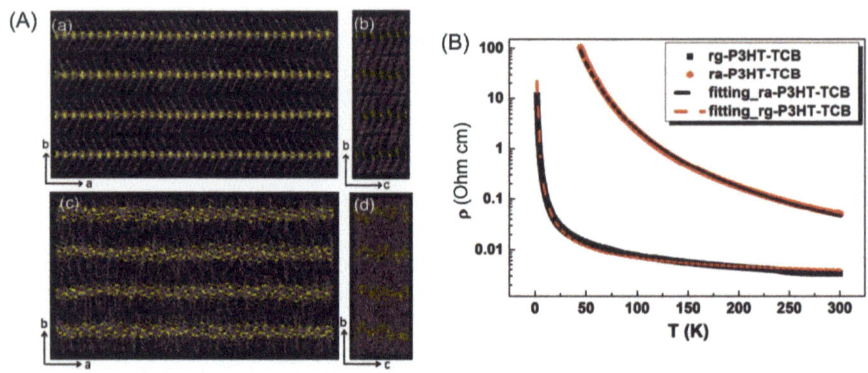


FIGURE 6.18 (A) Top view and side view of the simulation snapshots for rg-P3HT and ra-P3HT chains; (B) the resistivity of ra-P3HT-TCB fits well with the variable-range hopping conduction model, while the resistivity of rg-P3HT-TCB fits well with the quasi-1D heterogeneous hopping conduction model.

properties of doped highly oriented P3HT films with different side-chain regioregularity. They revealed that the regularity of side chains affects seriously both the carrier transport edge and the dimensionality of the transport paths and therefore the carrier mobility. Combining molecular dynamics simulations and experiments, they showed that regular side-chain structure favors one-dimensional (1D) transport along the backbone chain direction, while the irregular side-chain structure presents the three-dimensional (3D) electron hopping behavior (Fig. 6.18). As a consequence, the regular side-chain P3HT samples (rg-P3HT) demonstrated high carrier mobility of 2.9 $cm^2/Vs$, which is more than one order of magnitude higher than that in irregular (random) side chain P3HT films (ra-P3HT).

## 6.3.3 PEDOT

PEDOT is another type of polythiophene derivatives with smaller energy gap, which generally has high electrical conductivity and good chemical stability. PEDOT can be synthesized by oxidizing EDOT using different oxidants such as ammonium persulfate (APS) and iron salts, which can release counter-ions to neutralize the positive charge of the backbone. The structure of the counter-ions has great impact on the carrier mobility and electrical conductivity of the host PEDOT. According to the counter-ions used in polymerization process, there are generally two types of PEDOTs, PEDOT:polystyrenesulfonate (PEDOT:PSS) and small-sized anion-doped PEDOT (S-PEDOT). PEDOT:PSS is commercially available. PEDOT:PSS is doped with hydrophilic polyanion PSS and therefore is water-soluble. S-PEDOT is usually obtained via in situ polymerization or electrochemical polymerization and is completely insoluble in all solvents. Either type of PEDOT has its distinctive advantage/disadvantage as thermoelectric materials. Due to the difference in counter-ion structure and polymerization process, different approaches can be applied to optimize their thermoelectric performance.

### *6.3.3.1 PEDOT:PSS*

PEDOT:PSS is formed by the oxidation of EDOT using ammonium persulfate (APS) in PSS aqueous solution [22]. After ion-exchange, the resulted solution can be used directly for film printing. Jiang et al. firstly reported the thermoelectric performance of DMSO-treated PEDOT:PSS with a $ZT$ value of $1.75 \times 10^{-3}$ for the cold-pressed pellet [23]. A year later, Chang et al. reported the thermoelectric performance of PEDOT:PSS film with a power factor of $4.78\ \mu W/mK^2$ [24]. After then, numerous studies have been conducted focusing on the optimization of electrical conductivity of PEDOT:PSS films using commercial products [25]. It is demonstrated that higher degree of polymerization favors to improving the electrical performance [26].

As mentioned above, the ordering degree of molecular structure and molecular conformation is also believed as one of the important factors to affect the carrier transportation. Excess PSS as secondary phase always exists in PEDOT:PSS and is considered being responsible for the structural disorder as well as the unexpected interfacial boundaries, which results in severely reduced $\sigma$ and $S$. Eliminating PSS content in the PEDOT:PSS has been tried to improve the structure ordering and carrier mobility. Kim et al. realized the significant removal of PSS by immersing PEDOT:PSS in dimethyl sulfoxide (DMSO) or ethylene glycol (EG) for a certain period, accompanied with a decrease in film thickness. After removing the excess PSS, both $\sigma$ and $S$ increase and an amazingly high $ZT$ value of 0.42 is reported [27]. Besides solvent treatments, Lee et al. developed chemical dedoping (hydrazine) [28] and ultrafiltration [29] approaches for effectively removing PSS. Acid treatment is also used to remove excess PSS. The counter-ion substitution through acid solution/vapor treatment (e.g., $H_2SO_4$ [30], oxalic acid [31], formic acid [32], HI [33]) results in greatly enhanced $\sigma$ while almost unchanged $S$.

#### 6.3.3.2 *Small-sized anion-doped PEDOT*

S-PEDOT is a completely insoluble polymer and usually can be synthesized via the chemical oxidation of EDOT with iron salts. At early stage, iron salt oxidant was directly mixed with EDOT monomer to produce S-PEDOT. The oxidation progressing rate in the direct mixing process is difficult to be controlled due to the catalyzation effect of proton, which is in situ generated during the polymerization. Introducing base inhibitors such as pyridine into the reaction system is found effective in controlling the reaction rate [34]. The resulted PEDOT:Tos via base-inhibited polymerization exhibits a high $\sigma \sim 1000$ S/cm. In vapor-phase polymerization (VPP) method, EDOT monomer is vaporized and is gradually absorbed by the oxidant mixture. The lower monomer concentration in the reaction system results in good controllability of reaction rate [35]. However, the PEDOT film synthesized by VPP method often exhibits poor $\sigma$ because of the existence of pin-hole defects caused by the precipitation of oxidant during reaction (Fig. 6.19).

To further stabilize the reaction and suppress the unexpected crystallization of oxidant, introducing inhibitor agent to the reaction is proposed. For example, when copolymer PEG-ran-PPG is added to the oxidant prior to the VPP process, the crystallization of oxidant is prevented and the $\sigma$ of the product PEDOT film exceeds 700 S/cm [36].

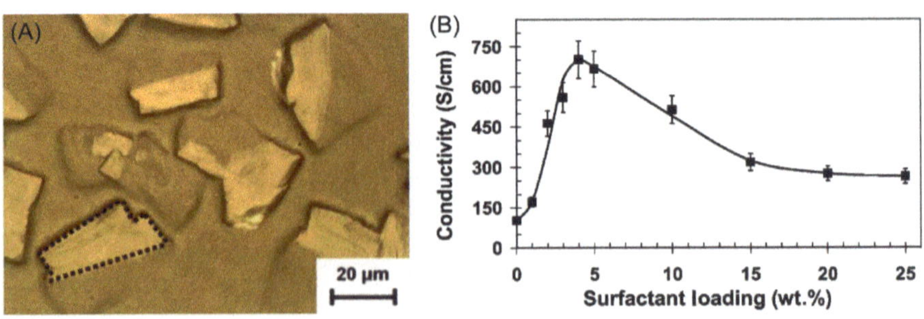


FIGURE 6.19 (A) Crystallized iron oxidant. (B) The influence of copolymer loading on the electrical conductivity.

Park et al. successfully synthesized high-performance PEDOT film ($\sigma$ is 1354 S/cm [37]) via a solution casting polymerization inhibited with PEPG/pyridine. However, the introduced copolymer inhibitors cannot be removed thoroughly by normal washing process and therefore overloaded inhibitor would lead to the decrease of electrical conductivity as shown in Fig. 6.19B.

Basic anion is found having self-inhibiting function in the PEDOT polymerization. When basic anion exists, the electrical performance of PEDOT:CSA can be improved by an order of magnitude [38]. Shi et al. systematically studied the role of anion in self-inhibited polymerization [39] and found that the anions with medium basicity and high solubility (e.g., DBSA) may serve as both the base inhibitor and anticrystallization inhibitor. They achieved a high PF of 69.6 $\mu W/mK^2$ in S-PEDOT film by adding DBSA as bifunctional inhibitors. Furthermore, the high oxidant concentration also makes it possible to prepare micrometer-thick films without the negative effect of inhibitors.

There are also other methods to synthesize S-PEDOT such as solution polymerization and electrochemical polymerization. For example, Zhang et al. synthesized PEDOT nanowires using solution oxidation polymerization stabilized by SDS [40]. The film acquired by direct filtering exhibits high PF of 35.8 $\mu W/mK^2$. Electrochemical polymerization can be used to synthesize PEDOT with unique morphologies for some potential applications. For example, Taggart et al. polymerized PEDOT nanowire array on lithographically patterned Ni substrate via electrochemical deposition (Fig. 6.20) [41]. The role of anion in electrochemical polymerization was studied by Culebras et al. [42]. The relatively large anion can prevent the polymer chain from being coiled. The TFSI-doped PEDOT (using $BMIM^+TFSI^-$ solution as electrolyte) exhibits a high PF of 147.2 $\mu W/mK^2$ after dedoping treatment.

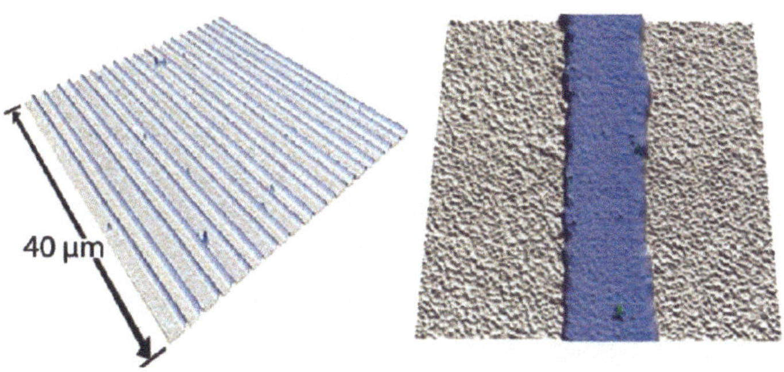


**FIGURE 6.20** S-PEDOT nanowire array produced by electrochemical deposition.

## 6.3.4 Other organic thermoelectric materials

In addition to the well-studied thermoelectric polymers (PANI, P3HT, and PEDOT), the exploration of organic thermoelectric materials has also been gradually extended to lots of other polymers. Typical representatives include poly(thiophene) derivatives, donor-acceptor (D-A)-type polymers and metal-organic complex. Fig. 6.21 shows the molecular structures of typical polymers as possible thermoelectric materials.

### *6.3.4.1 Poly(thiophene) derivatives*

Study on poly(3-hexylthiophene) (P3HT) as thermoelectric materials has been majorly focused on the chemical/structure modification and thermoelectric properties. Shannon Yee et al. found that the identity of heteroatom in polyheterocycles strongly influences the doping susceptibility and the thermoelectric properties of P3HT film [43]. The optical bandgap of polymer and the chemical potential can be adjusted by substituting S atom in thiophene ring with Te or Se. At low doping concentrations, poly(3-alkyltellurophene) (P3RTe) and poly(3-alkylselenophene) (P3RSe) can achieve the power factors over 10 $\mu W/m/K^2$, which is comparable to highly doped polythiophenes. The temperature-dependency of electrical conductivity shows that the charge transport in these films around room temperature is consistent with Mott's polaron hopping model. The activation energies decrease as the heteroatom is changed from sulfur to selenium and then to tellurium for a given dopant concentration. Poly(bisdodecylquaterthiophene) (PQT12) is another

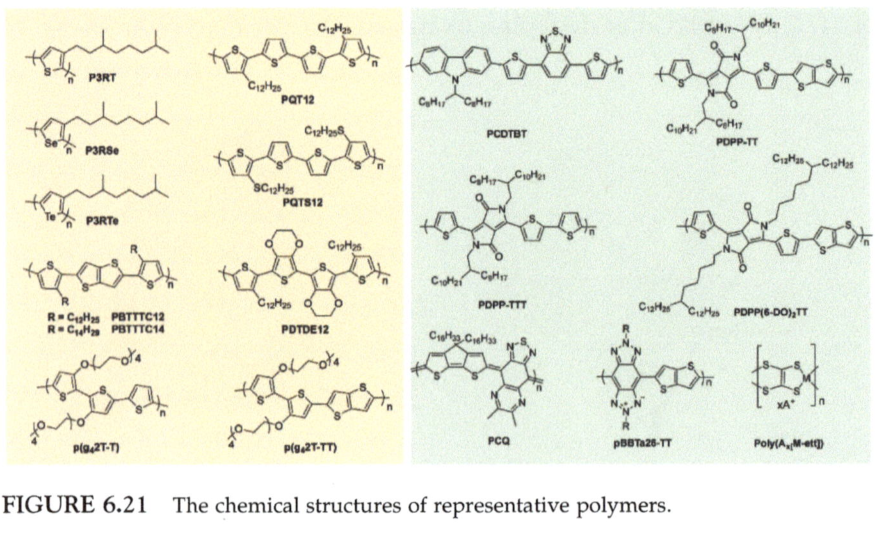


FIGURE 6.21 The chemical structures of representative polymers.

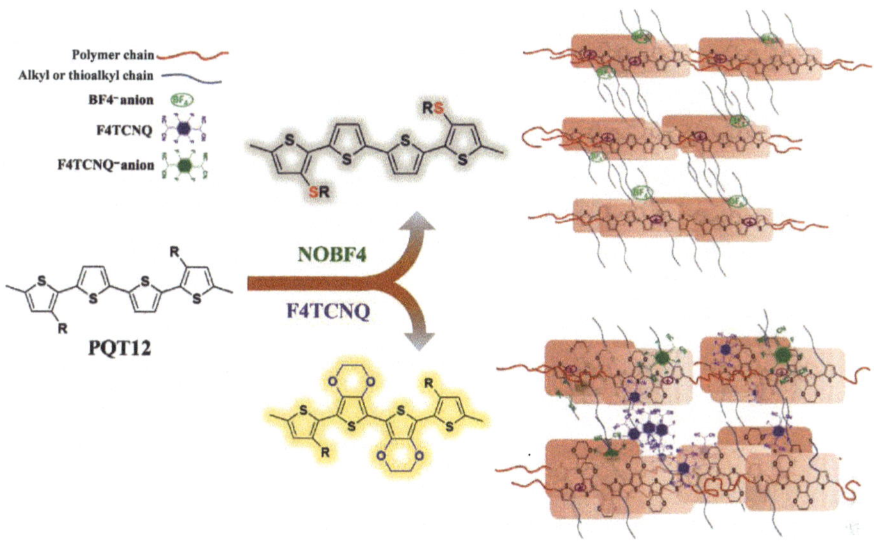


FIGURE 6.22 Chemical structures of conductive polymers and the packing models in films.

prototypical polythiophene used in thin-film electronic devices [44–46]. In order to improve its thermoelectric performance, Howard Katz et al. introduced sulfur atoms into the side chains to obtain poly(bisdodecylthioquaterthiophene) (PQTS12) and also introduced an ethylenedioxy group into the backbone to obtain biEDOT copolymers [47]. These modifications raise the HOMO levels and make host molecules more easily to be doped. An electrical conductivity of 350 S/cm is obtained by doping PQTS12 with NOBF4, and a high electrical conductivity of 140 S/cm is obtained by doping PDTDE12 with F4TCNQ (the chemical structures are shown in Fig. 6.22). There is no ionic contribution for charge transport; therefore the doped film exhibits good stability in air in the scale of month. Based on the experimental observation of film crystallinity and morphology, they argued that the electrical conductivity mainly depends on the bulk molecular packing status that is greatly affected by the dopant structure.

Poly(2,5-bis(3-alkylthiophen-2-yl)thieno[3,2-b]thiophene) (PBTTT) has high mobility and is widely used in OFET [48,49]. Daoben Zhu et al. proposed that its high structural ordering is in charge of the high bulk mobility and the enhanced power factor [50]. Henning Sirringhaus et al. demonstrated that the dopant molecules, F4TCNQ, can be incorporated into PBTTT film without disrupting the conjugated layers by solid-state diffusion (Fig. 6.23A) [51]. Therefore 2D coherent charge transport in highly ordered conducting polymers allows a high

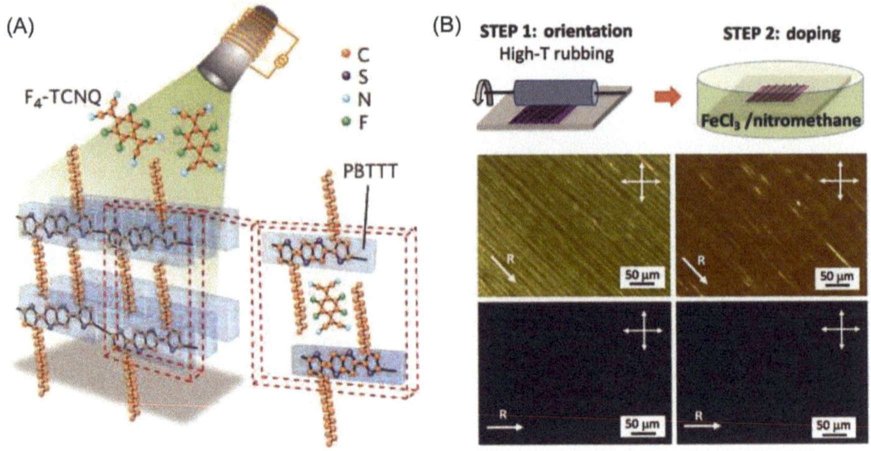


FIGURE 6.23 (A) Schematic diagram of the F4TCNQ evaporation procedure. (B) The two-step process to prepare oriented PBTTTC12 thin films; the polarized optical microscopy images under crossed polarizers for the as-rubbed (left) and doped (right) C12-PBTTT films.

electrical conductivity up to 248 S/cm. Furthermore, Chabinyc et al. prepared PBTTT polymer doped by (tridecafluoro-1,1,2,2-tetrahydrooctyl) trichlorosilane (FTS) using vapor deposition method and achieved a high electrical conductivity (1000 S/cm) with a Seebeck coefficient of 33 μV/K, in turn, a large power factor of ~100 μW/m/K$^2$ [52]. The oriented PBTTT thin film is also prepared by combining high-temperature rubbing and subsequent doping in a solution of $FeCl_3$ in nitromethane by Martin Brinkmann et al. (Fig. 6.23B) [53]. The control of crystallization and orientation of PBTTT-C12 prior to doping gives a unique handle over the charge transport and TE properties. As a result, the highest conductivity of $2.2 \times 10^5$ S/cm is achieved, which surpasses the value for stretch-oriented polyacetylene doped with iodine. The outstanding power factor of 2 mW/m/K$^2$ along the direction parallel to molecular chain is obtained. These findings demonstrate that chemical doping without interference of polymer packing would favor the charge transport and therefore improve electrical conductivity and power factor.

As one of the core issues to tune electrical conductivity of poly(thiophene)s, the realization of efficient chemical doping is still a challenge due to the poor solution processability. To solve this problem, a crucial progress is made by introducing polar side chains into poly(thiophene)s. For instance, using polar oligoethylene glycol as side chains, polymer p($g_4$2T-T) shows an increased ionization energy (IE) compared to P3HT and has better miscibility with dopants [54]. These characters allow a maximum electrical conductivity of up to 100 S/cm for p($g_4$2T-T) doped

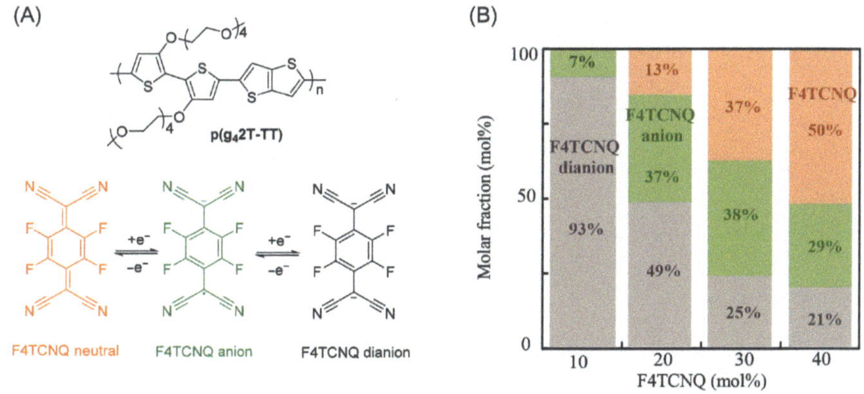


**FIGURE 6.24** (A) Chemical structure of polymer and F4TCNQ anion and dianion. (B) The molar fractions of neutral F4TCNQ, anion, and dianion in doped film with various amounts of F4TCNQ derived by fitting of the absorption coefficients.

with F4TCNQ or a less oxidant dopant, DDQ, via solution coprocessing. Unexpectedly, the existence of polar side chains favors the thermal stability of doped film up to 150°C, which is of particular interest for organic thermoelectrics. Christian Müller et al. found that double doping of conjugated polymers can be realized in F4TCNQ-doped bithiophene-thienothiophene copolymer with tetraethylene glycol side chains (p($g_4$2T-TT)), leading to an ionization efficiency of up to 200% (Fig. 6.24) [55]. They argued that forming dianions is still energetically favorable as long as the IE of the host polymer is smaller than the electron affinity of the already negatively charged dopant. In other words, the use of stronger dopants (with higher electron affinity) with the ability to capture two electrons from polymers would lead to a wide class of highly conducting polymers with high doping efficiency.

### *6.3.4.2 Donor-acceptor type polymers*

The thermoelectric performance of organic thin films is strongly dependent on the charge carrier mobility [56,57]. The D-A polymer semiconductors present a large space to tune their electrical properties, because the alternative D-A structure would be favorable for strong intermolecular interactions and electronic coupling [58], leading to efficient charge transport by hopping. The tunability of electrical conductivity of D-A type polymers has been successfully proved in transistors and organic solar cell devices [59]. Some D-A type polymers, such as carbazole-based polymers [60], diketopyrrolopyrrole (DPP)-based polymers [61–63], and D-A polymer with quinoidal character [64,65], have been investigated by focusing on the structure dependency of electrical

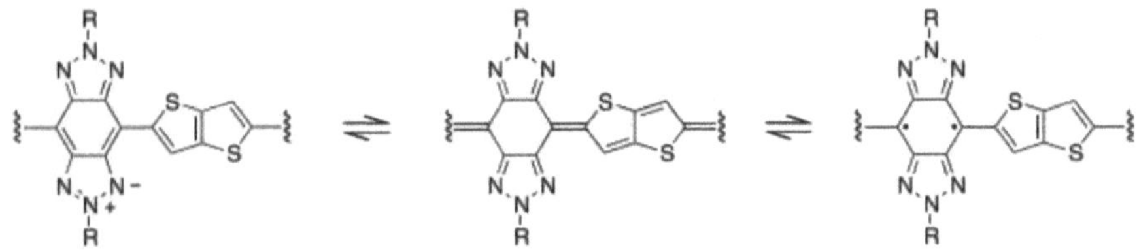

FIGURE 6.25 Doping-induced chemical structure transformation among zwitterionic (left), quinoidal (middle), and biradical (right) structures in pBBTa26-TT polymer.

conductivity. For example, the existence of the quinoidal character in pBBTa26-TT backbone will lead to the intrinsically conductive and contribute to the highly delocalized polarons upon doping (Fig. 6.25). Although a number of D-A polymers show high mobility, most of them have not yet been reported possessing good thermoelectric performance. One of the difficulties is that the acceptor moiety in D-A structure is usually inactive for p-type doping [65]. Nevertheless, rational design of polymer structure and exploring new dopants with larger molecular size may be the next research direction to overcome this limitation.

#### 6.3.4.3 *Metal-organic complex*

Metal-organic complexes are expected as the doping-free thermoelectric materials. For instance, the thermoelectric properties of poly (nickel-ethylenetetrathiolate) and its analogs have been studied experimentally [66,67] and theoretically [68]. The intrinsically prominent power factors would be over 103 $\mu W/m/K^2$ based on the density functional calculations. High electrical conductivity in metal-organic frameworks (MOFs) has also been successfully demonstrated [69]. Kuang-Lieh Lu et al. reported a reasonably high electrical conductivity (10.96 S/cm) with a low activation energy (6 meV) in the single crystal of copper-sulfur-based MOFs [70]. It may provide a new roadmap for designing highly conductive MOFs for thermoelectric applications.

## 6.4 Polymer-based thermoelectric composites

Similar to inorganic thermoelectric materials, incorporating nanosized secondary phase into polymer matrix to form organic/inorganic composites and/or hybrids has also been widely used to improve the thermoelectric performance of polymers. For example, enhancement of *ZT* has been frequently reported in PANI and PEDOT-based composites, in which the commonly used inorganic nanoparticles are carbon nanotubes (CNTs), graphene, inorganic semiconductors, and metals. Obviously, according to the general mixture rule, neither the electrical conductivity nor the *ZT* of the composite could exceed the largest one of either individual component. There is still contentious debate over the exact mechanism how hybrid/composite materials are able to outperform their individual components. The proposed possible mechanisms for enhancing their thermoelectric performance include the interface-induced ordering of molecular structure/conformation, strengthened phonon scattering, and the bridging function as well. In this section, the fabrication and performance optimization of typical polymer-based composites are highlighted and the possible mechanisms are also briefly reviewed.

## 6.4.1 Interface-induced ordering of molecular chain arrangement

The carrier transport of conducting polymers is principally controlled by the interchain and intra-chain hopping processes. Increasing the degree of ordering of molecular arrangement can decrease the hoping barrier and therefore increase the carrier mobility. CNTs not only have 1D ordered structure, but also have a large number of $\pi$ bonds on the tube surface, which may form $\pi-\pi$ conjugation with conducting polymers. The polymer molecules can grow along the surface of CNTs to form an ordered chain arrangement, which may increase the carrier mobility.

In 2010 Yao et al. developed an in situ polymerization technique to prepare SWNT/PANI composites (Fig. 6.26B) [71]. The obtained composite exhibits high carrier mobility, electrical conductivity, and Seebeck coefficient as compared with the values calculated by the mixed rule as shown in Fig. 6.26A. This is explained by the formation of an ordered PANI chain structure on the surface of CNTs induced by the strong $\pi-\pi$ conjugated interactions as schematically shown in Fig. 6.27B. They further prepared CNT/PANI composite fibers with CNTs arranged in the direction of electric field by electrospinning. In this electrospun CNT/PANI composite, the PANI backbone chains are orientated along the CNT axis and therefore along the fiber axis due to the $\pi-\pi$ conjugation interactions between PANI and CNTs, which resulted in significant anisotropy of thermoelectric properties, especially with remarkable increase of both electrical conductivity and Seebeck coefficient in the fiber axial direction.

The conformation ordering process of PANI chain can be assisted by introducing m-cresol solvent into the in situ polymerization process. In this m-cresol-assisted polymerization process, m-cresol makes the coiled PANI molecular chains straighten in the solution due to the chemical

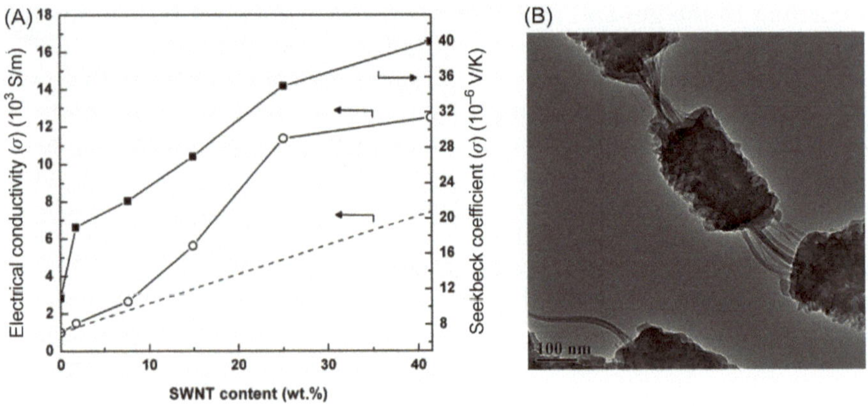


FIGURE 6.26 (A) The electrical conductivity and Seebeck coefficient of SWNT/PANI composites with different SWNT contents; (B) the TEM image of SWNT/PANI composites.

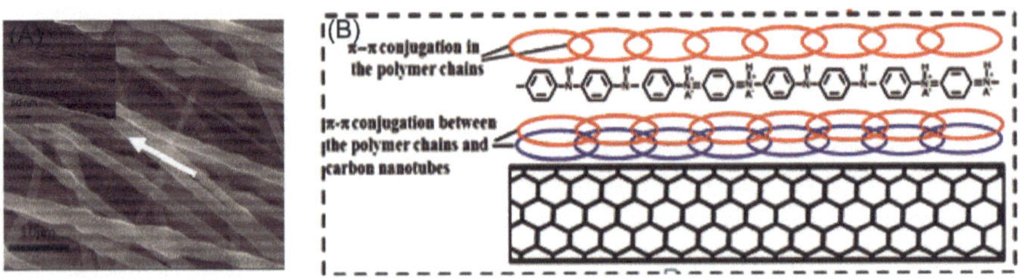


FIGURE 6.27 (A) SEM and TEM (the inset) images of the CNT/PANI composite fibers in the direction parallel to the electric field; (B) schematic diagram of the π−π conjugation interactions between PANI and CNTs.

interaction between PANI molecules and m-cresol. Morphologically, the expanded PANI molecules are easier to form π−π conjugation with CNTs over a wide area, which shall raise the volume ratio of the ordered conformation. The PANI/CNTs nanocomposite films prepared by the m-cresol assisted in situ polymerization exhibit large in-plane electrical conductivity up to $1.44 \times 10^3$ S/cm and power factor of 217 μW/m/$K^2$ (Fig. 6.28A) [72]. The existence of ordered regions of PANI molecular structure in the composite films is experimentally confirmed by XRD and conductive atomic force microscopy (CAFM) (Fig. 6.28B). Similar results have been also observed in PANI/graphene composites, because graphenes possess similar surface chemical structure (π bonds) to CNTs [73–74].

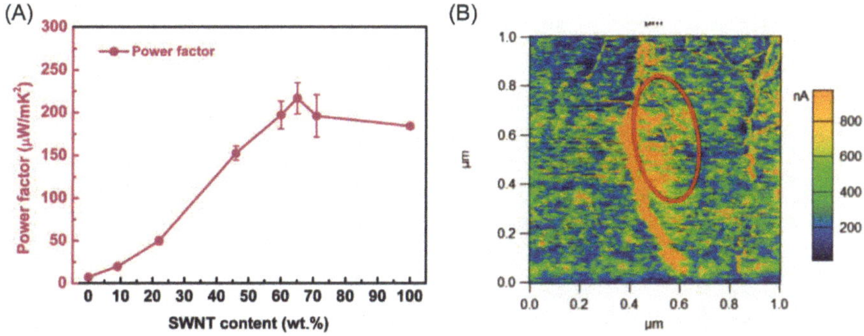


FIGURE 6.28 (A) Power factor at room temperature for SWNTs/PANI composite films with different SWNT contents; (B) surface current image of 1 wt.% SWNTs/PANI composite film on glass substrates.

## 6.4.2 Interfacial scattering to phonons and electrons

The existence of large quantity of nanointerfaces in polymer/inorganic composites can effectively scatter electrons or phonons and significantly change electron and phonon transport properties. The energy-filtering effect has been extensively studied in inorganic thermoelectric materials and is widely considered to independently promote Seebeck coefficient without obvious suppression of electrical conductivity [75]. For organic/inorganic composites with intimate contacts between polymer and nanoparticles, the energy-filtering effect is also expected by appropriately designing the band structures of interfaces [76–77]. As shown in Fig. 6.29, the existence of an energy offset between the transport bands of the filler and polymer would induce an energy barrier ($\Delta E_g$) at the interface. Only the carriers with energy greater than $\Delta E_g$ can pass the barrier while those with energy less than $\Delta E_g$ shall be scattered, which may enhance the Seebeck coefficients by increasing the average energy of the transported carriers.

In 2009 Meng et al. observed the enhanced Seebeck coefficients and power factors in PANI/CNT composites, which are several times higher than either of their individual components [78]. They ascribed this result to the size-dependent energy-filtering effect caused by the nanostructured PANI coating layer wrapped around the CNTs. Subsequently, the increase of Seebeck coefficient has been found in a variety of polymer-based composites with inorganic nanoparticles as dispersing phase. For example, He et al. found that the Seebeck coefficient of $Bi_2Te_3$ nanowires–dispersed P3HT/$Bi_2Te_3$ composites is several times higher than that of pure P3HT (Fig. 6.30), but the electrical conductivity did not decrease significantly [77]. Through experimental measurements and theoretical analysis, they proposed that the energy filtering effect is the dominant

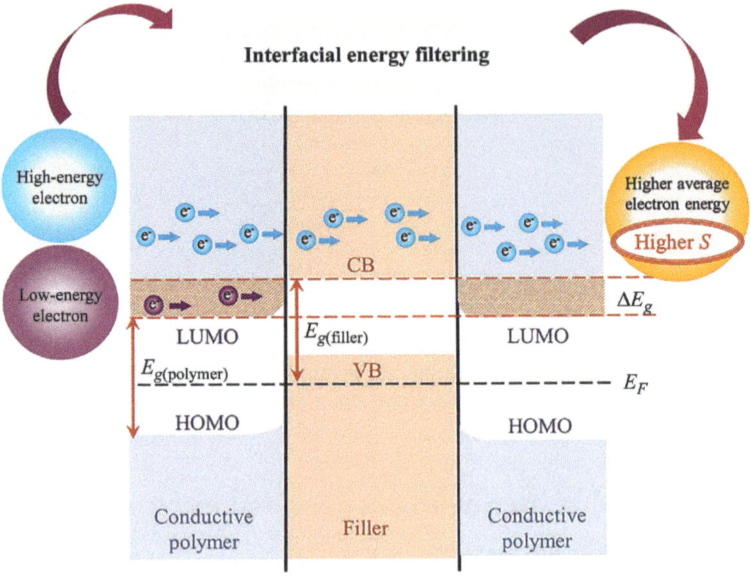


FIGURE 6.29 Schematic illustration of the energy-filtering effect at the interface between the conductive polymer and the filler.

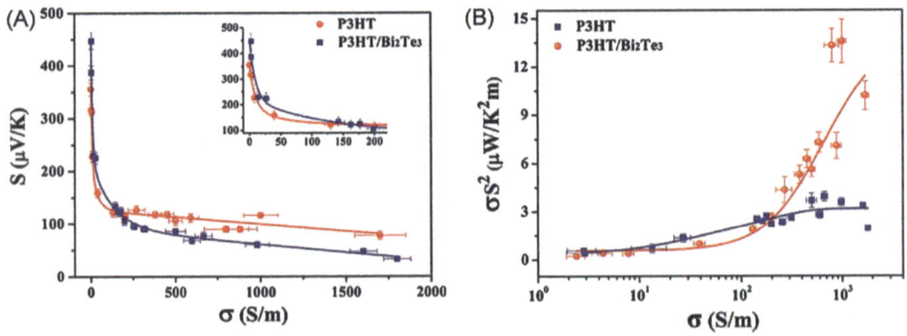


FIGURE 6.30 (A) The correlation between the Seebeck coefficient S and the electrical conductivity $\sigma$ in P3HT and P3HT-$Bi_2Te_3$ nanocomposites; the inset shows the close-up in the range of low electrical conductivity; (B) the correlation between the power factor and the electrical conductivity $\sigma$ in P3HT and P3HT-$Bi_2Te_3$ nanocomposites.

reason for the enhanced power factor. In the heavily doped P3HT/$Bi_2Te_3$ composite films, the moderate interface barrier between P3HT and $Bi_2Te_3$ selectively scatters the low-energy carriers. In 2017 Wang et al. fabricated PANI/SWNT/Te hybrid films and reported that the maximum power factor of the ternary nanocomposite films is much higher than those of the individual components (PANI, SWNTs, and

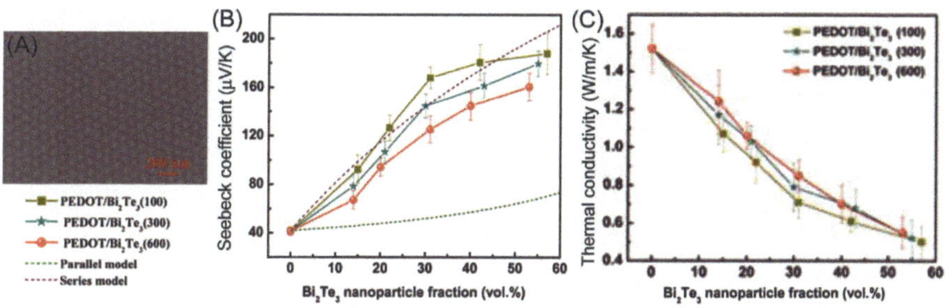


**FIGURE 6.31** (A) SEM images of PEDOT/$Bi_2Te_3$ hybrid films. (B) Seebeck coefficient of PEDOT/$Bi_2Te_3$ films as a function of $Bi_2Te_3$ nanoparticle fraction. (C) Thermal conductivity of PEDOT/$Bi_2Te_3$ films as a function of $Bi_2Te_3$ nanoparticle fraction.

Te nanorods) [79]. They claimed that the two interfaces, at PANI/SWNTs and PANI/Te with controlled interfacial energy barriers, may act as energy filters to scatter low-energy carriers. They further fabricated the flexible organic/inorganic hybrids where size-tunable $Bi_2Te_3$ nanoparticles are discontinuously monodispersed in the continuous conductive polymer phase [80]. The PEDOT-$Bi_2Te_3$ nanointerfaces in hybrid films are believed to provide the sites for selectively scattering both low-energy carriers and phonons and contribute to the large Seebeck coefficient and low thermal conductivity, as shown in Fig. 6.31. Consequently, figure of merit (*ZT*) of 0.58 is obtained in the PEDOT-$Bi_2Te_3$ film at room temperature, outperforming other reported organic materials and organic/inorganic hybrids.

Dispersing quantum dots into organic materials has also been reported as an effective way to improve the Seebeck coefficient (*S*) through phonon drag effect. Shi et al. reported a one-step synthesis of PEDOT/Te composite films using novel $Te^{IV}$-based oxidants [81], in which the Te quantum dots with the size less than 5 nm are dispersed homogeneously. The dispersion of Te quantum dots in a very low content brings significant enhancement of *S*. As a result, a high power factor about 100 $\mu W/m/K^2$ is achieved for the composite with only 2.1–5.8 wt.% Te, which is 50% higher than that of pure PEDOT film, as shown in Fig. 6.32. It provides an encouraging direction for exploring high-performance organic nanocomposites, though the physical mechanism is waiting to be revealed.

### 6.4.3 Organic/inorganic nanointercalated superlattice

Another strategy to realize high-performance organic/inorganic hybrids is to form superlattice structures by stacking metallic (or semiconducting) inorganic materials and organic molecules layer by layer

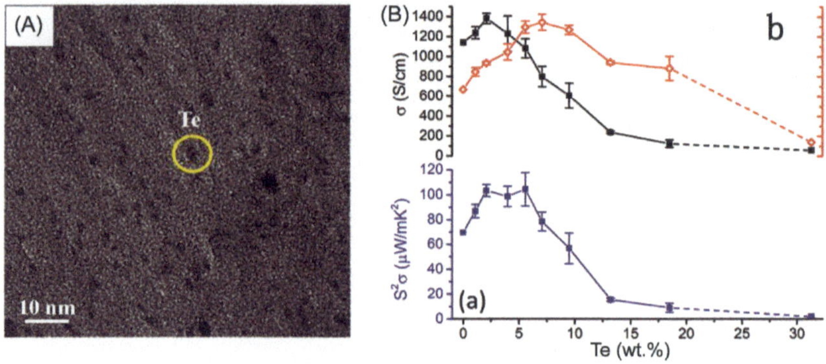


FIGURE 6.32 (A) The SEM image of PEDOT/Te hybrid films with Te quantum dots dispersed; (B) thermoelectric properties of PEDOT/Te films with different Te content.

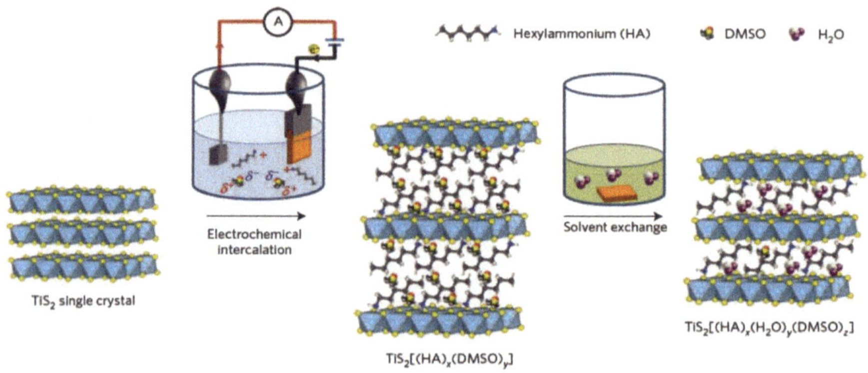


FIGURE 6.33 Schematic diagram of structure and preparation process of $TiS_2$ organic/inorganic hybrid superlattices.

[82]. In such superlattice structure, high $S^2\sigma$ can be secured via the inorganic atomic layer, while low $\kappa$ can be contributed by the organic molecular layer. In this multilayer structure, extra charge carries can be generated from intercalation complexes that are formed by chemical interactions between organic molecules and inorganic layers.

Wan et al. intercalated the organic cations in electrolyte into the van der Waals gap of $TiS_2$ single crystals through a facile electrochemical process and then synthesized a hybrid superlattice of alternating inorganic $TiS_2$ monolayers and organic cations [82] (Fig. 6.33). Electrons are externally injected into the inorganic layers and are then stabilized by organic cations, providing n-type carriers. The inorganic component $TiS_2$ leads to a high power factor and the organic intercalation significantly reduced the thermal conductivity. As a result, the electrical conductivity and power factor of hybrid superlattice of $TiS_2/[(\text{hexylammonium})_x(H_2O)_y(DMSO)_z]$

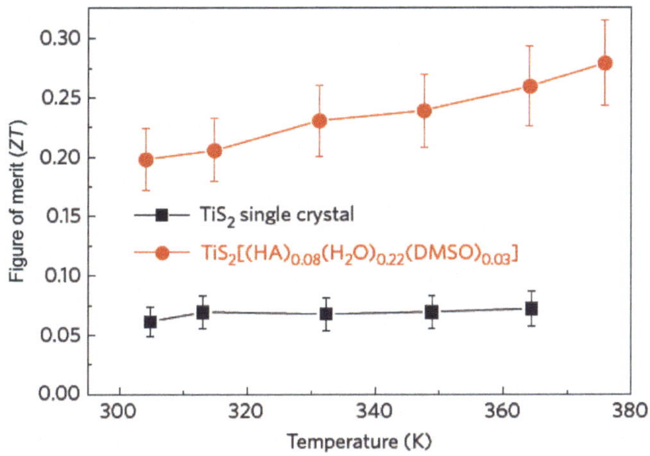


**FIGURE 6.34** Temperature-dependent *ZT* of $TiS_2$ organic/inorganic hybrid superlattices.

is up to 790 S/cm and 450 μW/m/$K^2$, with a thermal conductivity of 0.12 W/m/K. Finally, the maximum *ZT* reaches 0.28 at 373K (Fig. 6.34). Moreover, the carrier concentrations of hybrid superlattices can be optimized by electively de-intercalating organic cations on the basis of the different boiling points of the organic molecules. For example, the organic layers composed of a random mixture of tetrabutylammonium (TBA) and hexylammonium (HA) molecules are intercalated into the $TiS_2$ superlattice. By heating the hybrid materials in vacuum under an intermediate temperature, the HA molecules with a lower boiling point are selectively de-intercalated, which reduced the electron density, while the TBA molecules with a higher boiling point remained. Thus the carrier concentrations can be effectively reduced, resulting in a remarkably high power factor of 904 μW/m/$K^2$ at 300K [83].

The layer-by-layer self-assembly structure of nanocarbon materials and conducting polymers are also prepared by the $\pi-\pi$ conjugation interaction between the two components for adjusting the thermoelectric performance of organic/inorganic composite. Grunlan et al. prepared PANI/GP/PANI/DWCNT multilayer composite films (Fig. 6.35) through sequential exposure of a substrate to aqueous solutions containing PANI, PSS-stabilized graphene (Gp), and SDBS-stabilized double-walled carbon nanotubes (DWNT) [84]. The PANI, Gp, and DWNT are sequentially assembled into a 3D film via the $\pi-\pi$ conjugation and electrostatic interactions between layers. Polyaniline chains are adsorbed on the surface of CNTs forming an ordered chain structure. PANI-covered DWNT bridges the gaps between graphene sheets, resulting in enhanced carrier mobility and therefore increased electrical conductivity and Seebeck coefficient. Moreover, the phonon scattering effect at the nanointerfaces could also

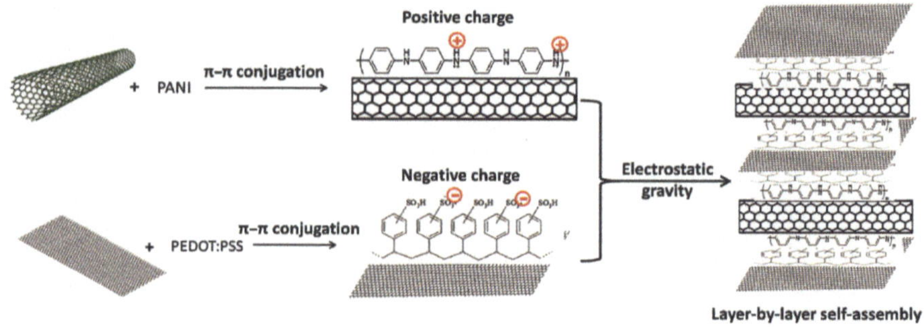


FIGURE 6.35 Schematic diagram of structure and preparation process of PANI/GP/PANI/DWCNT multilayer composite films.

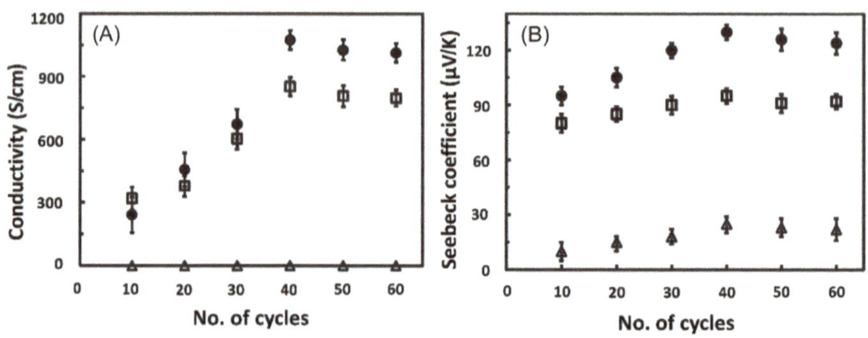


FIGURE 6.36 (A) Electrical conductivity and (B) Seebeck coefficient of PANI/GP/PANI/DWCNT multilayer composite films as a function of film layers.

reduce the thermal conductivity. The electrical conductivity and Seebeck coefficient of the composite reach 1100 S/cm and 129 μV/K, respectively. The TE power factor of the films reaches 1825 $\mu W/m/K^2$ (Fig. 6.36). By replacing insulating stabilizers (PSS and SDBS) with water-soluble intrinsically conductive polymer (PEDOT:PSS), the electrical conductivity and Seebeck coefficient are further improved. The synthesized PANI/graphene-PEDOT:PSS/PANI/DWNT-PEDOT:PSS multilayer composite film exhibits large power factor of 2710 $\mu W/m/K^2$, which is competitive with commercially available bismuth tellurides [85].

## 6.4.4 Charge transfer by the junctions

For the abnormal improvement of electrical conductivity in organic/inorganic nanocomposites, an initial explanation is the bridging function for charge transfer generated by inorganic nanowires. Kim et al.

reported the formation of a 3D network structure of CNTs in PVAC/CNT composite and claimed that such network enhances the electrical conductivity by the CNT interconnecting (bridging) function (Fig. 6.37) [86]. They used pure single-walled carbon nanotubes (SWNTs) with higher conductivities to replace the mixed CNTs and obtained a high electrical conductivity up to $10^5$ S/m [87]. The power factor of the obtained composite materials reaches 160 $\mu$W/m/K$^2$ (Fig. 6.38). Basically, this bridging model in the composite is similar to the percolation model, that is, highly conducting dispersed phases forming the conducting network to improve the electrical transport properties.

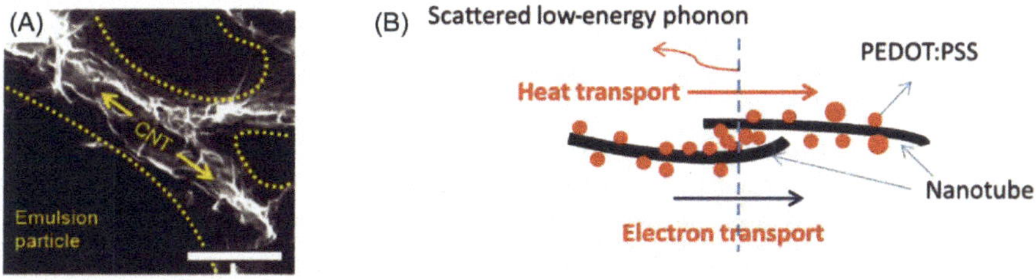


**FIGURE 6.37** (A) SEM image of three-dimensional network structure of CNT/PVAC composites; (B) schematic diagram of charge transfer between nanotube-PEDOT:PSS-nanotube junctions in the composites.

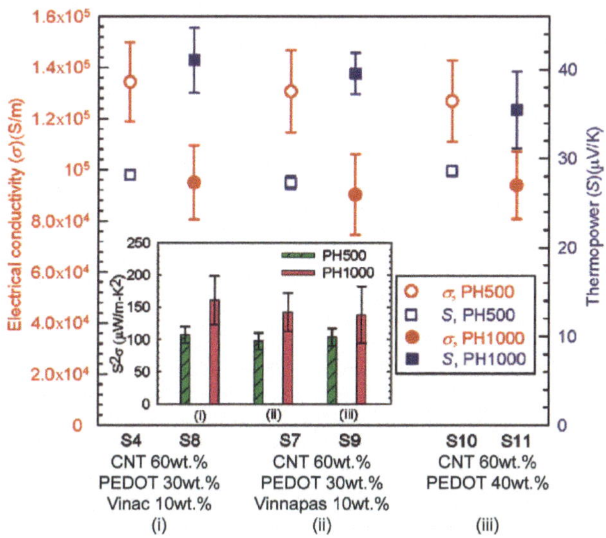


**FIGURE 6.38** The electrical conductivity and Seebeck coefficient of CNT/PEDOT:PSS composites. The inset is the power factor for the composites.

## 6.5 Summary

Although organic conductors as a new category of thermoelectric materials have received wide attention and various attempts have been delivered on modulating the thermoelectric transport properties, the progress on organic thermoelectrics is not so significant as that on inorganic thermoelectric materials. So far, lots of remarkable achievements on improving electrical conductivity, through chemical doping, structure design, forming composites or hybrids, have been developed for couple of conducting polymers. However, all of these recognized conducting polymers exhibit relatively lower thermopower as compared with the state-of-the-art inorganic thermoelectric materials. Enhancing the thermopower without significantly sacrificing their electrical conductivity is still a great challenge. The lack of the in-depth understanding of transport physics and doping mechanism in organic materials leaves the structure design and performance optimization of organic thermoelectric materials almost in guideless. The advances with new molecular designs and synthetic endeavors are needed toward a better understanding of structure-property relation. Typically, it is an urge task to establish the theoretical scheme of thermoelectric transport physics associated with the molecular structures and chain conformation for organic polymers.

## References

[1] B. Russ, A. Glaudell, J.J. Urban, et al., Organic thermoelectric materials for energy harvesting and temperature control, Nat. Rev. Mater. 1 (10) (2016) 16050.
[2] C. Liao, M. Zhang, M.Y. Yao, et al., Flexible organic electronics in biology: materials and devices, Adv. Mater. 27 (46) (2015) 7493–7527.
[3] B. Lüssem, C.-M. Keum, D. Kasemann, et al., Doped organic transistors, Chem. Rev. 116 (22) (2016) 13714–13751.
[4] A.P. Kulkarni, C.J. Tonzola, A. Babel, et al., Electron transport materials for organic light-emitting diodes, Chem. Mater. 16 (23) (2004) 4556–4573.
[5] A.J. Heeger, Bulk heterojunction solar cells with internal quantum efficiency approaching 100%, Nat. Photonics 3 (5) (2009) 297–302.
[6] O. Bubnova, X. Crispin, Towards polymer-based organic thermoelectric generators, Energy Environ. Sci. 5 (11) (2012) 9345–9362.
[7] N. Lu, L. Li, M. Liu, Universal carrier thermoelectric-transport model based on percolation theory in organic semiconductors, Phys. Rev. B 91 (19) (2015) 195205.
[8] M.C.J.M. Vissenberg, M. Matters, Theory of the field-effect mobility in amorphous organic transistors, Phys. Rev. B 57 (20) (1998) 12964–12967.
[9] V. Ambegaokar, B.I. Halperin, J.S. Langer, Hopping conductivity in disordered systems, Phys. Rev. B 4 (8) (1971) 2612–2620.
[10] N. Mateeva, H. Niculescu, J. Schlenoff, et al., Correlation of Seebeck coefficient and electric conductivity in polyaniline and polypyrrole, J. Appl. Phys. 83 (6) (1998) 3111–3117.

[11] N. Toshima, Conductive polymers as a new type of thermoelectric materials, Macromol. Symp. 186 (2002) 81–86.
[12] H. Yan, T. Ohta, N. Toshima, Stretched polyaniline films doped by (±)-10-camphorsulfonic acid: anisotropy and improvement of thermoelectric properties, Macromol. Mater. Eng. 286 (2001) 139–142.
[13] Q. Yao, Q. Wang, L. Wang, Y. Wang, J. Sun, H. Zeng, et al., The synergic regulation of conductivity and Seebeck coefficient in pure polyaniline by chemically changing the ordered degree of molecular chains, J. Mater. Chem. A 2 (8) (2014) 2634–2640.
[14] L.M. Wang, Q. Yao, J.X. Xiao, et al., Enhanced thermoelectric properties of polyaniline nanofilms induced by self-assembled supramoleculars, Chem. Asian J. 11 (2016) 1955–1962.
[15] H. Lee, G. Anoop, H. Lee, et al., Enhanced thermoelectric performance of PEDOT:PSS/PANI-CSA polymer multilayer structures, Energy Environ. Sci. 9 (2016) 2806–2811.
[16] Y. Xuan, X. Liu, S. Desbief, P. Leclère, X. Crispin, et al., Thermoelectric properties of conducting polymers: the case of poly(3-hexylthiophene), Phys. Rev. B 82 (2010) 115454.
[17] E. Lim, K.A. Peterson, G.M. Su, M.L. Chabinyc, et al., Thermoelectric properties of poly(3-hexylthiophene) (P3HT) doped with 2,3,5,6-tetrafluoro-7,7,8,8- tetracyanoquinodimethane (F4TCNQ) by vapor-phase infiltration, Chem. Mater. 30 (2018) 998–1010.
[18] S.Y. Qu, Q. Yao, W. Shi, L.D. Chen, et al., The influence of molecular configuration on the thermoelectrical properties of poly(3-hexylthiophene), J. Electron. Mater. 45 (3) (2016) 1389–1396.
[19] S.Y. Qu, Q. Yao, B. Yu, K. Zeng, L.D. Chen, et al., Optimizing the thermoelectric performance of poly(3-hexylthiophene) through molecular-weight engineering, Chem. Asian J. 13 (2018) 3246–3253.
[20] S.Y. Qu, Q. Yao, L.M. Wang, C. Uher, L.D. Chen, et al., Highly anisotropic P3HT films with enhanced thermoelectric performance via organic small molecule epitaxy, NPG Asia Mater. 8 (2016) e292.
[21] S.Y. Qu, C. Ming, Q. Yao, L.D. Chen, et al., Understanding the intrinsic carrier transport in highly oriented poly(3-hexylthiophene): effect of side chain regioregularity, Polymers 10 (2018) 815.
[22] Q. Wei, M. Mukaida, K. Kirihara, et al., Thermoelectric power enhancement of PEDOT:PSS in high-humidity conditions, Appl. Phys. Express 7 (3) (2014) 031601.
[23] F. Jiang, J. Xu, B. Lu, et al., Thermoelectric performance of poly(3,4-ethylenedioxythiophene):poly(styrenesulfonate), Chin. Phys. Lett. 25 (6) (2008) 2202–2205.
[24] K. Chang, M. Jeng, C. Yang, et al., The thermoelectric performance of poly(3,4-ethylenedi oxythiophene)/poly(4-styrenesulfonate) thin films, J. Electron. Mater. 38 (7) (2009) 1182–1188.
[25] B. Zhang, J. Sun, H.E. Katz, et al., Promising thermoelectric properties of commercial PEDOT:PSS materials and their $Bi_2Te_3$ powder composites, ACS Appl. Mater. Interfaces 2 (11) (2010) 3170–3178.
[26] Z. Fan, D. Du, H. Yao, et al., Higher PEDOT molecular weight giving rise to higher thermoelectric property of PEDOT:PSS: a comparative study of Clevios P and Clevios PH1000, ACS Appl. Mater. Interfaces 9 (13) (2017) 11732–11738.
[27] G.H. Kim, L. Shao, K. Zhang, et al., Engineered doping of organic semiconductors for enhanced thermoelectric efficiency, Nat. Mater. 12 (8) (2013) 719–723.
[28] S.H. Lee, H. Park, S. Kim, et al., Transparent and flexible organic semiconductor nanofilms with enhanced thermoelectric efficiency, J. Mater. Chem. A 2 (20) (2014) 7288.
[29] S.H. Lee, H. Park, W. Son, et al., Novel solution-processable, dedoped semiconductors for application in thermoelectric devices, J. Mater. Chem. A 2 (33) (2014) 13380.

[30] J. Kim, J.G. Jang, J.-I. Hong, et al., Sulfuric acid vapor treatment for enhancing the thermoelectric properties of PEDOT:PSS thin-films, J. Mater. Sci. Mater. Electron. 27 (6) (2016) 6122–6127.
[31] C. Liu, H. Shi, J. Xu, et al., Improved thermoelectric properties of PEDOT:PSS nano-films Treated with oxalic acid, J. Electron. Mater. 44 (6) (2015) 1791–1795.
[32] D.A. Mengistie, C.H. Chen, K.M. Boopathi, et al., Enhanced thermoelectric performance of PEDOT:PSS flexible bulky papers by treatment with secondary dopants, ACS Appl. Mater. Interfaces 7 (1) (2015) 94–100.
[33] J. Wang, K. Cai, H. Song, et al., Simultaneously enhanced electrical conductivity and Seebeck coefficient in poly (3,4-ethylenedioxythiophene) films treated with hydroiodic acid, Synth. Met. 220 (2016) 585–590.
[34] B. Winther-Jensen, D.W. Breiby, K. West, Base inhibited oxidative polymerization of 3,4-ethylenedioxythiophene with iron(III)tosylate, Synth. Met. 152 (1–3) (2005) 1–4.
[35] J. Kim, E. Kim, Y. Won, et al., The preparation and characteristics of conductive poly (3,4-ethylenedioxythiophene) thin film by vapor-phase polymerization, Synth. Met. 139 (2) (2003) 485–489.
[36] M. Fabretto, K. Zuber, C. Hall, et al., The role of water in the synthesis and performance of vapour phase polymerised PEDOT electrochromic devices, J. Mater. Chem. 19 (42) (2009) 7871.
[37] T. Park, C. Park, B. Kim, et al., Flexible PEDOT electrodes with large thermoelectric power factors to generate electricity by the touch of fingertips, Energy Environ. Sci. 6 (3) (2013) 788.
[38] A. Yoshida, N. Toshima, Gold nanoparticle and gold nanorod embedded PEDOT:PSS thin films as organic thermoelectric materials, J. Electron. Mater. 43 (6) (2014) 1492–1497.
[39] Y. Shiraishi, S. Hata, Y. Okawauchi, et al., Improved thermoelectric behavior of poly (3,4-ethylenedioxythiophene)-poly(styrenesulfonate) using poly(N-vinyl-2-pyrrolidone)-coated $GeO_2$ nanoparticles, Chem. Lett. 46 (7) (2017) 933–936.
[40] J. Zhang, K. Zhang, F. Xu, et al., Thermoelectric transport in ultrathin poly(3,4-ethylenedioxythiophene) nanowire assembly, Compos. Part B Eng. 136 (2018) 234–240.
[41] D.K. Taggart, Y. Yang, S.C. Kung, et al., Enhanced thermoelectric metrics in ultra-long electrodeposited PEDOT nanowires, Nano Lett. 11 (1) (2011) 125–131.
[42] M. Culebras, C.M. Gómez, A. Cantarero, Enhanced thermoelectric performance of PEDOT with different counter-ions optimized by chemical reduction, J. Mater. Chem. A 2 (26) (2014) 10109.
[43] S.A. Gregory, A.K. Menon, S. Ye, et al., Effect of heteroatom and doping on the thermoelectric properties of poly(3-alkylchalcogenophenes), Adv. Energy Mater. 8 (34) (2018) 1802419.
[44] B.S. Ong, Y. Wu, P. Liu, et al., High-performance semiconducting polythiophenes for organic thin-film transistors, J. Am. Chem. Soc. 126 (11) (2004) 3378–3379.
[45] H. Li, J. Dailey, T. Kale, et al., Sensitive and selective $NO_2$ sensing based on alkyl- and alkylthio-thiophene polymer conductance and conductance ratio changes from differential chemical doping, ACS Appl. Mater. Interfaces 9 (24) (2017) 20501–20507.
[46] K. Besar, S. Yang, X. Guo, et al., Printable ammonia sensor based on organic field effect transistor, Org. Electron. 15 (11) (2014) 3221–3230.
[47] H. Li, M.E. DeCoster, R.M. Ireland, et al., Modification of the poly(bisdodecylquaterthiophene) structure for high and predominantly nonionic conductivity with matched dopants, J. Am. Chem. Soc. 139 (32) (2017) 11149–11157.
[48] Z. Fei, P. Pattanasattayavong, Y. Han, et al., Influence of side-chain regiochemistry on the transistor performance of high-mobility, all-donor polymers, J. Am. Chem. Soc. 136 (43) (2014) 15154–15157.

[49] I. McCulloch, M. Heeney, C. Bailey, et al., Liquid-crystalline semiconducting polymers with high charge-carrier mobility, Nat. Mater. 5 (4) (2006) 328–333.
[50] Q. Zhang, Y. Sun, F. Jiao, et al., Effects of structural order in the pristine state on the thermoelectric power-factor of doped PBTTT films, Synth. Met. 162 (9–10) (2012) 788–793.
[51] K. Kang, S. Watanabe, K. Broch, et al., 2D coherent charge transport in highly ordered conducting polymers doped by solid state diffusion, Nat. Mater. 15 (8) (2016) 896–902.
[52] S.N. Patel, A.M. Glaudell, D. Kiefer, et al., Increasing the thermoelectric power factor of a semiconducting polymer by doping from the vapor phase, ACS Macro Lett. 5 (3) (2016) 268–272.
[53] V. Vijayakumar, Y. Zhong, V. Untilova, et al., Bringing conducting polymers to high order: toward conductivities beyond 105 S $cm^{-1}$ and thermoelectric power factors of 2 mW $m^{-1}$ $K^{-2}$, Adv. Energy Mater. 9 (24) (2019) 1900266.
[54] R. Kroon, D. Kiefer, D. Stegerer, et al., Polar side chains enhance processability, electrical conductivity, and thermal stability of a molecularly p-doped polythiophene, Adv. Mater. 29 (24) (2017) 1700930.
[55] D. Kiefer, R. Kroon, A.I. Hofmann, et al., Double doping of conjugated polymers with monomer molecular dopants, Nat. Mater. 18 (2) (2019) 149–155.
[56] W. Shi, D. Wang, Z. Shuai, High-performance organic thermoelectric materials: theoretical insights and computational design, Adv. Electron. Mater. 5 (11) (2019) 1800882.
[57] Y. Lu, J.-Y. Wang, J. Pei, Strategies to enhance the conductivity of n-type polymer thermoelectric materials, Chem. Mater. 31 (17) (2019) 6412–6423.
[58] J. Shaw, H. Zhong, C.P. Yau, et al., Alternating copolymers incorporating dithienogemolodithiophene for field-effect transistor applications, Macromolecules 47 (24) (2014) 8602–8610.
[59] L. Pandey, C. Risko, J.E. Norton, et al., Donor-acceptor copolymers of relevance for organic photovoltaics: a theoretical investigation of the impact of chemical structure modifications on the electronic and optical properties, Macromolecules 45 (16) (2012) 6405–6414.
[60] R.B. Aïch, N. Blouin, A. Bouchard, et al., Electrical and thermoelectric properties of poly(2,7-carbazole) derivatives, Chem. Mater. 21 (4) (2009) 751–757.
[61] K. Broch, D. Venkateshvaran, V. Lemaur, et al., Measurements of ambipolar Seebeck coefficients in high-mobility diketopyrrolopyrrole donor–acceptor copolymers, Adv. Electron. Mater. 3 (11) (2017) 1700225.
[62] J. Saska, G. Gonel, Z.I. Bedolla-Valdez, et al., A freely soluble, high electron affinity molecular dopant for solution processing of organic semiconductors, Chem. Mater. 31 (5) (2019) 1500–1506.
[63] Y. Karpov, T. Erdmann, I. Raguzin, et al., High conductivity in molecularly p-doped diketopyrrolopyrrole-based polymer: the impact of a high dopant strength and good structural order, Adv. Mater. 28 (28) (2016) 6003–6010.
[64] T.L. Dexter Tam, C.K. Ng, S.L. Lim, et al., Proquinoidal-conjugated polymer as an effective strategy for the enhancement of electrical conductivity and thermoelectric properties, Chem. Mater. 31 (20) (2019) 8543–8550.
[65] D. Di Nuzzo, C. Fontanesi, R. Jones, et al., How intermolecular geometrical disorder affects the molecular doping of donor–acceptor copolymers, Nat. Commun. 6 (2015) 6460.
[66] Y. Sun, L. Qiu, L. Tang, et al., Flexible n-type high-performance thermoelectric thin films of poly(nickel-ethylenetetrathiolate) prepared by an electrochemical method, Adv. Mater. 28 (17) (2016) 3351–3358.

[67] Y. Sun, P. Sheng, C. Di, et al., Organic thermoelectric materials and devices based on p- and n-type poly(metal 1,1,2,2-ethenetetrathiolate)s, Adv. Mater. 24 (7) (2012) 932–937.
[68] W. Shi, G. Wu, K. Hippalgaonkar, et al., Poly(nickel-ethylenetetrathiolate) and its analogs: theoretical prediction of high-performance doping-free thermoelectric polymers, J. Am. Chem. Soc. 140 (41) (2018) 13200–13204.
[69] A.A. Talin, A. Centrone, A.C. Ford, et al., Tunable electrical conductivity in metal-organic framework thin-film devices, Science 343 (6166) (2014) 66–69.
[70] A. Pathak, J.-W. Shen, M. Usman, et al., Integration of a $(-Cu-S-)_n$ plane in a metal–organic framework affords high electrical conductivity, Nat. Commun. 10 (1) (2019) 1721.
[71] Q. Yao, L.D. Chen, W.Q. Zhang, et al., Enhanced thermoelectric performance of single-walled carbon nanotubes/polyaniline hybrid nanocomposites, ACS Nano 4 (4) (2010) 2445–2451.
[72] L.M. Wang, Q. Yao, J.X. Xiao, et al., Engineered molecular chain ordering in single-walled carbon nanotubes/polyaniline composite films for high-performance organic thermoelectric materials, Chem. Asian J. 11 (2016) 1804–1810.
[73] L.M. Wang, Q. Yao, H. Bi, et al., Large thermoelectric power factor in polyaniline/graphene nanocomposite films prepared by solution-assistant dispersing method, J. Mater. Chem. A 2 (2014) 11107–11113.
[74] L.M. Wang, Q. Yao, H. Bi, et al., PANI/graphene nanocomposite films with high thermoelectric properties by enhanced molecular ordering, J. Mater. Chem. A 3 (2015) 7086–7092.
[75] A.J. Minnich, M.S. Dresselhaus, Z.F. Ren, et al., Bulk nanostructure thermoelectric materials: current research and future prospects, Energy Environ. Sci. 2 (2009) 466–479.
[76] K.C. See, J.P. Feser, C.E. Chen, et al., Water-processable polymer–nanocrystal hybrids for thermoelectrics, Nano Lett. 10 (2010) 4664–4667.
[77] M. He, J. Ge, Z. Lin, et al., Thermopower enhancement in conducting polymer nanocomposites via carrier energy scattering at the organic-inorganic semiconductor interface, Energy Environ. Sci. 5 (2012) 8351.
[78] C.Z. Meng, C.H. Liu, S.S. Fan, A promising approach to enhanced thermoelectric properties using carbon nanotube networks, Adv. Mater. 22 (4) (2010) 535–539.
[79] L.M. Wang, Q. Yao, W. Shi, et al., Engineering carrier scattering at the interfaces in polyaniline based nanocomposites for high thermoelectric performances, Mater. Chem. Front. 1 (2017) 741–748.
[80] L.M. Wang, Z.M. Zhang, Y.C. Liu, et al., Exceptional thermoelectric properties of flexible organic – inorganic hybrids with monodispersed and periodic nanophase, Nat. Commun. 9 (2018) 3817–3824.
[81] W. Shi, S.Y. Qu, H.Y. Chen, et al., One-step synthesis and enhanced thermoelectric properties of polymer-quantum dot composite films, Angew. Chem. Int. Ed. 57 (27) (2018) 8037–8042.
[82] C.L. Wan, X.K. Gu, F. Dang, et al., Flexible n-type thermoelectric materials by organic intercalation of layered transition metal dichalcogenide $TiS_2$, Nat. Mater. 14 (6) (2015) 622–627.
[83] C. Wan, R. Tian, M. Kondou, et al., Ultrahigh thermoelectric power factor in flexible hybrid inorganic-organic superlattice, Nat. Commun. 8 (1-9) (2017) 1024.
[84] C. Cho, B. Stevens, J.-H. Hsu, et al., Completely organic multilayer thin film with thermoelectric power factor rivaling inorganic tellurides, Adv. Mater. 27 (19) (2015) 2996–3001.

[85] C. Cho, K.L. Wallace, P. Tzeng, et al., Outstanding low temperature thermoelectric power factor from completely organic thin films enabled by multidimensional conjugated nanomaterials, Adv. Energy Mater. 6 (2016) 1502168.
[86] D. Kim, Y. Kim, K. Choi, et al., Improved thermoelectric behavior of nanotube-filled polymer composites with poly(3,4-ethylenedioxythiophene) poly(styrenesulfonate), ACS Nano 4 (1) (2010) 513–523.
[87] C. Yu, K. Choi, L. Yin, et al., Light-weight flexible carbon nanotube based organic composites with large thermoelectric power factors, ACS Nano 5 (10) (2011) 7885–7892.

CHAPTER

# 7

# Design and fabrication of thermoelectric devices

## 7.1 Introduction

As the fundamental unit to realize thermoelectric (TE) energy conversion, a TE device generally consists of p- and n-type TE materials wired electrically in series (or partly parallel) and thermally in parallel, which are sandwiched between two insulator plates (usually polymers or ceramics). Although the maximum energy conversion efficiency of a device is determined by the TE properties of the components of TE materials, the practical conversion efficiency and the output performance (power density, service durability, etc.) are seriously affected by the device engineering schemes, including the geometry of TE legs, connecting schemes, various interfaces, and so on. Furthermore, a variety of applications puts forward diverse requirements on device structure and performance. For example, as for the power generation applications, the complex and harsh service conditions such as high temperature, large temperature difference, and thermal-mechanical coupling raise strict requirements especially on the reliability of TE devices. As for the environmental energy harvesting applications, the small temperature difference in service environment requires the devices to have sensitive response to the tiny temperature change. Furthermore, for the wearable self-supply energy devices, good flexibility, long-term stability, and environmental suitability are required. Therefore in order to maximize the practical conversion efficiency and to synergistically meet the diverse requirements on their functions and structures for various practical applications, comprehensive approaches to the device technology should be applied to deal with these complex scientific and technical issues. This chapter will focus on the basic structures and design

*Thermoelectric Materials and Devices*
DOI: https://doi.org/10.1016/B978-0-12-818413-4.00007-7

principles, key integration technologies, testing methods, and service behavior of TE devices, aiming at providing readers the fundamental knowledge and solutions for device design and manufacture technology.

## 7.2 Structures of thermoelectric devices

A TE uni-couple is made up of one n-type and one p-type TE materials connected by metal electrodes. The shaped TE materials soldered with electrode are usually called TE elements or legs. A practical TE module is fabricated by connecting many uni-couples to realize satisfied power output or cooling ability. The structure of a conventional TE module follows the π-shaped configuration shown in Fig. 7.1 [1], in which the column-like or cube-like n- and p-type elements are connected electrically in series and thermally in parallel. The module is electrically insulated by polymer or ceramic plates on the top and bottom ends, which contact with heat source and heat sink, respectively. Through structure design, it is relatively easy to maximize the conversion efficiency approaching the theoretical value on the basis of materials' TE properties. The thermal stress in the device working at large

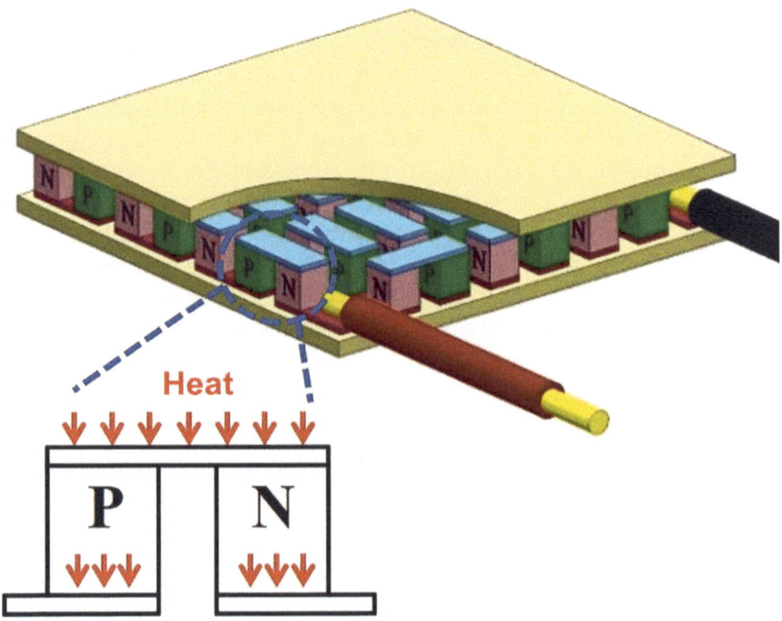


FIGURE 7.1 Structural diagram of the π-shaped TE module.

temperature gradient is inevitable, and therefore the thermal stress relieving design becomes one of the key issues, mainly through matching the coefficient of thermal expansion (CTE) of different parts including TE materials, conductive electrodes, and insulation substrates.

In addition to the conventional π-shaped configuration, some alternative structures have been proposed to meet the diverse requirements of the applications with special service environments. The typical ones are the tube-shaped module and Y-shaped module. In 2002 Weinberg presented a tube-shaped TE module, in which a number of n- and p-type ring-shaped TE elements alternately arranged around the cylindrical heat source [2], as shown in Fig. 7.2. A ring-shaped insulating plate is placed between the adjoined n- and p-type elements to realize the electrical insulation. In addition, short-tubular metal electrodes are connected onto the outer and inner walls of the TE elements in an alternate arrangement along the axial direction in order to realize the electric connection in series. In principle, both the inner surface and the outer surface are electrically insulated by cylindrical liners, whose functions are the same as those of the upper and bottom ceramic plates used in the π-shaped module. The tubular module is suitable for power generation utilizing nonplate-like heat sources or fluid medium heat sources. However, the radial transmission characteristics of heat flow and current bring difficulties to the optimization design of temperature field and electric field. In addition, the difficulty in machining and welding of curved TE materials and electrodes raises the manufacturing cost.

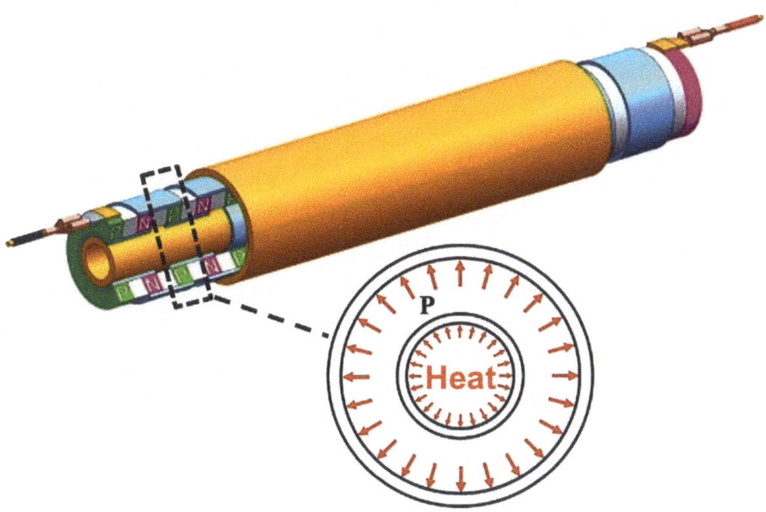


FIGURE 7.2 Structural diagram of the tube-shaped TE module.

Since the power generation devices generally work under large temperature differences, thermal stresses within device are inevitable. Especially, most TE materials are brittle, and stress damage becomes one of the most prominent problems in the service and fabrication process of TE modules. Therefore relieving stress or designing stress-free structure becomes one of the important goals in device design. Fig. 7.3 shows the novel Y-shaped configuration presented by Bell in 2006 [3]. This Y-shaped structure can more readily accommodate TE elements with different thickness, cross-sectional area, and CTE. The p- and n-type TE elements are arranged alternately with inserting an electrode between the neighboring elements. The electrodes also facilitate the thermal transport from the heat source to the side TE elements in a cornering transfer trajectory. This Y-shaped structure allows integration of TE elements with unconstrained geometries, that is, the p- and n-type elements can be designed in different shapes and sizes. Furthermore, the TE materials are stress-free, because TE elements remain relatively independent and the stresses are applied on the electrodes and insulation components. Each TE element can be optimized independently or partially independently, which is particularly favorable for the segmented module. Specifically, each p- and n-type leg can have a different cross-sectional area and/or thickness with each layer optimized to reach

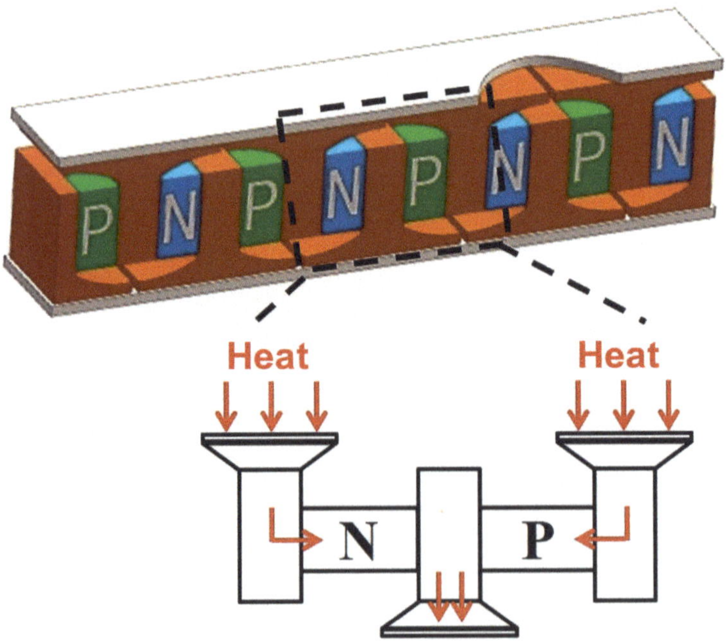


FIGURE 7.3 Structural diagram of the Y-shaped TE module.

the highest $ZT$ in their particular temperature range. This character combats negative influences of the mismatch between the optimum power outputs of the different TE segments that achieve their maximum efficiency at significantly different current densities. However, the heat flux and current density through TE materials in the Y-shaped module are not uniform, which will bring deterioration on the maximum performance to some extent. Moreover, the heat loss caused by the electrode connectors that bound the TE elements also degrades the device efficiency.

## 7.3 Fabrication and evaluation technologies of thermoelectric devices

### 7.3.1 Manufacturing process

Soldering techniques have been widely used in the fabrication of commercial TE devices. Fig. 7.4 illustrates the process flow of the π-shaped bismuth telluride (BT) module prepared by soldering method. The p- and n-type TE materials are bonded with electrodes (usually thin copper plate) by soldering. Aluminum oxide or aluminum nitride ceramic plates are usually used as the electrical insulating substrates. The copper plate (or layer) is patterned onto the ceramic substrates by

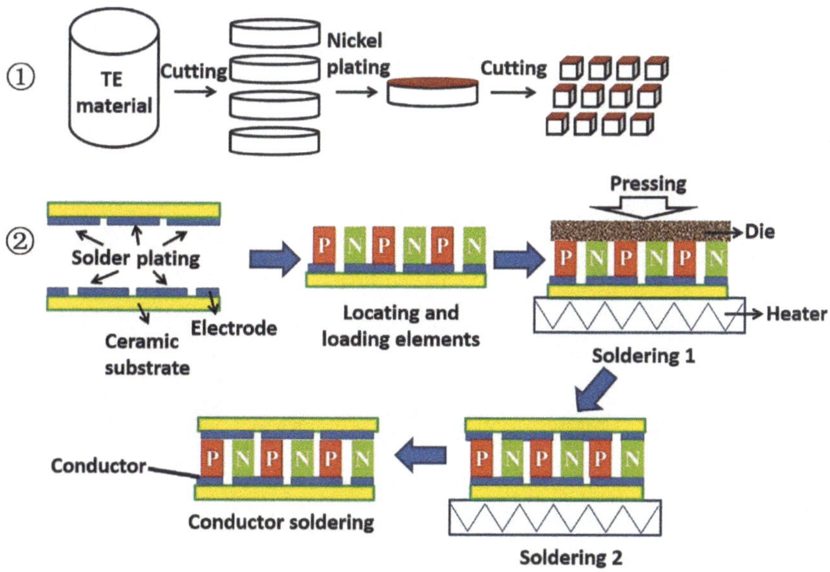


FIGURE 7.4 Process flow diagram of π-shaped bismuth telluride module using soldering method.

means of direct bonding technique. The n- and p-type BT elements and the ceramic plates are joined together by soldering to form a sandwich structure, in which the p- and n-type BT elements are connected alternately in series. In order to improve the wettability of solder on TE elements, nickel is usually plated in advance on the top and down surfaces of BT elements. Although the soldering technique is widely adopted in industrial production, its inherent weakness (such as relatively low melting point of the solder, severe diffusion of the solder component elements at high temperatures) restricts the module operation temperature usually below 200°C. In addition, the sandwiched structure with brittle ceramic plates as upper and lower insulating substrates endows the module with poor mechanical shock resistance.

To meet the requirement of power generation modules operating at high temperatures, a number of new bonding techniques have been developed. Among them, arc spraying, diffusion welding, and one-step spark plasma sintering (SPS) are well established. The process flow diagram of arc spraying technique is shown in Fig. 7.5. In this process, the electrode materials (e.g., Mo and Al) are melted by arc and then sprayed directly onto the p- and n-type TE legs that are mounted in a thermal-endurance frame made of high-temperature resistant ceramic or polymer [4,5]. Because the electrodes are directly bonded with TE legs by spraying the metal melts onto the TE materials without using any solders, the operating temperature of arc-sprayed module will not be restricted by the solders. Surface engineering (e.g., Ni plating) can be accomplished before arc spray in order to improve the adhesion

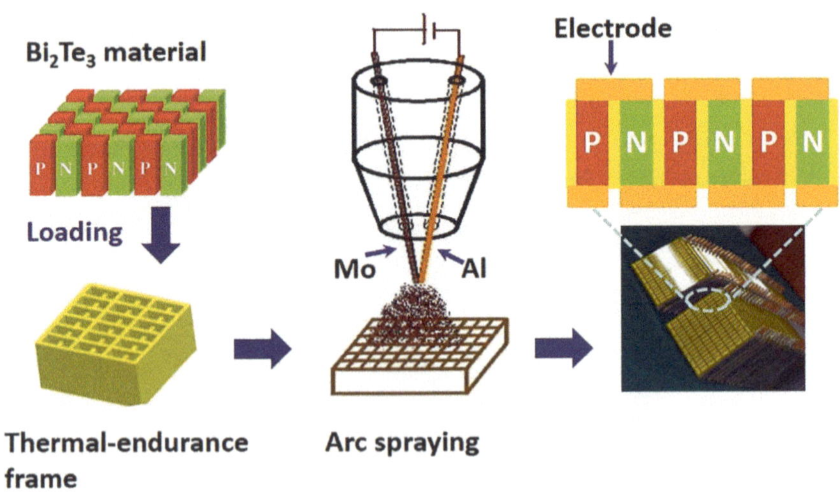


FIGURE 7.5 Process flow diagram of π-shaped bismuth telluride module using arc spraying.

between the electrode and TE materials. Comparing with the traditional soldered module, the arc-sprayed module can operate at higher temperatures with higher mechanical strength, but the framework usually causes extra heat losses and therefore deteriorates the conversion efficiency.

In the practical application of TE device, the hot-side electrodes endure harsher conditions than the cold side electrodes. The combination of high-temperature bonding technique and low-temperature soldering technique is a wise way to conveniently realize the fabrication of high-temperature devices. The SPS technique is used to bond and connect the hot-side electrode to p- and n-type TE materials to prepare π-shaped uni-couples. And then, a series of π-shaped uni-couples are connected into a module by soldering their cold-side electrodes. The process flow using one-step SPS is shown in Fig. 7.6. In this process, the densification of TE materials (p- and n-type) and the bonding of the hot-side electrode are simultaneously accomplished by one batch of SPS or hot-pressing (HP) sintering. Interface layers can be introduced between the electrode and TE materials to tune the bonding strength and interfacial resistances. This one-step sintering technique has been applied to the fabrication of skutterudite (SKD) modules and PbTe modules [6–8].

## 7.3.2 Electrodes and interfacial engineering

The characteristics of the conductive electrodes and the interfaces between electrodes and TE materials play an important role not only in the conversion efficiency but also in service performance (reliability and durability) of TE device. The match of physical properties between

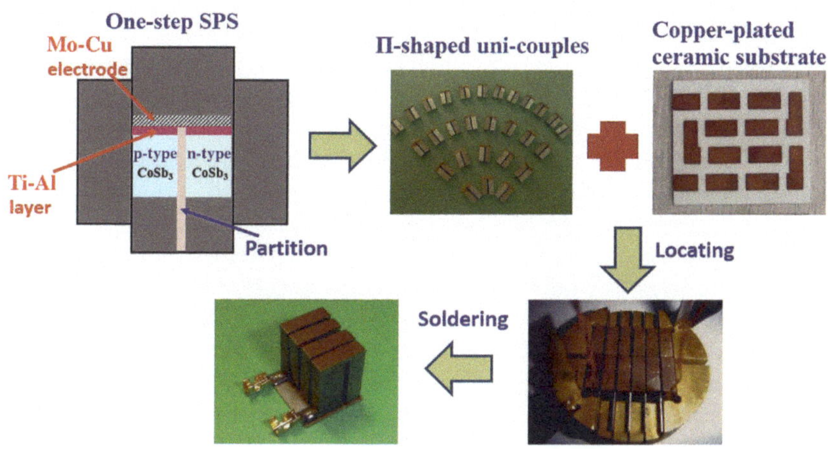


FIGURE 7.6 Process flow diagram of TE module using one-step SPS.

electrode and TE material is the first priority in selecting electrode materials. The major criteria for selecting electrode materials include: (1) high electrical and thermal conductivities to eliminate energy dissipation; (2) matched CTE with TE materials (both p- and n-legs) to minimize the thermal stress; (3) good wettability to TE materials to provide high bonding strength and low interfacial thermal/electrical resistances; (4) chemical inert, say, no serious interdiffusion or chemical reactions with TE materials at the operating temperatures; (5) thermally stable at high temperatures with good oxidation resistance; (6) excellent processability and easy to bond with TE materials. Usually, the common conductive electrode materials, such as Cu, Ag, Au, Al, and their alloys, cannot meet all the requirements at the same time. It is difficult to synchronously realize high bonding strength, low electrical contact resistance, and high refractory stability by directly bonding an electrode to TE materials. Introducing transition layer (barrier layer and/or bonding layer) between the electrodes and TE materials is commonly used to assist the realization of these contradict requirements simultaneously.

Typical candidate electrodes and the corresponding bonding technologies for the representative TE materials are listed in Table 7.1. The fabrication technology of $Bi_2Te_3$-based refrigeration module is well established. Copper or nickel is usually used as the electrode material. Arc spraying is also used to fabricate Al electrode onto $Bi_2Te_3$ alloy [4,9]. The arc-sprayed $Bi_2Te_3$ module possesses better durability and can be used for power generation up to 300°C. Carbon, tungsten, and Mo-Si alloys are used as the electrode materials for SiGe-based module. Because of the poor weldability of carbon to SiGe, spring pressuring was used to connect C electrode to SiGe. But the electrical/thermal contact resistances in the spring-pressed elements are usually larger than those in the metallurgically bonded elements. Hot isostatic pressing, HP, and SPS can be used to metallurgically bond tungsten or Mo-Si alloy electrode with SiGe materials [10,11]. Fe and Ni are commonly used as electrode materials for PbTe-based module. However, severe diffusion of Ni or Fe during service at high temperatures leads to the degradation of performance [12,13]. As for the TE materials of oxide compounds, it is difficult to metallurgically bond the oxides with metals, and silver pasting has been successfully adopted for connecting silver electrodes onto oxide TE materials. Nevertheless, the high volatility of silver pastes at high temperature may accelerate the failure of the module [14]. Sustained efforts have been made on the development of $CoSb_3$-based skutterudite module focusing on the exploration of electrodes and bonding technologies. In early stage, NASA-JPL presented the spring pressure contact method to connect electrodes with SKD materials [15]. However, the inherent high electrical/thermal contact resistances depress the conversion efficiency. Fan et al. firstly realized the

**TABLE 7.1** Typical candidate electrodes and corresponding bonding technologies.

| TE material | Electrode | Bonding technology | Characteristics |
|---|---|---|---|
| $Bi_2Te_3$ | Cu | Soldering | Matched CTE, low thermal and electrical contact resistances, but poor stability |
| | Al | Plasma spraying | High cost and easy oxidation |
| SiGe-based | C | Spring pressure contact | Mismatched CTE, but high thermal and electrical contact resistances |
| | W | HP or SPS | Matched CTE, but high thermal and electrical contact resistances |
| | Mo-Si alloy | HP or SPS | Matched CTE, low thermal and electrical contact resistances |
| PbTe-based | Fe or Ni | HP or SPS | Strong interface bonding, but mismatched CTE |
| Oxide thermoelectric materials | Ag | Silver paste | Low thermal and electrical contact resistances, but low stability |
| $CoSb_3$-based filled skutterudite | Mo | SPS | Mismatched CTE |
| | Cu | Spring pressure contact | High thermal and electrical contact resistances |
| | Mo-Cu alloys | SPS | Matched CTE, low thermal and electrical contact resistances |
| Half-Heusler | Cu | Brazing | Low thermal and electrical contact resistances, but low stability |

rigid bonding of SKD to Mo by inserting Ti as transition layer using SPS method [16]. However, the large mismatch of CTEs between Mo and $CoSb_3$ generates high residual stress, resulting in cracking at bonding interface and leading to the degradation of module. By using Mo-Cu alloy with optimized composition, the CTEs can be tuned to match the SKD materials, which significantly reduces the residual stress and increases the bonding strength [8,17].

The transition layers are designed to play multiple functions simultaneously or individually, such as serving as a barrier to block the element interdiffusion and/or interfacial reaction at high temperature, or as a binder to enhance the adhesion between TE materials and electrode, or as a buffer layer to relax the mismatch of physical properties between electrode and TE materials (CTE, Young's modulus, etc.), or as a surface modification layer to improve the wettability or to realize Ohmic contact. For the commercial bismuth telluride modules used for refrigeration, nickel plating on $Bi_2Te_3$ surface was commonly employed to improve the wettability of tin solder, which can enhance the bonding and also reduce the electrical/thermal contact resistances. However, as for the generation devices, the interfacial diffusion rates and differences of physical properties between the TE material and electrode may be boosted by orders of magnitude with elevating temperatures, which makes the selection and integration of transition layers more challenging.

$CoSb_3$-based skutterudite (SKD) is one of the most promising materials for power generation in the middle temperature range. The major challenge for SKD module is how to deal with the severe diffusion and high reactivity of Sb element. At high temperature, Sb element reacts with most of the candidate electrode materials, such as Cu, Ni, Mo. El-Genk et al. proved the ineffectiveness of copper as electrode by demonstrating the great increase of contact resistance and the sharp depression of the output power of SKD module consisting of Cu electrode after aging at 600°C for 150 days [18]. Zhao et al. fabricated $CoSb_3$-based SKD module via one-step SPS bonding technique by employing Cu-Mo or Cu-W alloys as electrodes and Ti as transition layer. Solid reaction and interdiffusion take place easily between the highly active Ti and SKD material or electrode alloys during SPS process, which leads to high bonding strength and low electrical contact resistivity [19]. However, sustained occurrence of severe interdiffusion and reaction at the Ti/SKD interface under high temperature (aging at 550°C) leads to the continuous growth of the multiple interfacial layers: TiSb, $TiSb_2$, and TiCoSb (Fig. 7.7). Especially, the growth of intermetallic compound (TiCoSb) layer causes the linear decrease of the adhesion strength and increase of the electrical contact resistance.

Gu et al. invented a new transition layer of Ti-Al alloy [20]. They employed mixed powder of Ti and Al instead of Ti powder to fabricate Ti-Al/SKD elements. In the one-step SPS, solid reaction between Ti and Al particles easily takes place forming Ti-Al intermetallic layer on the surface of Ti particles (Fig. 7.8A). Because the Ti-Al intermetallic compounds are good conductors with high melting point, the Ti-Al/Ti shell-core structure prevents the further diffusion between Ti and Sb during service, therefore suppressing the growth of diffusion layer (Fig. 7.8B). Such composite transition layer also enables low interfacial resistance as demonstrated in Fig. 7.8C.

SiGe TE devices have served for a long term in the radioisotope thermoelectric generator (RTG) powered for space missions. Various electrodes and bonding technologies have been developed for fabricating SiGe devices. Table 7.2 summarizes the electrode structure of SiGe-based TE

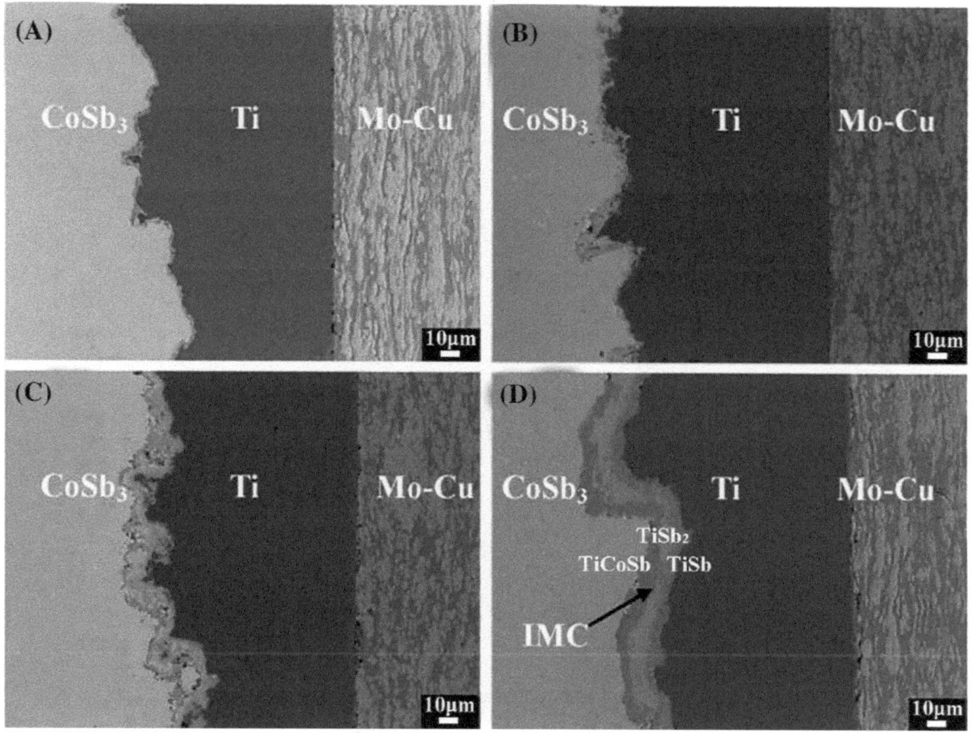


**FIGURE 7.7** SEM images of $CoSb_3$/Ti/Mo-Cu interface aged at 550°C for different aging time: (A) 0 day, (B) 8 days, (C) 20 days, and (d) 30 days.

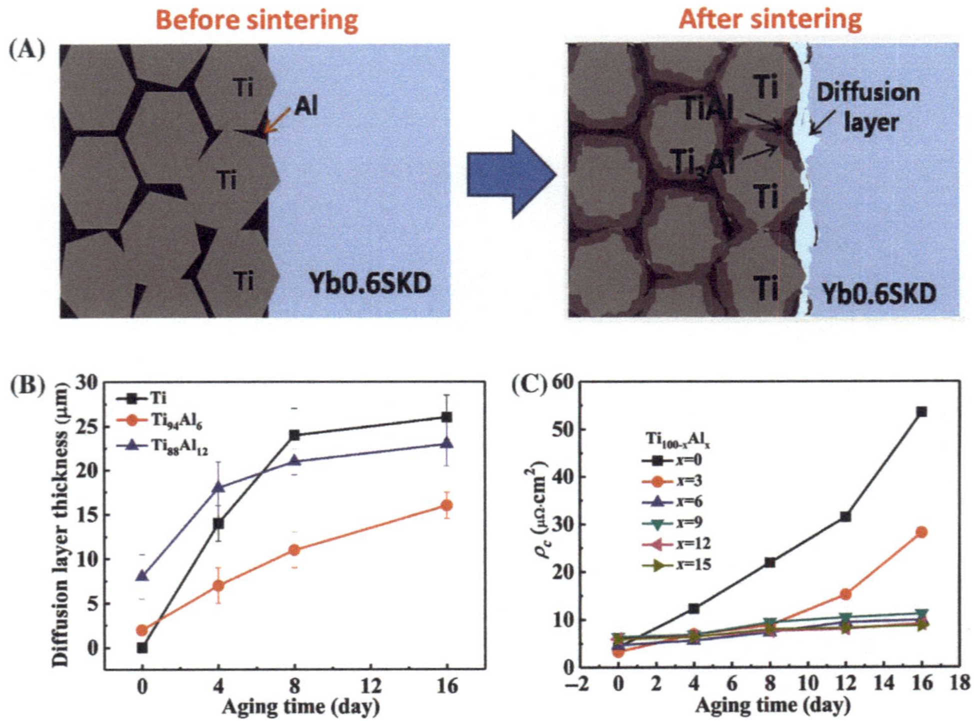


**FIGURE 7.8** (A) Schematic diagram of evolution of $Ti_{100-x}Al_x/Yb_{0.6}Co_4Sb_{12}$ interface structure. (B) Variation of diffusion layer thickness with the aging time for different Ti-Al/Yb-SKD joints aged at 600°C. (C) Variation of contact resistivity with the aging time for the Ti-Al/Yb-SKD joints aged at 600°C.

module and the corresponding contact resistivity [11,21–29]. Besides the spring pressuring method, metallurgical bonding techniques have also been used to fabricate the electrodes. C and Mo-Si can be bonded with SiGe alloys without transition layer. Tungsten (W) possesses high electrical conductivity and good stability at high temperatures, and its CTE can be tuned by forming composite with $Si_3N_4$ to match the CTE of SiGe. C, Ti, $MoSi_2 + Si_3N_4$ composite, and $TiB_2 + Si_3N_4$ composite can be used as the transition layer for SiGe/W and SiGe/($W + Si_3N_4$) joints. The high electrical conductivity and relatively high activity of $MoSi_2$ and $TiB_2$ make them suitable as transition layer to realize strong adhesion and low contact resistance. For example, Yang et al. [30] prepared (W-$Si_3N_4$)/($TiB_2$-$Si_3N_4$)/p-SiGe element *via* one-step SPS using 70 vol.% W + 30 vol.% $Si_3N_4$ composite as electrode, and 80 vol.% $TiB_2$ + 20 vol.% $Si_3N_4$ composite as barrier layer. The $TiB_2$-$Si_3N_4$ composite barrier layer not only guarantees the as-prepared elements having a low electric contact resistivity of 15 $\mu\Omega \cdot cm^2$, but also enables reasonably good thermal

**TABLE 7.2** The hot-side structure of SiGe-based TE module and the corresponding contact resistivity.

| Hot side structure (electrode/barrier layer/TE materials) | Joint stability | Contact resistivity ($\mu\Omega \cdot cm^2$) |
|---|---|---|
| W/C/SiGe | Good (aging condition: 1000°C, 140 h) | 300 |
| CVD-W/Sealed-C/SiGe | Good (aging condition: 1000°C, 1000 h) | <100 |
| (Si-Mo)/SiGe | Good | <100 |
| W/Ti/n-SiGe | Holes are formed within the barrier layer (before aging) | – |
| C/Ti/n-SiGe | Holes are formed within the barrier layer (aging condition: 1100°C, 300 h) | – |
| (Si-$MoSi_2$)/SiGe | Good (before aging) | >2000 |
| (W-$Si_3N_4$)/($TiB_2$-$Si_3N_4$)/n-SiGe | Good (before aging) | ~200 |
| (W-$Si_3N_4$)/($TiB_2$-$Si_3N_4$)/p-SiGe | Good (before aging) | ~100 |
| ($W + Si_3N_4$)/($MoSi_2 + Si_3N_4$)/n-SiGe | Good (before aging) | ~150 |
| ($W + Si_3N_4$)/($MoSi_2 + Si_3N_4$)/p-SiGe | Good (before aging) | ~300 |
| ($W + Si_3N_4$)/($TiB_2 + Si_3N_4$)/p-SiGe | Good (aging condition: 1000°C, 120 h) | <100 |

stability by maintaining a sufficiently small contact resistivity of about $75\ \mu\Omega \cdot cm^2$ after aging at 1000°C for 120 hours.

### 7.3.3 Measurement of electrical and thermal contact resistances

In principle, the conventional four-probe method used in the resistance measurement of bulk materials can be modified for measuring contact resistance. The schematic measurement arrangement of electrical contact resistance is shown in Fig. 7.9A. Probes A and D are used as current terminates and probes B and C are used as voltage contacts. During measurement, probes A, B, and D are fixed and only probe C moves along the A–D line. When probe C scans on the surface of material-I or material-II, the potential drop between B and C presents a linear change, and the resistivity of the bulk material (I or II) can be obtained from the potential drop, current and dimensions. When probe C scans across the interface, the potential drop may present a sudden change and the contact resistance ($R_C$) is obtained from the applied current and the total potential drop across the interface. According to the definition, the electrical contact resistivity is the product of contact resistance ($R_C$) and sample's cross-sectional area, namely $\rho_c = R_C \times A$. In practical experiments, in order to realize homogeneous electric field, probe A and D are reasonably replaced by metal blocks (Fig. 7.9B), and a rectangular sample with two parallel end surfaces is sandwiched by these two blocks.

As an example, Fig. 7.10 shows a measurement result of electrical contact resistivity of $Yb_{0.3}Co_4Sb_{12}/Ti/Yb_{0.3}Co_4Sb_{12}$ joint, tested by the method shown in Fig. 7.9B. The thickness of Ti barrier layer is about $50\ \mu m$, which is regarded to the nearly flat part in the voltage-displacement curve. It contains two $Ti/Yb_{0.3}Co_4Sb_{12}$ interfaces, at the left and right sides of Ti layer, respectively. Across these two interfaces, changes in voltage (0.004 and 0.006 mV) are observed. The applied current for the measurement is set up as a constant value of 100 mA. Using

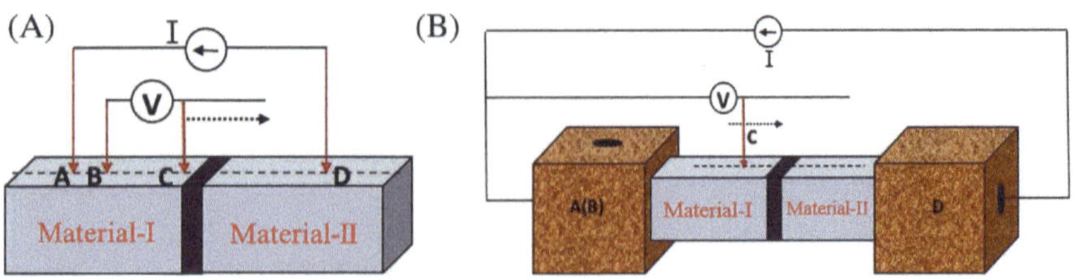


**FIGURE 7.9** (A) Schematic diagram of four-probe method. (B) Schematic diagram of four-probe method in reality.

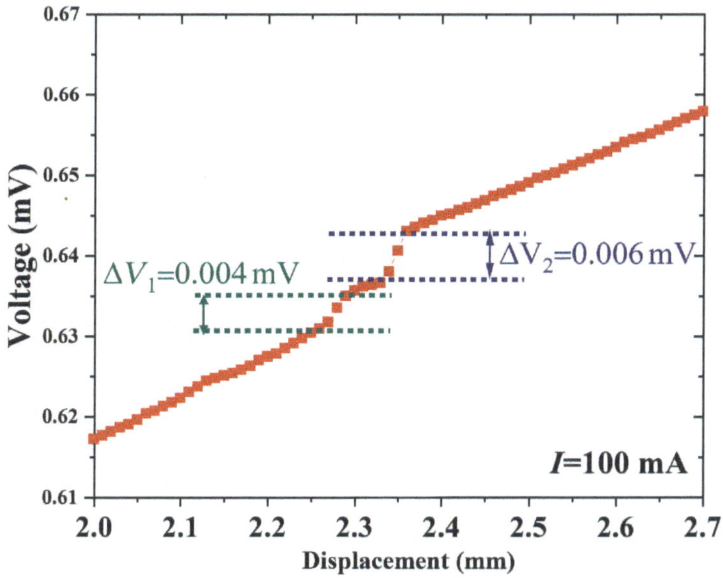


**FIGURE 7.10** Measurement result of contact resistivity of $Yb_{0.3}Co_4Sb_{12}$/Ti/$Yb_{0.3}Co_4Sb_{12}$ joint prepared by HP.

the cross-sectional area of $3 \times 3\ mm^2$, the average contact resistivity of Ti/$Yb_{0.3}Co_4Sb_{12}$ interface is calculated as $4.5\ \mu\Omega \cdot cm^2$.

The measurement of thermal contact resistance is more difficult than that of electrical contact resistance. Nowadays, there are no reliable methods to directly measure the interfacial thermal resistance. By combining the conventional multilayer heat conduction equation and precise measurement of thermal conductivity of bulk material, one can find a convenient way to estimate the thermal contact resistance [7]. Fig. 7.11 shows a multiple layer–plate model to demonstrate the estimation of interfacial thermal resistance through back-deduction operation. This multilayer plate is composed of TE material, barrier layer, and electrode. According to the conduction equation in multilayer structure, the total thermal resistance of the plate is the sum of the thermal resistances of these three components, if there is no extra thermal resistance caused by the interfaces. Accordingly, by taking into account the contribution of the interfacial thermal resistance, the total thermal resistivity (the reciprocal of the thermal conductivity $\kappa$) shall be given as:

$$\frac{l}{\kappa} = \frac{l_1}{\kappa_1} + \frac{l_2}{\kappa_2} + \frac{l_3}{\kappa_3} + \delta \tag{7.1}$$

where, $l$ is the total thickness of the plate, $l_1$, $l_2$, and $l_3$ are thicknesses of the three different layers, while $\kappa_1$, $\kappa_2$, and $\kappa_3$ are the corresponding

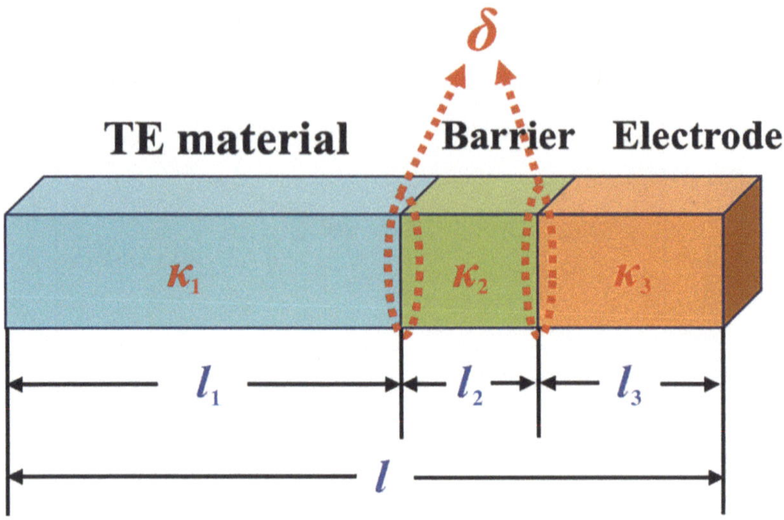


FIGURE 7.11 Schematic of multiple layer–plate model for estimating extra interfacial thermal resistance.

thermal conductivities. $\delta$ is the additional thermal resistance caused by the interfaces. Therefore it is convenient to estimate the interfacial thermal resistances by independently measuring the thermal conductivities of the bulk multilayer plate and the individual material of three components.

### 7.3.4 Evaluation of conversion efficiency and output power

The maximum conversion efficiency ($\eta_{max}$) and maximum output power ($P_{max}$) are two main indexes for the evaluation of TE device performance. Fig. 7.12A shows the schematic diagram of measurement arrangement for evaluating the output performance of TE power generation module. During the measurement, a temperature difference (hot-end temperature $T_h$, cold end temperature $T_c$) is firstly established, and then the load resistance is changed under the given temperature difference. By measuring the voltage drop ($V_s$) on the standard resistance ($R_s$), the output current ($I_{out}$) can be accurately obtained by $I_{out} = V_s/R_s$. By changing load resistance and measuring the corresponding output voltage ($V_{out}$) and output current ($I_{out}$), one can obtain $I_{out}$–$V_{out}$ and $I_{out}$–$P_{out}$ plots as shown in Fig. 7.12B. The linear fitting of $I_{out}$–$V_{out}$ produces $R_{in}$ (line slope) and $V_{oc}$ (intercept). $P_{max}$ is obtained when the total load resistance is equal to the device internal resistance ($R_{in}$).

The released heat from the cold side ($Q_{out}$) can be measured by using a copper block with known thermal conductivity as the heat

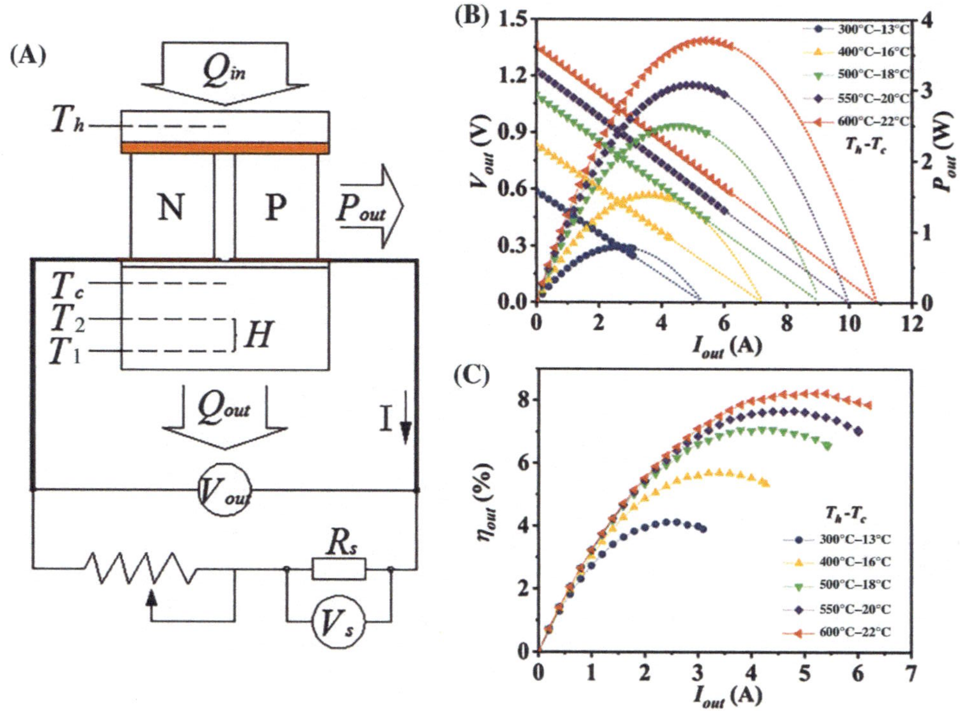


**FIGURE 7.12** (A) Schematic diagram of measurement arrangement for TE device output performance. (B) $I_{out}-V_{out}$ and $I_{out}-P_{out}$ curves of a TE device under different temperature drops. (C) Conversion efficiency ($\eta$) versus output current ($I_{out}$).

flow meter. It can be calculated by accurately measuring the temperature $T_1$ and $T_2$ (as shown in Fig. 7.12A) and the geometric dimension of the copper block:

$$Q_{out} = k \times \frac{T_1 - T_2}{H} \cdot (L \times W) \tag{7.2}$$

where $k$ is the thermal conductivity of the copper block, $L$ and $W$ are the side lengths, and $H$ is the perpendicular distance between $T_1$ and $T_2$.

According to the definition, the conversion efficiency can be calculated by

$$\eta = \frac{P_{out}}{Q_{in}} \times 100\% = \frac{P_{out}}{P_{out} + Q_{out}} \times 100\% \tag{7.3}$$

Finally, the current-dependent conversion efficiency under different temperature drops can be obtained as shown in Fig. 7.12C.

The inevitable measurement errors of various parameters and the difficulty of controlling heat transfer and temperature field complexly affect the accuracy of evaluation of device performance. At first, because the Peltier effect is sensitive to current, the current change by varying load resistance during the measurements will disturb the heat balance and cause the variation of temperatures ($T_h$ and $T_c$). Waiting enough time is necessary to reach the thermal equilibrium after each measuring operation. Secondly, extra heat exchanges (radiation and convection) between the TE legs and surrounding mediums, especially under large temperature differences, are inevitable and will cause additional errors for the measured heat flow. Furthermore, the cross-sectional temperature distribution shall also affect the device output performance. In the practical measurement, setting heat-shields around the devices and heat flow meter is commonly taken as effective way to eliminate the side heat exchanges and maintain homogenous temperature distribution in the cross-sectional plane.

### 7.3.5 Harman method

As described in Chapter 1, General Principles of Thermoelectric Technology, the figure of merit $Z$ of a TE material can be found from the independently measured thermal conductivity, electrical conductivity, and Seebeck coefficient according to the definition of $Z = S^2\sigma/\kappa$. Theoretically, $Z$ value can also be obtained by directly evaluating the performance of TE device. According to Chapter 1, General Principles of Thermoelectric Technology, the maximum temperature difference

$\Delta T_{max}$ produced by a TE cooling device under an adiabatic condition is related to material's $Z$ value and temperature by the equation of

$$Z = \frac{2\Delta T_{max}}{T_c^2} \tag{7.4}$$

where $T_c$ is the temperature at the cooled side. Since the temperature difference produced by the Peltier effect varies parabolically with the current passing through the TE device [see Eq. (1.43), the maximum temperature difference of the device can be easily determined from $\Delta T$–$I$ plot. And then, it is easy to estimate $Z$ using Eq. (7.4)]. Obviously, this is only a rough method to estimate the average $Z$ (or $ZT$) in the temperature range between $T_c$ and $T_h$. Nowadays, the state-of-the-art TE devices can produce temperature difference over 60K–70K, even approach 90K. This is really a wide range within which the materials' parameters ($S$, $\sigma$, $\kappa$ and therefore $Z$) shall present great changes. Furthermore, the influence of the device topology structure and contact resistances on the module performance is overlooked in this approach.

As early as in 1950s, Harman proposed a method and demonstrated the principle of direct measurement of material's $ZT$ [31]. It is thus named as his name. The schematic diagram of measurement arrangement is shown in Fig. 7.13. When applying a DC current $I$ to the sample, a temperature difference ($\Delta T$) shall be produced between the two ends of the sample due to the Peltier effect. At the same time, between the two ends of the sample, in addition to the resistance-induced voltage drop ($V_R = IR$), a Seebeck voltage ($V_\alpha$) shall be also produced accompanied with the temperature difference. Therefore, the total voltage drop ($V_\tau$) shall be the sum of Seebeck voltage and resistance voltage, $V_\tau = V_\alpha + V_R$. It is reasonable to ignore the joule heat for the time being. The absorbed/released heat by the Peltier effect, the heat conduction

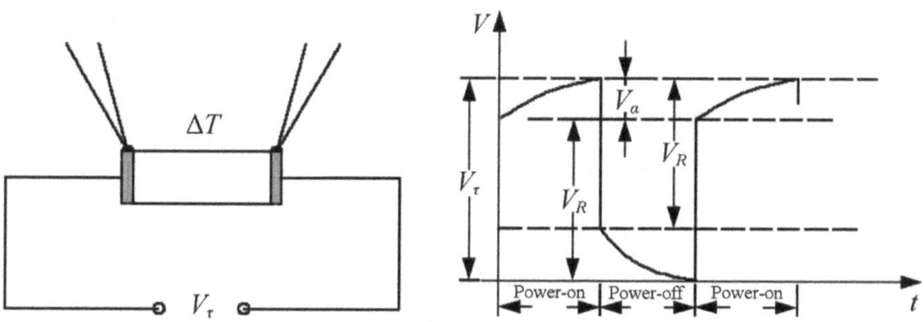


FIGURE 7.13 Schematic diagram of Harman method.

driven by the temperature difference, Seebeck voltage, and electrical current are correlated with each other as described in Chapter 1, General Principles of Thermoelectric Technology. Under adiabatic conditions, according to the basic equations described in Chapter 1, General Principles of Thermoelectric Technology, one can find the correlation between $ZT$ and voltages as

$$ZT = \frac{\dot{V}_{\tau}}{V_{R}} - 1 \tag{7.5}$$

Therefore, by carefully measuring voltage curve and determining the component voltages ($V_{\tau}$ and $V_{R}$) under adiabatic conditions, $ZT$ can be directly estimated. The adiabatic condition can be approximately realized by employing vacuum condition, designing appropriate sample size, and using tiny probes. Finite increment method or steady-state method can be used to measure $V_{\tau}$ and $V_{R}$. The details of the measuring technique can be found in ref. [1].

## 7.4 Modeling and structure design of thermoelectric devices

### 7.4.1 Modeling approaches

In Chapter 1, General Principles of Thermoelectric Technology, the output performance (conversion efficiency, power output, cooling ability, etc.) of a π-shaped uni-couple was conceptually described with correlation to the TE properties of the constituent materials. The derivation of these relations are made based on a series of simplifications and assumptions, majorly including: (1) the uni-couple is an ideal device without considering the electrical and thermal contact resistances between the electrodes and TE materials; (2) the TE properties of the p- and n-type materials are temperature independent, and the Thomson effect is absent; (3) only the heat that transfers through the TE legs is taken into account, while heat exchanges through convection and radiation between the TE legs and surrounding mediums are ignored; (4) the generated Joule heat is averagely transferred to the hot and cold sides. Based on the above simplifications and assumptions, the maximum conversion efficiency can be expressed as

$$\eta_{max} = \frac{T_h - T_c}{T_h} \cdot \frac{\sqrt{1 + Z_{pn}T_{ave}} - 1}{\sqrt{1 + Z_{pn}T_{ave}} + T_c/T_h} \tag{7.6}$$

where $Z_{pn}$ is the figure of merit of the uni-couple, $T_h$ is the hot-side temperature, $T_c$ is the cold-side temperature, and $T_{ave}$ is the average temperature of the hot and cold sides, respectively.

The maximum power output of such a π-shaped TE uni-couple is defined as the power output generated when the load resistance $R_L$ matches the uni-couple resistance $R_{pn}$, which is expressed as

$$P_{max} = \frac{S_{pn}^2 \Delta T^2}{4R_{pn}} \tag{7.7}$$

where $\Delta T$ is the temperature difference between the hot and cold sides, and $S_{pn} = S_p - S_n$. It is especially noteworthy that, $Z_{pn}$ parameter is a characteristic not of a pair of materials but rather, of a particular uni-couple, since it includes terms that involve the relative dimensions of the TE elements [Eqs. (1.18) and (1.19)]. Only in special cases can this conceptual value $Z_{pn}$ be accurately related to the true figure of merit of TE material. One such case is when the p- and n-type materials are exactly equivalent to one another apart from the sign of the Seebeck coefficient. Another case is when one leg is the superconductor, whose Seebeck coefficient and electrical resistivity are effectively zero.

In practice, the TE properties of materials are temperature-dependent, and the p- and n-type legs are not equivalent in most cases. The Thomson effect cannot be ignored especially when the temperature gradient and/or current are very large. Moreover, considerable electrical and thermal resistances in the interfaces and the thermal radiation and convection shall result in a great loss in conversion efficiency. Therefore Eqs. (7.6) and (7.7) are not applicable for predicting the performance of a practical TE device. In order to accurately describe the heat transfer and energy conversion process in practical TE devices, a variety of theoretical modeling approaches and practical methods have been proposed. By using these models and methods, one can further make clear the influence of structural factors on the device performance and then approach the optimization of device performance.

#### *7.4.1.1 Global energy balance model*

It is most essential to establish heat balance equations (algebraic equation) at both the hot side and cold side of the TE device. As described in Section 1.3, based on a global balance of heat transfer and TE effects, it is easy to obtain the analytical solution expression describing the device performance on the assumptions that the TE properties of materials are independent on or linearly correlated to temperature. In this model, the maximization of $Z_{pn}$ is geometrically satisfied by

$$\gamma_{opt} = \sqrt{\frac{\kappa_n \rho_p}{\kappa_p \rho_n}} \tag{7.8}$$

Here, $\gamma$ is the ratio of the aspect ratios of p-leg to n-leg, $\gamma = A_p H_n / A_n H_p$, where $A$ and $H$ stand for the cross-sectional area and the height of TE legs, respectively, so that

$$Z_{pn}^{max} = \frac{(S_p - S_n)^2}{\left(\sqrt{\kappa_p \rho_p} + \sqrt{\kappa_n \rho_n}\right)^2} \quad (7.9)$$

Therefore, when the p- and n-type materials have different electrical resistivities and thermal conductivities, and $\rho_p \kappa_n / \rho_n \kappa_p \neq 1$, the aspect ratio of the p- and n-leg needs to be optimized according to Eq. (7.8).

The global energy balance model is widely used for determining the size of TE legs because of its easiness and convenience. However, its accuracy is limited due to the simplifying assumptions. Some improved versions have been proposed by considering Thomson effect [32,33], or thermal conduction resistance and convection heat transfer [34,35]. All these modifications are still based on using temperature-independent heat transfer coefficients and linear dependency of TE properties on temperature. The global energy balance model is well applicable to the devices working under low temperature and small temperature difference, in which both the temperature distribution and potential distribution can be linearly approximated. However, for estimating the performance of generation devices working at high temperatures and under large temperature difference, the nonlinear temperature distribution and nonlinear potential distribution will lead to substantial errors by using this model.

#### 7.4.1.2 One-dimensional local energy balance model

In order to quantitatively evaluate the influence of device geometry on output performance, one-dimensional heat conduction models have been proposed to establish differential equations describing thermal-to-electrical energy conversion process [36–39]. In this model, Fourier heat conduction equation (differential equation) is established by taking into account both the input heat sources and the internal heat sources (Joule heat and Thomson heat). In the transient mode, the one-dimensional local energy balance equation in the TE leg is [33]

$$\rho \cdot C_p \cdot \frac{\partial T(x,t)}{\partial t} = k \cdot \frac{\partial}{\partial x}\left(\frac{\partial T(x,t)}{\partial x}\right) + \frac{1}{\sigma} \cdot \left(\frac{I}{A}\right)^2 - \beta \cdot \frac{I}{A} \cdot \frac{\partial T(x,t)}{\partial x} \quad (7.10)$$

Here, $\rho$, $\beta$, and $A$ are the density, Thomson coefficient, and leg sectional area, respectively. Others represent the same meaning as those mentioned in Chapter 1, General Principles of Thermoelectric Technology. At two ends of the TE leg, that is, $x = 0$, $T = T_c$, and $x = H$, $T = T_h$, the heat flows shall be associated with Seebeck effect and heat conduction as

$$Q(x=0,t) = S \cdot I \cdot T_c - k \cdot A \cdot \left.\frac{\partial T}{\partial x}\right|_{(x=0,t)} \tag{7.11}$$

$$Q(x=H,t) = S \cdot I \cdot T_h - k \cdot A \cdot \left.\frac{\partial T}{\partial x}\right|_{(x=H,t)} \tag{7.12}$$

The analytical solutions of heat flow, temperature distribution can be derived from the Eqs. (7.10)–(7.12) through integral operation by giving temperature-dependent TE parameters and assuming a constant cross-sectional area of TE leg. Similarly, the potential distribution can be easily computed by using the obtained temperature profile and temperature-dependent electrical properties. Compared with the global energy balance model, more precise description of distribution of heat flow is performed and the temperature-dependency of TE parameters can be involved in the one-dimensional local energy balance model. However, due to the assumption of one-dimensional flow, say, electrical current and heat flux are parallel, the side heat exchanges by radiation and convection are not involved, while these "ineffective" heat exchanges shall cause energy loss and deteriorate the conversion efficiency. Furthermore, thermal and electric contact resistances at junctions are not included in this model.

#### 7.4.1.3 Electrical analogy method

Electrical analogy method was proposed to simulate the process of TE energy conversion to deal with the complex correlations and boundary conditions, such as the temperature-dependent TE parameters, transient mode, or variable section of the TE legs [40]. In this method, a TE leg is equivalently described as an analogical circuit composed of a series of discretized nodes. In each individual node, the local energy balance Eq. (7.10) is tenable. Fig. 7.14 shows the analogical scheme of the thermal transfer in a TE leg, in which the discretized heat transfer

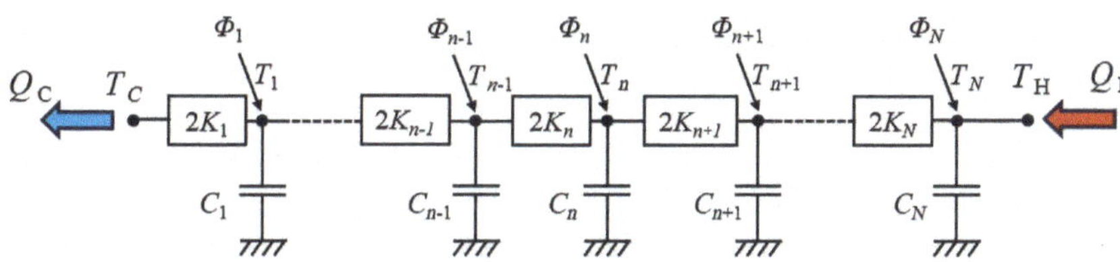


FIGURE 7.14 Analogical scheme of the heat transfer in a TE leg. The TE leg is discretized into $N$ pieces of heat transfer cells. The heat capacitance and thermal conductance of heat transfer cell $n$ are $C_n$ $\left(C_n = \rho_n \cdot Cp_n \cdot \frac{H}{N} \cdot A\right)$ and $K_n$ $\left(K_n = \frac{k_n \cdot A}{H/N}\right)$, respectively [40].

cells are interconnected in series, and heat capacitors are combined in parallel [40].

In steady-state mode, the discretization of the balance equation at each node can be expressed as

$$\frac{T_{n-1} - T_n}{\frac{1}{2} \cdot \left(\frac{1}{K_{n-1}} + \frac{1}{K_n}\right)} + \frac{T_{n+1} - T_n}{\frac{1}{2} \cdot \left(\frac{1}{K_{n+1}} + \frac{1}{K_n}\right)} + \Phi_n = 0 \tag{7.13}$$

Here, the first and second terms refer to the input and output heat flow in node $n$. $\Phi_n$ is the "extra" heat flow caused by Joule effect and Thomson effect in addition to the heat flow through conduction. $\Phi_n$ is given by

$$\Phi_n = R_n \cdot I^2 - \beta_n \cdot I \cdot \frac{T_{n+1} - T_{n-1}}{2} \tag{7.14}$$

Taking the Eqs. (7.11) and (7.12) as the boundary conditions, thermal conductance $K_n$ and electrical resistance $R_n$ in each element can be estimated based on the local temperature and leg section ($A_n$ can vary in this model) [40]. Thanks to discretizing the TE leg into multiple nodes, this analogical model is appropriate for describing both steady state and transient state for geometries with varied cross-sectional area. Besides, lateral heat losses (convective and radiative exchanges) can also be added to the elementary cell. The details on the modeling operation and calculation can be referred to the original paper [40].

Compared with the global energy balance model, prediction accuracy is obviously improved in the local energy balance model and electrical analogy model, since the temperature dependence of the TE properties and the contribution of Thomson effect can be involved into the latter two. However, due to the ignorance of the energy loss caused by assembly factors such as the heterogeneous interface between electrodes and TE materials, there is still irreparable limitation for these two approaches. For a practical TE module, all the causes of energy losses, such as assembling factors and ineffective heat transfer (radiation, convection), are interconnected with each other. Therefore the synchronous optimization of multiple parameters is vitally important to realize the optimal device performance.

#### *7.4.1.4 Three-dimensional finite element method*

The numerical analysis approach based on finite element method (FEM) can be applied to deal with the complex TE phenomena in a three-dimensional geometric configuration by fully coupling the electrical and thermal governing equations. Under a steady state, the basic vector equations of energy conservation and continuity of electric charge are [41]

$$\nabla \cdot \vec{q} = \dot{q} \tag{7.15}$$

$$\nabla \cdot \vec{J} = 0 \tag{7.16}$$

and the TE constitutive equations are:

$$\vec{q} = [S]T\vec{J} - [\kappa]\nabla T \tag{7.17}$$

$$\vec{J} = -[\sigma]\nabla V - [\sigma][S]\nabla T \tag{7.18}$$

Combing the above vector equations, the governing equations describing the distribution of temperature and electric potential can be derived as

$$\nabla([\kappa]\nabla T) + \frac{\vec{J}^2}{\sigma} - T\vec{J} \cdot \left[\left(\frac{\partial S}{\partial T}\right)\nabla T + (\nabla S)_{\mathrm{T}}\right] = 0 \tag{7.19}$$

$$\nabla([\sigma]\nabla V + [\sigma][S]\nabla T) = 0 \tag{7.20}$$

where vectors $\vec{j}$ and $\vec{q}$ represents the current density and the heat flux in three dimensions, respectively, $[\kappa]$ is the thermal conductivity matrix, $[\sigma]$ stands for the electrical conductivity matrix, $[S]$ is the Seebeck coefficient matrix, $V$ is the electrostatic potential, and $T$ is the absolute temperature.

When setting the value of current, by solving the Eqs. (7.19) and (7.20), the 3D temperature distribution and electric potential can be obtained and then transformed into output voltage, output power, absorbed heat, and conversion efficiency by mathematical operation. However, due to the temperature dependence of the characteristic parameters of TE materials, the above governing equations are strongly nonlinear. Furthermore, for some special operating environments, the boundary conditions may be complex functions of temperature and position. Therefore it is difficult to obtain accurate solutions of the equations, and FEM can be used to obtain numerical solution of the governing equations [42]. It regards the solution domain as being composed of many subdomains called finite elements, which are interconnected by nodes. The analysis scheme is to give a piecewise approximate solution of the basic equations. Since TE legs can be divided into numbers of subdomains with any shape and sizes, complex geometric boundaries and complex continuum problems can be easily dealt with by the three-dimensional FEM method.

The device performance parameters (temperature distribution, potential distribution, output power, conversion efficiency, etc.) under actual boundary conditions can be numerically acquired by taking into account all the influence factors (interfacial electrical and thermal resistances, geometry, electrode, etc.). Fig. 7.15 demonstrates the logical framework for the full-parameter optimization of a TE power generation module, which could be realized by using the three-dimensional FEM simulation.

## 7.4.2 Examples of module design by three-dimensional finite element method

The conversion efficiency of TE power generation devices is not only related to the TE materials' performance and the temperatures at hot and cold sides, but also greatly affected by the geometrical dimensions of TE legs. When the constituent n- and p-type materials have different TE properties, the two TE legs should be designed into different sizes for maximizing the performance. This section takes the one-stage skutterudite (SKD) module and skutterudite/bismuth telluride (SKD/BT) segmented module as examples to demonstrate the optimization design

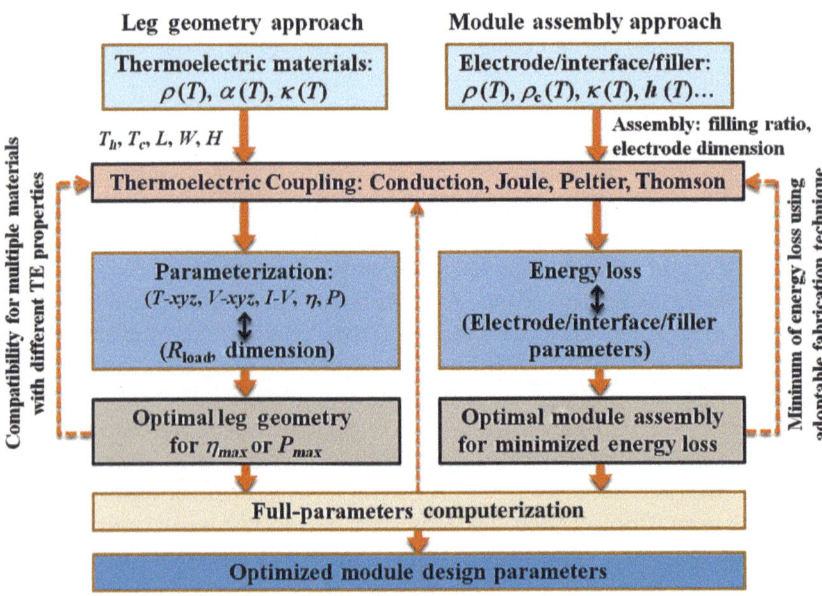


FIGURE 7.15 Logical framework for the full-parameter optimization of a thermoelectric power generation module [42].

of the device structure and to reveal the influence of structural and interfacial parameters on the device performance.

A three-dimensional geometrical model of one-stage π-shaped SKD uni-couple is shown in Fig. 7.16A. The meshing result obtained by using the ANSYS Workbench platform is shown in Fig. 7.16B. After assigning the material properties (shown in Fig. 7.17 and Table 7.3) and setting the boundary conditions to the uni-couple [for example, the temperature of the cold source $T_{cooler} = 25°C$, the temperature of the heat source $T_{heater} = 600°C$, the heat transfer coefficient between the heat source and TE module is set as $6 \times 10^3$ W/(m$^2$ · K), and heat transfer coefficient between the cold source and TE module is set as $1.2 \times 10^4$ W/(m$^2$ · K)], the three-dimensional temperature distribution and potential distribution in different parts can be calculated by using FEM, as shown in Fig. 7.16C and D. By extracting the current, voltage, and heat flow values, the values of output power and conversion efficiency can be obtained according to their definitions. The calculated device output parameters (output power, conversion efficiency, exchanged heat, open circuit voltage, and internal resistance) are shown in Fig. 7.16E–G as functions of the structural parameters (the ratio of leg height and cross-sectional area, $H/A_{pn}$; the cross-sectional area ratio of p- and n-legs, $A_p/A_n$). As can be seen from Fig. 7.16E, with increasing $A_p/A_n$, both the maximum power output ($P_{max}$) and conversion efficiency ($\eta_{max}$) tend to rise first and then fall. They reach their peaks at different specific sizes, indicating that it is difficult to achieve $P_{max}$ and $\eta_{max}$ at the same $p/n$ sectional ratio. As can be seen from Fig. 7.16G, both $P_{max}$ and power density ($P_d$) decrease along with the increase of $H/A_{pn}$, while $\eta_{max}$ presents a reverse trend. This is because the increase of $H/A_{pn}$ leads to a linear increase of the internal resistance ($R_{in}$) but a parabola-like increase of open circuit voltage ($V_{OC}$) as shown in Fig. 7.16F. Therefore under the operation condition with a fixed temperature drop across the heat source and cooler source, one can set small $H/A_{pn}$ (with a reasonable range) to obtain high $P_{max}$ or $P_d$, but large $H/A_{pn}$ to obtain high $\eta_{max}$ according to the different requirements of practical applications.

Generally, for a given TE material, the optimal operating temperature is limited in a certain range, because the TE properties of a material are always dependent on temperature. This gives an up-limit of performance for a single-stage device. Cascaded or segmented structures are commonly used to transcend such limitation in single-stage device, especially when the devices operate under large temperature differences. This is to use different materials setting at different temperature ranges to match their peak $ZT$s so that to ensure the average $Z_{pn}T$ value of device within the whole temperature range larger than the single-stage device. A typical segmented TE uni-couple is demonstrated in

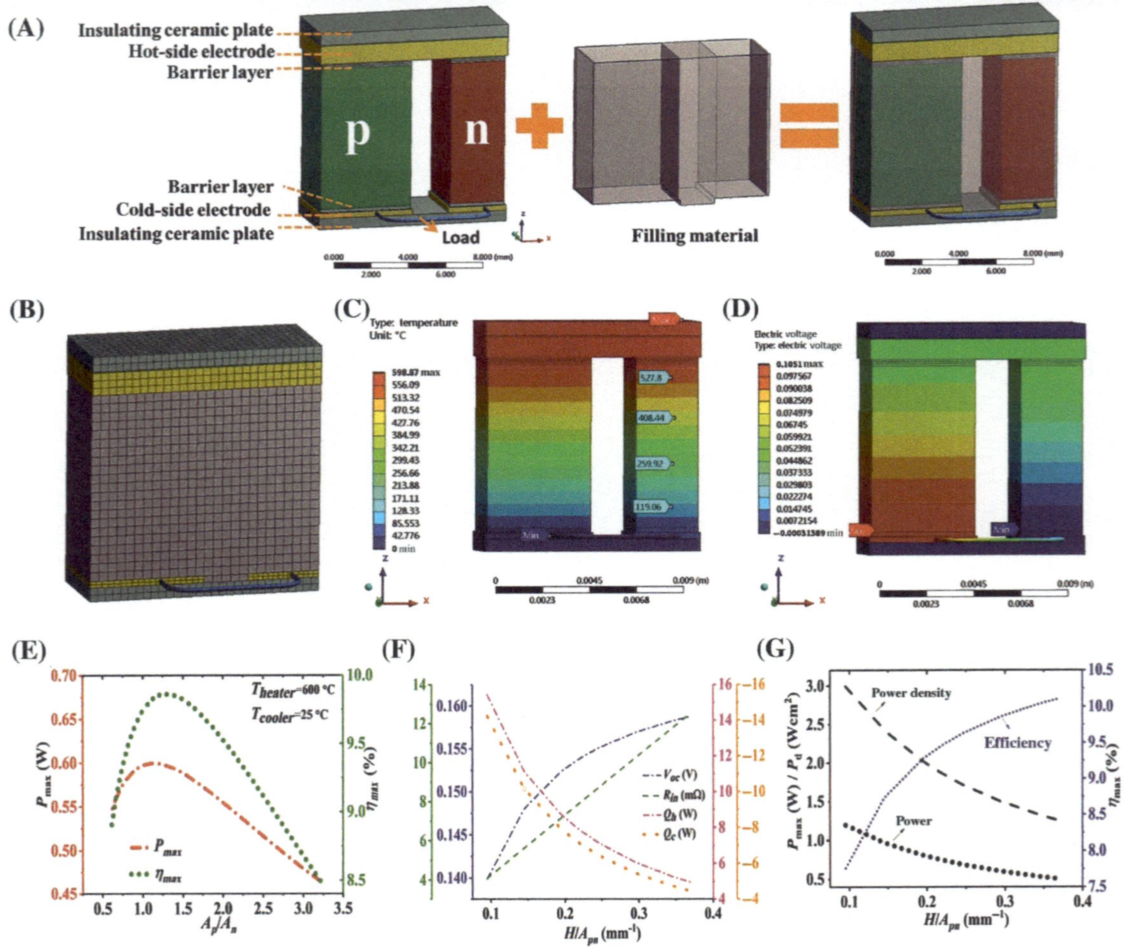


FIGURE 7.16 (A) The three-dimensional geometrical model of one-stage π-shaped SKD uni-couple. (B) The meshing result of one-stage SKD uni-couple obtained by using the ANSYS Workbench platform. (C) Three-dimensional temperature distribution and (D) potential distribution of the one-stage SKD uni-couple without fillers for clear view. (E) The maximum power output ($P_{max}$) and conversion efficiency ($\eta_{max}$) of SKD uni-couple as a function of the cross-sectional area ratio of p- and n-legs ($A_p/A_n$). (F) Open circuit voltage ($V_{OC}$), internal resistance ($R_{in}$), absorbed heat at hot side ($Q_h$) and released heat at cold side ($Q_c$) of the SKD uni-couple as a function of the ratio of leg height and cross-sectional area (H/

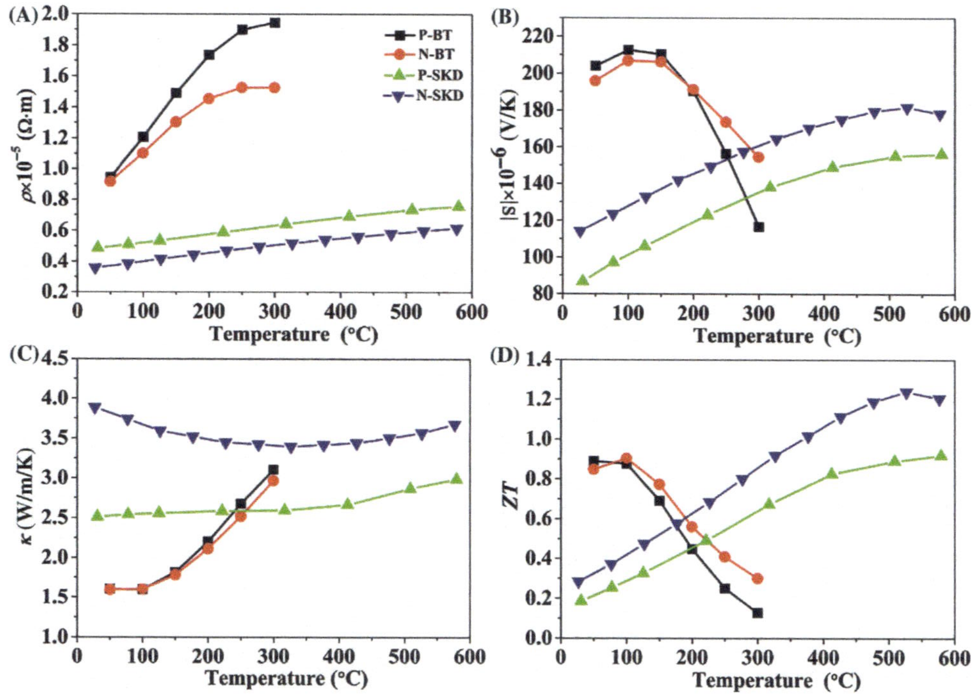


**FIGURE 7.17** (A) Electrical resistivity ($\rho$), (B) absolute Seebeck coefficient ($|S|$), (C) thermal conductivity ($\kappa$), (D) $ZT$s of BT (p-$Bi_{0.4}Sb_{1.6}Te_3$, n-$Bi_2Te_{2.5}Se_{0.5}$) and SKD (p-type $CeFe_{3.85}Mn_{0.15}Sb_{12}$, n-type $Yb_{0.3}Co_4Sb_{12}$) as functions of the absolute temperature [42].

TABLE 7.3 Material property parameters.

| Item | Thermal conductivity (W/m/K) | Electrical resistivity (ohm · m) | Function |
|---|---|---|---|
| AlN | 200 | – | Insulating ceramic plate |
| $Mo_{50}Cu_{50}$ | 250 | $2.67 \times 10^{-8}$ | Hot-side electrode |
| Ag-Cu-Zn | 401 | $1.68 \times 10^{-8}$ | Hot-side brazing solder |
| Ti-Al | 21.9 | $5.26 \times 10^{-7}$ | Barrier layer |
| SnSb solder | 55 | $1.14 \times 10^{-7}$ | Cold-side solder |
| Cu | 380–260 (3°C–150°C) | $1.68 \times 10^{-8}$ | Cold-side electrode and heat flow meter |
| $Al_2O_3$ | 30.3–9.1 (100°C–600°C) | – | Insulating ceramic plate |

Fig. 7.18A. In the segmented uni-couple, the constituent materials usually possess different temperature dependences of their TE parameters (thermal conductivity, electrical conductivity, and Seebeck coefficient). The coupling between the material properties and the leg geometry directly determines the temperature distribution, heat flux distribution, current density, and its distribution within the TE legs, and therefore affects the conversion efficiency.

Furthermore, the segmented structure possesses extra bonding junctions between different TE materials that not only complicate the fabrication process but also inevitably introduce extra electrical and thermal contact resistances. By using the three-dimensional full-parameter numerical method, the precise computerized design of SKD/BT segmented module can be performed [42]. The temperature dependency of materials' properties and various parasitic energy losses (caused by interface resistance, convection/radiation heat losses) are taken into account in the simulation and design optimization. Fig. 7.18 presents the predictive performance of BT/SKD segmented uni-couple at various structural and functional factors. Experimental measurements show a good agreement with simulation results (Fig. 7.18F). Unprecedentedly, the rational structure design and the advanced fabrication technology with extremely low thermal/electrical losses enable a device efficiency up to 96.9% of the theoretical efficiency based on the material's properties. In addition, by changing the geometric shape and boundary conditions, this three-dimensional full-parameter method can be extended to the design of TE devices with complex structures such as ring-shaped, Y-shaped and thin-film devices.

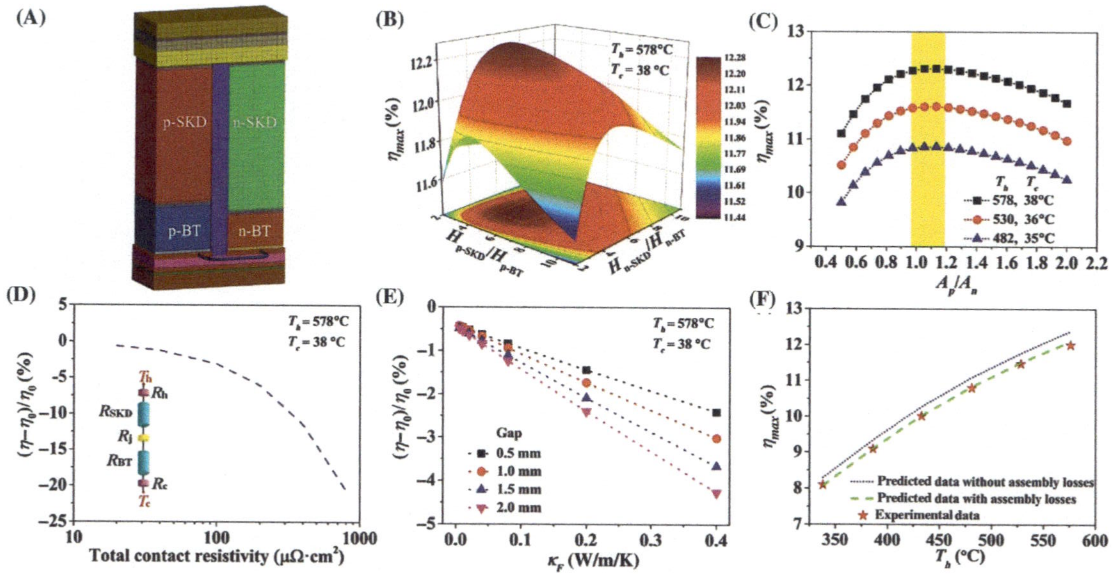


**FIGURE 7.18** (A) Schematic diagram of three-dimensional SKD/BT-segmented TE power generation uni-couple. (B) Maximum conversion efficiency ($\eta_{max}$) as a function of $H_{n\text{-SKD}}/H_{n\text{-BT}}$ and $H_{p\text{-SKD}}/H_{p\text{-BT}}$ at $T_h = 578°C$ and $T_c = 38°C$. (C) Maximum conversion efficiency ($\eta_{max}$) as a function of $A_p/A_n$ under different operating temperatures. (D) The effect of the total contact resistivity of a segmented SKD/BT uni-couple on the efficiency loss; $\eta_0$ represents the efficiency without considering the contact resistivity; the inset depicts the electrical resistance network for a segmented TE uni-couple. (E) The effect of the thermal conductivity of fillers ($k_F$) on the efficiency loss under different gap dimensions; $\eta_0$ represents the efficiency without considering fillers. (F) Comparison of the experimental $\eta_{max}$ with simulation.

## 7.5 Thermoelectric microdevices

The miniaturization of TE module will significantly extend the application of thermoelectric technology [43]. With the development of technology of integrated circuits, the power consumption of microelectronic system increases constantly. For example, the driving power of a quartz watch is 5 μW, while the power consumption is 50 μW for a cardiac pacemaker and 100 μW for a wireless sensor [44]. These microelectronic systems ask for micro-battery with high power density and long lifetime to replace chemical batteries. On the other hand, with the rapid development of high power density electronic device technology, there is an explosive demand for precise temperature control and active refrigeration in local area, because the high power density devices generate huge localized heat flow (called hotspot). For example, the localized heat flow density in CPU can be as high as 300 $W/cm^2$ [45,46], and LED produces heat density in the order of 100 $W/cm^2$ [47]. TE micromodules are highly expected for the applications as both in situ power generator and local area active refrigerator.

According to the heat flow direction in the TE legs, there are two typical schemes for structure design of TE micromodules, called the in-plane type and cross-plane type, as shown in Fig. 7.19. The in-plane type device has a thin-film structure where heat flow is parallel to film or substrate. This structure can build relatively large temperature difference. Almost all deposition technologies for preparing thin films can be easily adopted for the fabrication of in-plane micro-device. However, the large heat loss from substrate, small heat absorbing area, and high internal resistance endow the in-plane module with small output power density. The structure factors and performance of typical micro-generators and refrigerators are listed in Tables 7.4 and 7.5.

The structure of cross-plane device is similar to the conventional π-shaped device module. Compared with the in-plane structure, the cross-plane device is featured with large heat-absorbing area and large output power density. However, in the cross-plane device, the

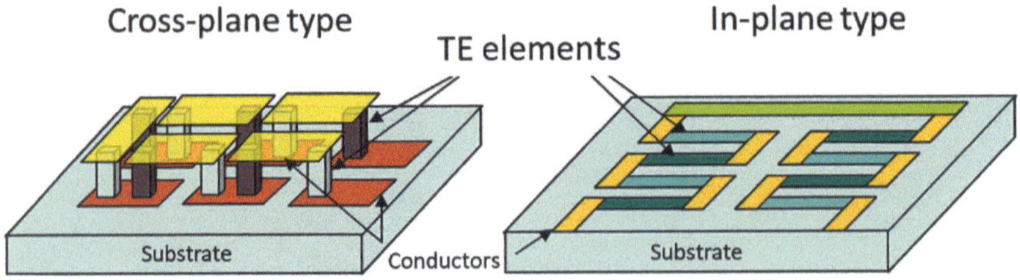


**FIGURE 7.19** Two typical structures of TE micromodule: cross-plane type (left) and in-plane type (right).

**TABLE 7.4** Structure factors and performance of micro TE power generators.

| Producer | Thermoelectric materials | Device structure | Area ($mm^2$) | Number of p-n couples | Height of TE leg (μm) | Temperature difference (K) | Open circuit voltage (V) | Output power density ($mW/cm^2$) |
|---|---|---|---|---|---|---|---|---|
| Infineon [47] | n/p-Poly Si films | In-plane | 6 | 59,400 | – | 5 | 0.66 | 0.00018 |
| DTS [48] | n-$Bi_2Te_3$ film p-$(BiSb)_2Te_3$ film | In-plane | 63.65 | 2250 | 50 | 5 | 2 | 0.0024 |
| JPL [49] | n-$Bi_2Te_3$ film p-$(BiSb)_2Te_3$ film | Cross-plane | 2.89 | 63 | 20 | 1.25 | 0.004 | 0.0346 |
| Micropelt [50] | n-$Bi_2Te_3$ film p-$(BiSb)_2Te_3$ film | Cross-plane | 25 | 1800 | 20 | 10 | 2.3 | 11.2 |
| RTI [51] | n/p-$Bi_2Te_3$ superlattice | Cross-plane | 460 | 512 | 5 | 5 | 1.247 | 6.74 |

**TABLE 7.5** Structure factors and performance of micro TE refrigerators.

| Producer | Thermoelectric materials | Device structure | Area ($mm^2$) | Number of p-n couples | Height of TE leg (μm) | Maximum cooling temperature difference (K) | Maximum cooling density ($mW/cm^2$) |
|---|---|---|---|---|---|---|---|
| Huang et al. [52] | n/p-Poly Si films | In-plane | 100 | 62,500 | – | 5.6 | – |
| JPL [49] | n-$Bi_2Te_3$ film p-$(BiSb)_2Te_3$ film | Cross-plane | 2.89 | 63 | 20 | 2 | – |
| Micropelt [50] | n-$Bi_2Te_3$ film p-$(BiSb)_2Te_3$ film | Cross-plane | 11.4 | – | 20 | 31 | 40 |
| RTI [53] | n-$Bi_2Te_3$ film n/p-$Bi_2Te_3$ superlattice | Cross-plane | – | 2 | 5 | 55 | 128 |
| Shakouri and Zhang [54] | p-Si/$Si_{0.8}Ge_{0.2}$ superlattice | Cross-plane | $0.04 \times 0.04$ | 1 | 3 | 2.5 | 600 |

interfacial electrical resistance accounts for larger proportion in the total inner resistance of the TE legs, which brings forward rigorous requirement for the electrode fabrication technology to minimize the interfacial resistances. Besides, it is difficult to generate a large temperature difference in the thickness direction of film. Various microprocessing technologies have been applied to integrate cross-plane TE device, such as the micromachining technology based on physical vapor deposition, reactive-ion etching, flip-chip bonding [55–57], integrated technology based on electrochemical deposition and welding [49,58,59], and screen printing technology using TE material paste [51]. Using electrochemical deposition and printing technology, thick TE legs more than 100 μm can be easily fabricated.

Compared with the fabrication of bulk TE device, there are more technique challenges in the fabrication of microdevice. The most difficult issues include the deposition of high-quality films with high TE performance, the fabrication of thin-film electrode with low contact resistance, and the integration of TE device fabrication process into the well-established microelectronic process. All the technique issues to affect the device performance of bulk TE device elaborated in the above sections of this chapter are equally applicable to microdevice. Topological structure, bonding scheme, heterojunction interface structures (TE/electrode, electrode/substrate) deliver great impact on the device output performance. Especially, the influence of the thermal contact resistance (called parasite resistance) in heterojunction is more significant and the thermal management becomes much more difficult.

The preparation of high-performance TE film is of course the first priority. Venkatasubramanian et al. prepared p-type $Bi_2Te_3/Sb_2Te_3$ superlattice thin film with $ZT_{max}$ of 2.4 and n-type $Bi_2Te_3/Bi_2Te_{2.83}Se_{0.17}$ superlattice thin film with $ZT_{max}$ of 1.4, and then integrated them into a cross-plane TE module [51,53]. They demonstrated very high refrigeration capability as 128 W/cm$^2$ and high output power density as 6.74 mW/cm$^2$ under the temperature difference of 5K. Shakouri and Zhang [54] fabricated superlattice TE refrigerator using p-type Si/$Si_{0.75}Ge_{0.25}$ superlattice thick film (200 layers and 3 μm in total thickness), and demonstrated a super large refrigeration power density of 600 W/cm$^2$, which enables a maximum refrigerated temperature difference of 3K. Although the superlattice thin films present great advantages for fabricating high-performance thin-film device, few practical products in market are available, because of the poor commercial viability and production repeatability.

Single-layer TE thin films have been prepared by various deposition techniques, such as physical vapor deposition, chemical vapor deposition, and electrochemical deposition as well. However, the reported TE performance is generally poor as compared with superlattice thin films.

Micropelt (a German company) fabricated n-type $Bi_2Te_3$ and p-type $(BiSb)_2Te_3$ thin films on metalized silicon substrates by sputtering and then integrated them into micromodule. The single-layer device demonstrated a refrigeration density of 40 $W/cm^2$ when operating as a refrigerator, and an output power of 11.2 $mW/cm^2$ as a generator [50]. Wang et al. [60] invented a unique method to prepare n- and p-type $Bi_2Te_3$ nanowire arrays on porous $Al_2O_3$ templates by electrochemical deposition. These nanowire arrays present relatively large Seebeck coefficient as $-188$ and 270 $\mu V/K$, respectively. But integrating such nanowire arrays into microdevice is still facing challenges.

## 7.6 Device service behavior

The TE generators are commonly requested to operate for a long time under high temperature and large temperature gradient. In many cases, the device may undergo harsh environment such as mechanical vibration, thermal shocks, high humidity or corrosivity atmosphere. Therefore it is inevitable for TE device to suffer performance degradation or failure during the long-term service. Essentially, the device service behavior is determined by the structure evolution of all the component parts under service conditions. There are three important aspects as basic scientific and engineering issues for TE degradation: (1) intrinsic stability of TE materials, (2) microstructure evolution of heterogeneous interfaces, and (3) dynamic stability of TE materials and device structure.

The TE transport properties of materials are sensitively affected by their microstructures and compositions. Basically, whatever structure changes occur during service, such as grain growth, phase transition, decomposition or localized precipitation, and element sublimation or surface regression, would cause changes of TE performance. For example, phase transition materials are usually excluded from the practical applications because phase transitions not only cause the changes of transport properties but also are frequently accompanied by volume change, leading to cracks or even structural failure. Grain growth shall occur in almost all materials serving at high temperatures, which may increase thermal conductivity. This would be especially serious for nanograined materials or nanocomposites. For example, PbTe-based materials with the dispersion of in situ precipitated nanoinclusions are reported to present grain coarsening and performance degradation after annealing at high temperature [61]. Basically, nano-inclusions are kinetically metastable and tend to grow at a rate that is determined by the diffusion constant of constituent elements. Element precipitation is another common reason for material degradation. Most of the TE materials are heavily doped semiconductors being intentionally supersaturated with dopants. Under most operating conditions (the operating temperature is usually lower than the material synthesis temperature), dopants tend to precipitate from solid solutions to form localized precipitates, causing the changes in the carrier concentrations. For example, precipitation of phosphorus in P-doped n-type SiGe alloy occurs between about 600K and 900K to decrease the carrier concentration and electrical conductivity [62]. The stability of TE materials can be evaluated by characterizing the changes of TE properties and microstructures when increasing the service time under operating conditions.

Another issue that might affect the stability of TE material and device is the mobile ions. In many Cu/Ag-containing materials and some of Zn/Mg-containing materials, the mobile ions ($Cu^{+}$, $Ag^{+}$, $Zn^{2+}$, $Mg^{2+}$) may migrate easily from one side to another side of the material and sometimes even deposit on the surface under an electric field and/or a thermal gradient [63,64]. Such metal deposition behavior would lead to the deterioration of material's TE performance and/or the formation of cracks at the device's boundaries during long-term service. Thus for quite a long time, the possibility of using these materials to fabricate stable device had been greatly concerned in TE community. In 2018 Qiu et al. proposed a parameter named as threshold voltage to judge the stability of the TE materials with mobile ions under external field [65]. The mobile ions will deposit from the material only when the voltage stressed on the material is higher than the critical voltage ($V_c$). Otherwise, the voltage-driven ion migration just results in the formation of gradient of the ion concentration inside the material without any metal deposition. In the latter case, these materials would demonstrate acceptable stability for TE devices. The threshold voltage is a material's intrinsic parameter and independent of the sample geometry. It was shown that the value of $V_c$ can be raised by creating vacancies at the sites of mobile ions or introducing "ion-blocking electron-conducting" interfaces (Fig. 7.20) [65–67]. Based on the understanding of the mechanism of ion migration and material's degradation, some preliminary strategies have been proposed to improve the service durability of the device made of $Cu_2X$ (X = S, Se, Te)-based liquid-like TE materials.

As one of the most promising power generation materials, the filled skutterudites are characterized with intrinsically high stability. The aging test shows that the TE properties of Yb-filled compound ($Yb_xCo_4Sb_{12}$) are very stable after a long-term aging at 923K as shown in Fig. 7.21. Besides the microstructure stability of bulk materials, the surface state of the TE legs should be also considered for the TE generators working at high temperature. It is found that the Sb sublimation in skutterudites is not ignorable at high temperature. After a long-term service, the Sb sublimation at surface may greatly affect the performance of

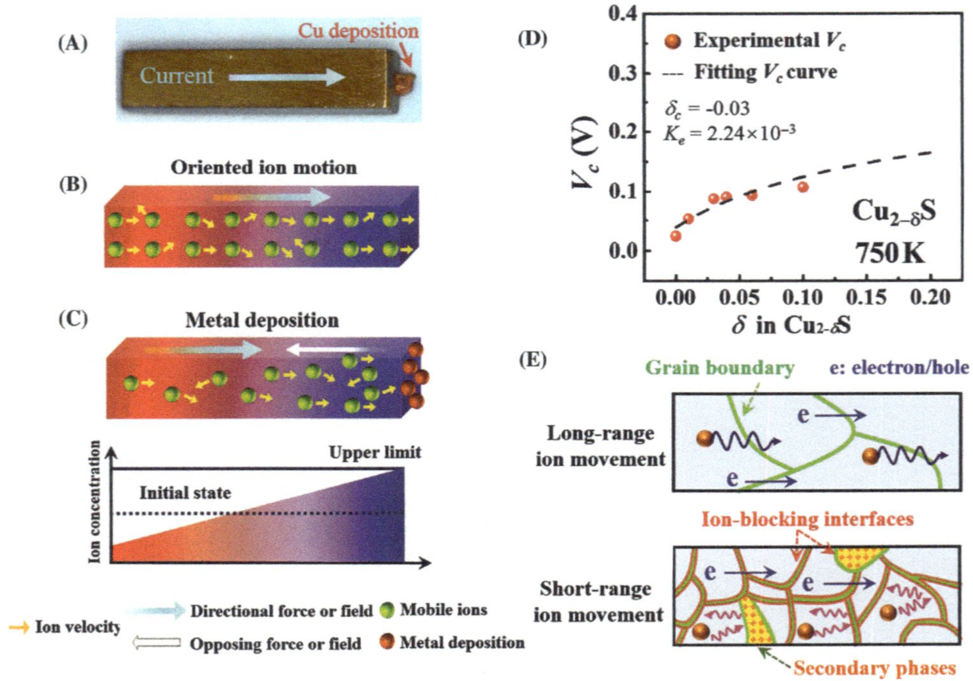


FIGURE 7.20 (A) Metallic Cu deposition on the surface of a $Cu_2S$ sample induced by a current. (B) Schematic of oriented ion motion in liquid-like thermoelectric materials under electric field. (C) Continuous metal deposition (or other decomposition) occurs when the local Cu ion concentration reaches an upper limit. (D) Critical voltage ($V_c$) as a function of Cu off-stoichiometry $\delta$ in the $Cu_{2-\delta}S$ ($\delta = 0$, 0.01, 0.03, 0.04, 0.06, and 0.1) samples at 750K. The symbols represent the measured $V_c$. The *dashed line* represents the theoretical $V_c$ curve. (E) Schematic for limiting the ion movement by including "electron-conducting and ion-blocking" thin interfaces: either grain boundaries (*red areas*) or a secondary phase (*yellow areas*).

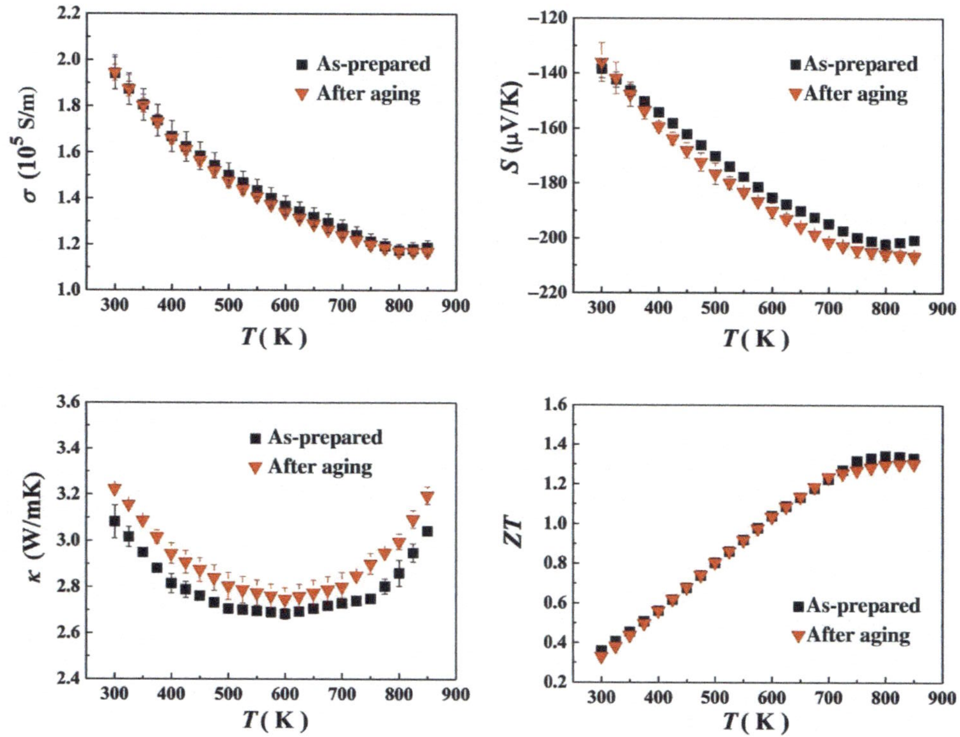


FIGURE 7.21 TE properties of n-type $Yb_xCo_4Sb_{12}$ after aging at 923K for 30 days.

device in many ways. First, the deposits of sublimated component(s) on the surroundings can result in either thermal or electrical shorts. Second, continued sublimation makes the TE legs slimmer and can eventually lead to structural failure. The sublimation behavior of several power generation materials has been characterized, and their sublimation rates in vacuum are shown in Fig. 7.22 [68]. Among them, SiGe shows the lowest sublimation rate ($\sim 4 \times 10^{-5}$ g/cm$^2$h), and the sublimation rates of TAGS ($AgSbTe_2$-GeTe), PbTe, and skutterudites are similar to the values above $\sim 1 \times 10^{-3}$ g/cm$^2$h. Fortunately, the sublimation of TE materials can be suppressed by surface coating, which is a well-established technology in chemical engineering industry. Caillat et al. characterized the open circuit voltages ($V_{OC}$) of metal coated and uncoated SKD legs in vacuum. Under the testing conditions (hot and cold junction temperatures are 892K and 316K, respectively), the $V_{OC}$ of uncoated SKD uni-couple decreased about 13% after 1200 hours operation, while the $V_{OC}$ of metal coated uni-couple decreased only 8% [18]. Xia et al. fabricated the $SiO_x$-coated SKD elements and also proved its effectiveness for the suppression of Sb sublimation and TE degradation [69].

As mentioned in Section 7.3, evolution of interfacial microstructure is unavoidable due to the interdiffusion and/or interreaction of elements at the boundaries under high temperature. The major damage caused by the interfacial structure evolution is the rise of the interfacial electrical resistance. Thus the interfacial influence on the TE degradation is commonly evaluated by measuring the changes of interfacial resistance vs time and temperature. The direct observation and semiquantitative

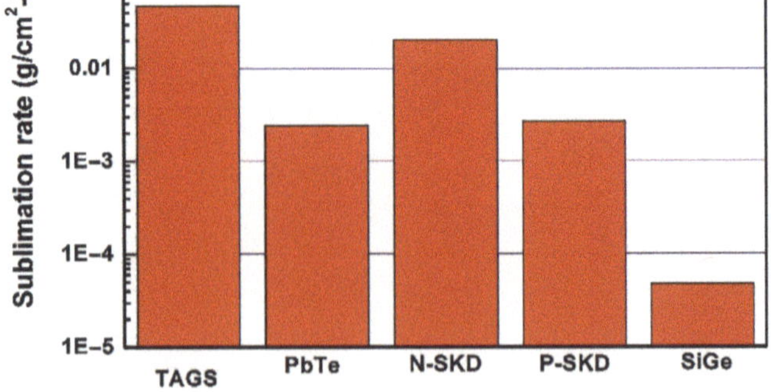


FIGURE 7.22 The sublimation rate of various TE materials at their maximum operating temperature under vacuum.

characterization of the corresponded microstructure are also helpful for predicting the degradation profile. For example, by combining the experimental observation and kinetic analysis, it is found that the variation of interfacial layer thickness can be described as a function of temperature and time by using the Arrhenius equation as shown in Fig. 7.23A [70]. Further investigation shows the variation of interfacial resistivity with time and temperature also obeys the Arrhenius relation (Fig. 7.23B). These preliminary results show the possibility to explain and predict the influence of interfacial microstructure evolution on the degradation of TE devices. The pioneer work on the full-life TE degradation of practical TE generators was performed by JPL group on the performance prediction of RTG. Hammel et al. [71] took the open circuit voltage and internal resistance as major evaluation indicators and collected numerous data from Pioneer, Viking, and other RTG data. They correlated the TE degradation with the hot-side temperature based on heritage model and described the degradation rate as time-independent percentage power loss per thousand hours. However, this heritage model did not include the effect of time. They further developed a hybrid model to predict the life-time power curves by combining the test data of MMRTG engineering unit with the heritage model. This hybrid model includes the effect of time and hot-side temperature and therefore provides a more reliable prediction for the degradation of TE generators.

The device reliability under dynamic service conditions (the typical ones include mechanical vibration and thermal cycling) is also important, especially for the applications in the recovery of automobile exhaust heat. In such dynamic service conditions, in addition to the influence of temperature, the thermal/mechanical stresses also become the important factors that may cause the device degradation or failure. The electrode joints are most likely to be suffered large stresses. In the practical operation, all the factors influencing the TE degradation will work together in TE device and generators, leading to a complicated description of the performance profile through the full life of device. Practically, it is necessary to establish a comprehensive modeling to predict the life-time power curves.

## 7.7 Summary

Since 1960s the TE generators have been successfully applied as irreplaceable power supply in space missions and other special power technology, and the TE refrigerators have been widely used in small-volume or localized temperature controlling with stably increasing markets. From 1990s because of the high expectation for efficiently utilizing

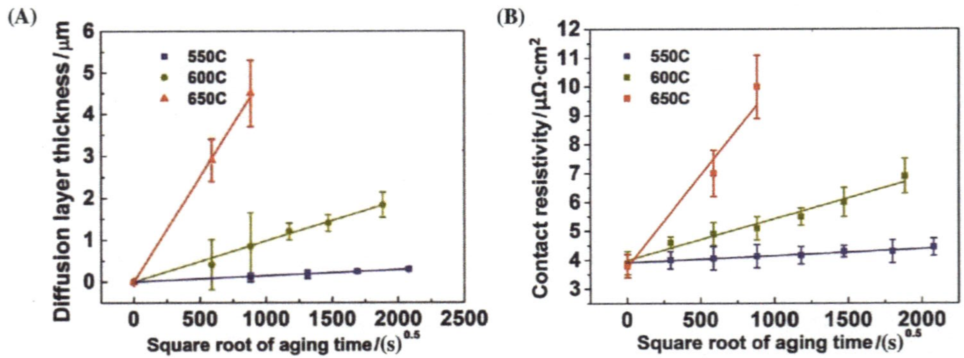


FIGURE 7.23 (A) Changes of interfacial reaction layer thickness and (B) changes of interfacial contact resistivity of Ti/Mo/$CoSb_3$ joint, after aging at different temperatures.

industrial waste heat and environmental thermal energy, thermoelectric technology has been especially spotlighted as a green energy technology. Especially, the achievements on basic science of TE transport theory have brought about the successful discovery of a series of new TE compounds and continual optimization of materials performance. Nevertheless, the TE applications are obviously lagging behind the astonishing progress in basic science on thermoelectrics. Typically, the TE generators have not been so widely commercialized as expected. The bottleneck of the large-scale application of power generation technology is the delayed development of TE device technology. Through continuous efforts in this field, the topological structure design method and integration technology with low parasitic energy losses have been preliminarily established, which enables the device output performance approaching the theoretical value on the basis of materials properties. However, service behavior for most devices is still not satisfied with the demands of long durability for industrial applications. Furthermore, cost-down is another burdensome task for thermoelectric engineers, including both the exploration of new TE materials composed of plenty and cheap elements, and the development of low-cost mass production technology. Finally, the thermoelectric technology is evidently featured with low power density, small volume, and no moving parts. Therefore, it is especially suitable for distributed or small-scale power generation, but less competitive for large-scale power generation. This should be also a guidance to follow in the development of practical thermoelectric technology.

## References

[1] D.M. Rowe, CRC Handbook of Thermoelectrics: Micro to Nano, CRC Press, Boca Raton, FL, 2006.

[2] F.J. Weinberg, D.M. Rowe, G. Min, Novel high performance small-scale thermoelectric power generation employing regenerative combustion systems, J. Phys. D Appl. Phys 35 (2002) L61–L63.

[3] D.T. Crane, L.E. Bell, Progress towards maximizing the performance of a thermoelectric power generator, in: Proceedings of 25th International Conference on Thermoelectrics, 6–10 August 2006.

[4] X.Y. Li, L.D. Chen, X.G. Xia, Y.S. Tang, S.Y. Tao, X.G. Xia, et al., Chinese Patent CN201408783Y, 2010.

[5] Y.Q. Wu, D.P. Lu, L.F. Jiang, L.G. Chen, L.H. F, X.Y. Li, et al., Chinese Patent CN101783386B, 2012.

[6] D.G. Zhao, X.Y. Li, W. Jiang, L.D. Chen, Fabrication of $CoSb_3$/MoCu thermoelectric joint by one-step SPS and evaluation, J. Inorg. Mater. 24 (2009) 545.

[7] D. Zhao, H. Geng, L. Chen, Microstructure contact studies for skutterudite thermoelectric devices, Int. J. Appl. Ceram. Technol. 9 (4) (2012) 733–741.

[8] D. Zhao, X. Li, L. He, W. Jianga, L. Chen, Interfacial evolution behavior and reliability evaluation of $CoSb_3$/Ti/Mo-Cu thermoelectric joints during accelerated thermal aging, J. Alloy. Compd. 477 (2009) 425–431.

[9] S.Q. Bai, F. Lei, T. Wu, X.Y. Huang, L.D. Chen, Chinese Patent CN103413889A, 2013.
[10] K. Hasezaki, H. Tsukuda, A. Yamada, S. Nakajima, Y. Kang, M. Niino, Thermoelectric semiconductor and electrode fabrication and evaluation of SiGe/electrode, in: The 16th International Conference on Thermoelectrics, Dresden, 1997, pp. 599–603.
[11] J.S. Lin, Y. Miyamoto, One-step sintering of SiGe thermoelectric conversion unit and its electrodes, J. Mater. Res. 15 (3) (2000) 647–652.
[12] V. Merges, Development of the radioisotope thermoelectric generator, in: The 2nd International Symposium on Power From Radioisolopes, Madrid, 1972, pp. 385–396.
[13] M. Orihashi, Y. Noda, L.D. Chen, Y. Kang, A. Moro, T. Hirai, Ni/PbTe and Ni/$Pb_{0.5}$Sn0.5Te ioining by plasma activated sintering, in: The 17th International Conference on Thermoelectrics, Nagoya, 1998, pp. 543–546.
[14] C.H. Lim, S.M. Choi, W.S. Seo, H.H. Park, A power-generation test for oxide-based thermoelectric modules using p-Type $Ca_3Co_4O_9$ and n-type $Ca_{0.9}Nd_{0.1}MnO_3$ Legs, J. Electron. Mater. 41 (2012) 1247–1255.
[15] M.S. El-Genk, H.H. Saber, T. Caillat, Efficient segmented thermoelectric unicouples for space power applications, Energy Convers. Manag. 4 (2003) 1755–1772.
[16] J.F. Fan, L.D. Chen, S.Q. Bai, X. Shi, Joining of Mo to $CoSb_3$ by spark plasma sintering by inserting a Ti interlayer, Mater. Lett. 58 (2004) 3876.
[17] Q. Zhang, X. Huang, S. Bai, X. Shi, C. Uher, L. Chen, Thermoelectric devices for power generation: recent progress and future challenges, Adv. Eng. Mater. 18 (2) (2016) 194–213.
[18] M.S. El-Genk, H.H. Saber, T. Caillat, J. Sakamoto, Tests results and performance comparisons of coated and un-coated skutterudite based segmented unicouples, Energy Convers. Manag. 47 (2006) 174–200.
[19] D. Zhao, X. Li, L. Wang, Y. Cai, W. Jiang, L. Chen, One-step sintering of $CoSb_3$ thermoelectric material and Cu-W alloy by spark plasma sintering, Mater. Sci. Forum 610 (2009) 389–393.
[20] M. Gu, X.G. Xia, X.Y. Li, X.Y. Huang, L.D. Chen, Microstructural evolution of the interfacial layer in the Ti-Al/$Yb_{0.6}$Co4$Sb_{12}$ thermoelectric joints at high temperature, J. Alloy. Compd. 610 (2014) 665–670.
[21] K. Hasezaki, H. Tsukuda, A. Yamada, S. Nakajima, Y. Kang, M. Niino, Thermoelectric semiconductor and electrode fabrication and evaluation of SiGe/electrode, in: Proceedings ICT'97—XVI International Conference on Thermoelectrics, 1997, pp. 599–602.
[22] G.L. Bennett, Space applications, in: D.M. Rowe (Ed.), CRC Handbook of Thermoelectrics, CRC Press, Boca Raton, FL, 1995.
[23] J.F. Mondt, SP-100 space subsystems, in: D.M. Rowe (Ed.), CRC Handbook of Thermoelectrics, CRC Press, Boca Raton, FL, 1995.
[24] R. Cockfield, Engineering development testing of the GPHS-RTG converter, in: Intersociety Energy Conversion Engineering Conference, 1981, pp. 321–325.
[25] G.L. Bennett, R.J. Hemler, A. Schock, Development and use of the Galileo and Ulysses power sources, Space Technol. Ind. Commer. Appl. 15 (3) (1994) 157–174.
[26] G.L. Bennett, J.J. Lombardo, R.J. Hemler, G. Silverman, C. Whitmore, W.R. Amos, et al., Mission of daring: the general-purpose heat source radioisotope thermoelectric generator, in: 4th International Energy Conversion Engineering Conference and Exhibit (IECEC), San Diego, CA, AIAA, 2006.
[27] G. Bennett, J. Lombardo, B. Rock, Power performance of the general-purpose heat source radioisotope thermoelectric generator, Space Nucl. Power Syst. (1987) 199–222.

[28] J.S. Lin, K. Tanihata, Y. Miyamoto, H. Kido, Microstructure and property of (Si-$MoSi_2$)/SiGe thermoelectric convertor unit, Funct. Graded Mater. 1997 (1996) 599–604.
[29] K. Hasezaki, H. Tsukuda, A. Yamada, S. Nakajima, Y. Kang, M. Niino, et al., SiGe/electrode response to long-term high-temperature exposure, in: XVII International Conference on Thermoelectrics, Proceedings of ICT'98, 1998, pp. 460–463.
[30] X.Y. Yang, J.H. Wu, M. Gu, X.G. Xia, L.D. Chen, Fabrication and contact resistivity of W-$Si_3N_4$/$TiB_2$-$Si_3N_4$/p-SiGe thermoelectric joints, Ceram. Int. 42 (7) (2016) 8044–8050.
[31] T.C. Harman, Special techniques for measurement of thermoelectric properties, J. Appl. Phys. 29 (1958) 1373.
[32] J. Chen, Z. Yan, L. Wu, The influence of Thomson effect on the maximum power output and maximum efficiency of a thermoelectric generator, J. Appl. Phys. 79 (11) (1996) 8823–8828.
[33] G. Fraisse, J. Ramousse, D. Sgorlon, et al., Comparison of different modeling approaches for thermoelectric elements, Energy Convers. Manag. 65 (1) (2013) 351–356.
[34] L. Chen, J. Gong, F. Sun, et al., Effect of heat transfer on the performance of thermoelectric generators, Int. J. Therm. Sci. 41 (1) (2002) 95–99.
[35] H.S. Lee, Optimal design of thermoelectric devices with dimensional analysis, Appl. Energy 106 (11) (2013) 79–88.
[36] A. Montecucco, J.R. Buckle, A.R. Knox, Solution to the 1-D unsteady heat conduction equation with internal Joule heat generation for thermoelectric devices, Appl. Therm. Eng. 35 (3) (2012) 177–184.
[37] K. Zabrocki, E. Muller, W. Seifert, One-dimensional modeling of thermogenerator elements with linear material profiles, J. Electron. Mater. 39 (9) (2010) 1724–1729.
[38] W.H. Chen, C.Y. Liao, C.I. Hung, A numerical study on the performance of miniature thermoelectric cooler affected by Thomson effect, Appl. Energy 89 (1) (2012) 464–473.
[39] S. Skipidarov, M. Nikitin (Eds.), Thermoelectrics for Power Generation: A Look at Trends in the Technology, IntechOpen, 2016.
[40] G. Fraisse, M. Lazard, C. Goupil, et al., Study of a thermoelement's behaviour through a modelling based on electrical analogy, Int. J. Heat. Mass Transf. 53 (17–18) (2010) 3503–3512.
[41] E.E. Antonova, D.C. Looman, Finite elements for thermoelectric device analysis in ANSYS, in: International Conference on Thermoelectrics, IEEE, 2005, pp. 215–218.
[42] Q.H. Zhang, J.C. Liao, Y.S. Tang, M. Gu, C. Ming, P.F. Qiu, et al., Realizing a thermoelectric conversion efficiency of 12% in bismuth telluride/skutterudite segmented modules through full-parameter optimization and energy-loss minimized integration, Energy Environ. Sci. 10 (2017) 956–963.
[43] S. Jun-Qiang, S. Xun, Z. Wen-Qing, C. Li-Dong, Heat conduction in thermoelectric materials and micro-devices, Physics 42 (2013) 112–123.
[44] R.J.M. Vullers, R. van Schaijk, I. Doms, C. Van Hoof, R. Mertens, Micropower energy harvesting, Solid-State Electron. 53 (2009) 684–693.
[45] M. Arik, J. Petroski, S. Weaver, Thermal challenges in the future generation solid state lighting applications: light emitting diode, in: Inter Society Conference on Thermal Phenomena, 2002, pp. 113–120.
[46] J. Sharp, J. Bierschenk, H.B. Lyon, Overview of solid-state thermoelectric refrigerators and possible applications to on-chip thermal management, Proc. IEEE 94 (2006) 1602–1612.
[47] M. Strasser, R. Aigner, M. Franosch, et al., Miniaturized thermoelectric generators based on poly-Si and poly-SiGe surface micromachining, Sens. Actuators A Phys. 97–98 (2002) 535.

[48] I. Stark, M. Stordeur, New micro thermoelectric devices based on bismuth telluride-type thin solid films, in: The 18th International Conference on Thermoelectrics. Hyatt Regency, Baltimore, MD, 1999, p. 465.
[49] G.J. Snyder, J.R. Lim, C.K. Huang, J.P. Fleurial, Thermoelectric microdevice fabricated by a MEMS-like electrochemical process, Nat. Mater. 2 (2003) 528.
[50] H. Bottner, J. Nurnus, A. Schubert, F. Volkert, New high density micro structured thermogenerators for stand alone sensor systems, in: The 26st International Conference on Thermoelectrics, IEEE, 2007, pp. 306–309.
[51] R. Venkatasubramanian, C. Watkins, C. Caylor, et al., Microscale thermoelectric devices for energy harvesing and thermal management, in: The Sixth International Workshop on Micro and Nanotechnology for Power Generation and Energy Conversion Applications, Berkeley, CA, 2006.
[52] I.Y. Huang, J.C. Lin, K.D. She, et al., Development of low-cost micro-thermoelectric coolers utilizing MEMS technology, Sens. Actuators A Phys 148 (2008) 176.
[53] G.E. Bulman, E. Siivola, B. Shen, R. Venkatasubramanian, Large external $\Delta T$ and cooling power densities in thin-film $Bi_2Te_3$-superlattice thermoelectric cooling devices, Appl. Phys. Lett. 89 (12) (2006) 122117.
[54] A. Shakouri, Y. Zhang, On-chip solid-state cooling for integrated circuits using thin-film microrefrigerators, IEEE Trans. Compon. Packag. Technol. 28 (1) (2005) 65–69.
[55] A. Watwe, R. Viswanath, Thermal implications of non-uniform die power map and CPU performance, in: Proceedings of InterPACK, 2003, p. 35044.
[56] H. Böttner, J. Nurnus, A. Gavrikov, G. Kühner, M. Jägle, C. Künzel, et al., New thermoelectric components using microsystem technologies, J. Microelectromech. Syst. 13 (3) (2004) 414–420.
[57] R. Venkatasubramanian, Thin-film thermoelectric device and fabrication method of same, US Patent No. 6300150B1, 2001.
[58] K.J. Shin, T.S. Oh, Micro-power generation characteristics of thermoelectric thin film devices processed by electrodeposition and flip-chip bonding, J. Electro. Mater. 44 (6) (2015) 1–8.
[59] G. Li, J.G. Fernandez, D.A.L. Ramos, V. Barati, N. Pérez, I. Soldatov, et al., Integrated microthermoelectric coolers with rapid response time and high device reliability, Nat. Electron. 1 (2018) 555–561.
[60] W. Wang, F. Jia, Q. Huang, J.Z. Zhang, A new type of low power thermoelectric micro-generator fabricated by nanowire array thermoelectric material, Microelectron. Eng. 77 (3–4) (2005) 223–229.
[61] H.C. Wang, J.H. Bahk, C.Y. Kang, J.P. Hwang, K.M. Kim, J.W. Kim, et al., Right sizes of nano- and microstructures for high-performance and rigid bulk thermoelectrics, Proc. Natl Acad. Sci. USA (111)(2014) 10949–10954.
[62] C.B. Vining, Silicon germanium, in: D.M. Rowe (Ed.), CRC Handbook of Thermoelectrics, CRC Press, Boca Raton, FL, 1995.
[63] D.R. Brown, T. Day, T. Caillat, G.J. Snyder, Chemical stability of $(Ag, Cu)_2Se$: a historical overview, J. Electron. Mater. 42 (2013) 2014–2019.
[64] G. Dennler, R. Chmielowski, S. Jacob, F. Capet, P. Roussel, S. Zastrow, et al., Are binary copper sulfides/selenides really new and promising thermoelectric materials? Adv. Energy Mater. 4 (2014) 1301581.
[65] P. Qiu, M.T. Agne, Y. Liu, Y. Zhu, H. Chen, T. Mao, et al., Suppression of atom motion and metal deposition in mixed ionic electronic conductors, Nat. Commun. 9 (2018) 2910.
[66] T. Mao, P. Qiu, P. Hu, X. Du, K. Zhao, T.-R. Wei, et al., Decoupling thermoelectric performance and stability in liquid-like thermoelectric materials, Adv. Sci. 7 (2020) 1901598.

[67] T. Mao, P. Qiu, X. Du, P. Hu, K. Zhao, J. Xiao, et al., Enhanced thermoelectric performance and service stability of $Cu_2Se$ Via tailoring chemical compositions at multiple atomic positions, Adv. Funct. Mater. 30 (2020) 1908315.
[68] T. Caillat, C.-K. Huang, B. Cheng, S.C. Chi, P. Gogna, J. Paik, et al., Advanced thermoelectric materials for radioisotope thermoelectric generators, in: 25th Symposium on Space Nuclear Power Albuquerque, NM, 2008.
[69] X. Xia, X. Huang, X. Li, M. Gu, P. Qiu, J. Liao, et al., Preparation and structural evolution of Mo/SiOx protective coating on $CoSb_3$-based filled skutterudite thermoelectric material, J. Alloy. Compd. 604 (2014) 94–99.
[70] M. Gu, S. Bai, X. Xia, X. Huang, X. Li, X. Shi, et al., Study on the high temperature interfacial stability of Ti/Mo/$Yb_{0.3}Co_4Sb_{12}$ thermoelectric joints, Appl. Sci. 7 (2017) 952.
[71] T.E. Hammel, R. Bennett, W. Otting, S. Fanale, Multi-mission radioisotope thermoelectric generator (MMRTG) and performance prediction model, in: 7th International Energy Conversion Engineering Conference, AIAA, 2009, 4576.

# Index

*Note*: Page numbers followed by "*f*" and "*t*" refer to figures and tables, respectively.

## N



## O



## P

## R



## S

FSC
www.fsc.org
MIX
Papier | Fördert
gute Waldnutzung
FSC® C083411

Kolibri 360 GmbH
Ferdinand-Jühlke-Straße 7
99095 Erfurt, Deutschland
produktsicherheit@kolibri360.de